D1271079

Reactive Hydrocarbons in the Atmosphere

Reactive Hydrocarbons in the Atmosphere

Edited by

C. Nicholas Hewitt

Institute of Environmental and Natural Sciences
Lancaster University
Lancaster, United Kingdom

ACADEMIC PRESS

San Diego   London   Boston   New York   Sydney   Tokyo   Toronto

This book is printed on acid-free paper. ∞

Academic Press
a division of Harcourt Brace & Company
525 B Street, Suite 1900, San Diego, California 92101-4495, USA
http://www.apnet.com

Academic Press
24-28 Oval Road, London NW1 7DX, UK
http://www.hbuk.co.uk/ap/

Library of Congress Catalog Card Number: 98-86876

International Standard Book Number: 0-12-346240-1

PRINTED IN THE UNITED STATES OF AMERICA
98 99 00 01 02 03 BB 9 8 7 6 5 4 3 2 1

CONTENTS

3 Modeling Biogenic Volatile Organic Compound Emissions to the Atmosphere

Alex Guenther

4 The Sampling and Analysis of Volatile Organic Compounds in the Atmosphere

Xu-Liang Cao and C. Nicholas Hewitt

5 Reactive Hydrocarbons in the Atmosphere at Urban and Regional Scales

Paolo Ciccioli, Enzo Brancaleoni, and Massimiliano Frattoni

CONTRIBUTORS

Numbers in parentheses indicate the pages on which the authors' contributions begin.

C. BOISSARD (209) LISA, Laboratoire Interuniversitaire des Systèmes Atmosphériques, Université Paris, Créteil, F-94010, France

B. BONSANG (209) Laboratoire des Sciences du Climat et de L'Environment CE Saclay, Gif-sur-Yvette Cedex, F-91191, France

ENZO BRANCALEONI (159) Instituto sull'Inquinam ento Atmosferico del CNR, Area della Ricerca di Roma, Monterotondo Scalo, I-00016, Italy

XU-LIANG CAO (119) Environmental Health Centre, Ottawa K1A O12, Canada

PAOLO CICCIOLI (159) Instituto sull'Inquinam ento Atmosferico del CNR, Area della Ricerca di Roma, Monterotondo Scalo, I-00016, Italy

R. G. DERWENT (267) Atmospheric Processes Research, Meteorological Office, Bracknell, RD12 2SZ, United Kingdom

RAY FALL (41) Department of Chemistry and Bioochemistry, Cooperative Institute for Research in Environmental Sciences, University of Colorado, Boulder, Colorado 80309

MASSIMILIANO FRATTONI (159) Instituto sull'Inquinam ento Atmosferico del CNR, Area della Ricerca di Roma, Monterotondo Scalo, I-00016, Italy

RANIER FRIEDRICH (1) Universitat Stuttgart, Institut fur Engiewirtschaft und Rationelle Energiewendung (IER), Stuttgart, D-70569, Germany

ALEX GUENTHER (97) Atmospheric Chemistry Division, National Center for Atmospheric Research, Boulder, Colorado 80303

C. NICHOLAS HEWITT (119) Institute of Environmental and Natural Sciences, Lancaster University, Lancaster LA1 4YQ, United Kingdom

ANDREAS OBERMEIER (1) Universitat Stuttgart, Institut fur Engiewirtschaft und Rationelle Energiewendung (IER), Stuttgart, D-70569, Germany

JOHN H. SEINFELD (293) California Institute of Technology, Pasadena, California 91125

PREFACE

The Earth's atmospheric environment is changing at an unprecedented rate. Emissions of trace gases due to Man's activities are causing perturbations in the chemical composition of the atmosphere, and these are beginning to have major quantifiable effects on the behavior of the Earth–atmosphere–biosphere system. For example, emissions of carbon dioxide, methane, nitrous oxide, and other radiatively active gases are causing changes to the Earth's energy balance, giving rise to increasing temperatures and other changes to climate. Emissions of ozone depleting substances, including the synthetic chlorofluorocarbons, are causing changes to stratospheric ozone concentrations, and hence to the amount of UV light reaching the Earth's surface. Emissions of reactive inorganic pollutants such as sulphur dioxide and nitrogen oxides are causing acidification of natural waters and soils at the regional scale. And finally, emissions of the whole vast family of reactive volatile organic compounds (or "reactive hydrocarbons" as used in this book's title) are causing changes to atmospheric chemistry—particularly the oxidizing capacity of the troposphere—over the whole range of spatial scales, from the localized urban scale through the regional scale and ultimately to the global scale.

This book seeks to describe the current state of knowledge of the emissions, chemistry, and effects of volatile organic compounds in the atmosphere. It covers compounds of both biogenic and anthropogenic origins, the former contributing to the background chemistry of the atmosphere, the latter causing the perturbations that are now so evident in many urban areas and, less obviously, in rural, semirural and suburban areas blighted by summertime photochemical smogs. The aim is to give an authoritative view of this rapidly changing subject, and I hope that each chapter, written by one or more acknowledged experts in the field, will provide a definitive and readable description of its topic and be of value to all those interested in understanding how fragile our atmospheric environment is.

My own interest in reactive hydrocarbons was engendered by a chance conversation as a young and impressionable scientist with Ralph Cicerone and Richard Wayne. Ray Fall provided the opportunity for me to begin my current work on biogenic hydrocarbons. As ever, Roy Harrison remains an inspiring mentor. I thank them all.

Nick Hewitt
Lancaster, United Kingdom

CHAPTER 1

Anthropogenic Emissions of Volatile Organic Compounds

RAINER FRIEDRICH* AND ANDREAS OBERMEIER

*Universität Stuttgart, Institut für Energiewirtschaft und Rationelle Energieanwendung (IER), Stuttgart, Germany

---

I. Introduction
II. Combustion Processes
    A. Exhaust Gas from Internal Combustion Engines
    B. Flue Gas Emissions from Combustion Installations
III. **Mining, Treatment, Storage, and Distribution of Fossil Fuels**
    A. Mining and Processing of Solid Fossil Fuels
    B. Production, Processing, Storage, and Distribution of Liquid Fossil Fuels
    C. Production, Processing, and Distribution of Gaseous Products
IV. **Solvent Use**
V. **Industrial Processes**
VI. **Biological Processes**
VII. **Modeling Anthropogenic Volatile Organic Compound Emissions**
    A. Requirements for Emission Data
    B. Source Classification
    C. Regionalization
    D. Temporal Resolution
VIII. **Emission Data**
IX. **Conclusions**
    References

The family of chemicals known as volatile organic compounds (VOCs) comprises a large variety of species, which are released into the atmosphere from numerous types of sources. Chapter 1 focuses on anthropogenic VOC emissions and begins with descriptions of important source categories (i.e., road transport, combustion of solid fuels in small furnaces, use of organic solvents, disposal of organic wastes on landfills, and ruminant husbandry). Parameters controlling the number of VOC emissions, information on the chemical composition of the emissions, and brief mention of the technical measures available for the abatement of VOC emissions are the major points mentioned in this context. Annual emission data for methane and nonmethane VOCs (NMVOCs) on global, European, and national scales are presented and discussed. Finally, the strong temporal variations and spatial distribution patterns of NMVOC emissions are outlined.

I. INTRODUCTION

The class of volatile organic compounds (VOCs) includes species with different physical and chemical behaviors. Pure hydrocarbons containing C and H as the only elements (e.g., alkanes, alkenes, alkynes, and aromatics) are important VOC classes. However, volatile organic compounds containing oxygen, chlorine, or other elements besides carbon and hydrogen are important, too. These latter classes include, for example, aldehydes, ethers, alcohols, ketones, esters, chlorinated alkanes and alkenes, chlorofluorocarbons (CFCs), and hydrochlorofluorocarbons (HCFCs).

In consideration of VOC emissions estimates, it is important to take all types of VOC compounds into account. One reason is that, in general, estimating total VOC emissions is the first step in preparing an emission inventory, followed by compound speciation according to additional information on the composition of VOC emissions expressed, for example, as percentage by mass of total emissions. Therefore, in principle, no VOC compounds are excluded from this chapter, although the focus is more on the reactive species.

Emissions of VOCs not only comprise a broad spectrum of species, but also are due to a wide variety of sources. Because of this, the number and variety of relevant source categories responsible for anthropogenic VOC emissions (and also biogenic emissions as outlined in Chapter 2) are rather large. They include:

- Combustion processes

- Production, treatment, storage, and distribution of fossil fuels
- Application of volatile organic solvents and solvent-containing products
- Industrial production processes
- Biological processes.

Combustion processes comprise, for example, internal combustion engine vehicles, combustion plants, and furnaces. Examples of solvent-containing products are paints and varnishes, metal degreasing agents, and adhesives. Biological processes include the digestive processes of ruminants, the handling of animal manure, and the disposal of organic wastes.

Within the first sections of this chapter on anthropogenic VOC emissions, the source categories mentioned earlier are described in more detail, focusing on:

- A systematic listing of important source types
- The explanation of decisive parameters controlling emissions
- A short description of available emission control techniques and measures.

The last sections of Chapter 1 are then dedicated to an overview of emission estimate procedures as well as available emission data and inventories on global, regional, and local scales.

II. COMBUSTION PROCESSES

The major products of a complete combustion of fossil fuels are carbon dioxide and water, following the overall reaction scheme

$$C_nH_m + (n + m/4)O_2 \longrightarrow nCO_2 + m/2\ H_2O,$$

where C_nH_m represents fossil fuels or other organic materials with a certain ratio of C/H, and $(n + m/4)\ O_2$ indicates the stoichiometric amount of oxygen that is theoretically required for complete combustion. However, in practice, technical combustion processes are more or less incomplete. One reason for this might be a lack of oxygen. In general, the amount of oxygen needed for the combustion process is higher than the stoichiometric amount, due to unavoidable inhomogeneities of the fuel–oxygen mixture. Other reasons for incomplete combustion might be combustion temperatures being low or the residence time of the fuel in the burner zone being too short.

Incomplete combustion of fossil fuel leads to emissions of CO (an intermediate product of the oxidation of carbon to CO_2) and VOCs. Hydrocarbons can enter the flue gas in a partly oxidized form (e.g., aldehydes) or even in an unoxidized form (e.g., alkanes and aromatics). Also the formation of new hydrocarbons (e.g., alkenes, alkynes, and aromatics) by radical reactions can be observed.

The more incomplete a reaction is, the greater the resultant VOC emissions. Relative to energy consumption, internal combustion engines are of major importance. As shown in Table 1; two-stroke engines of mopeds are characterized by the highest VOC emission factor, here expressed in kilograms of emitted VOCs per terajoule of fuel used. The exhaust gas of two-stroke engines contains not only noncombusted or partly combusted gasoline but also noncombusted or partly combusted lubricants in large amounts. VOC emission factors for four-stroke gasoline engines (without catalysts) are considerably lower, followed by diesel engines of heavy-duty vehicles, light-duty vehicles, and finally diesel engines of passenger cars with the lowest emission factors for engines without emission control. However, the lowest emission factors for mobile sources are from gasoline engines equipped with three-way catalysts.

Decisive emission parameters controlling emission rates from the combustion of fossil fuels in stationary plants and furnaces are fuel type and combustion technique, giving an indication of the completeness of combustion processes. By using sources in Germany as an example in Table 1, it is clear that the combustion of gaseous and liquid fuels, in general, leads to lower VOC emissions than the combustion of coal, wood, or other solids. The only exceptions are coal dust combustion installations, which emit less VOC than heavy fuel oil power plants.

Especially for solid fuels, the VOC emission factors for large combustion plants are by far lower than the emission factors for small furnaces used in households. The reason for this is that fossil fuel power plants are equipped with complex and costly combustion technologies, ensuring more or less complete combustion of specially prepared coal, whereas the construction of small furnaces in the household sector is relatively simple and designed more for easy operation than for really optimized combustion. Medium-sized industrial and commercial combustion installations have emission factors between the extreme values.

A. Exhaust Gas from Internal Combustion Engines

Exhaust gas emissions of VOCs from motor vehicles strongly depend on parameters such as vehicle speed and motor load. Typical variations of these parameters can be arranged in so-called driving modes or traffic situations.

TABLE 1 Total VOC Emission Factors of the Combustion of Fuels in Different Sectors in Germany[a,b]

Source type	Further description	Gas	Gasoline	Diesel	Light fuel oil	Heavy fuel oil	Coal	Wood, other solid fuels
Passenger car	Gasoline, without catalyst		473					
Passenger car	Gasoline, with three-way catalyst		17					
Passenger car	Diesel, before 1990			38				
Passenger car	Diesel, after 1996			33				
Light-duty vehicle	Gasoline, without catalyst		454					
Light-duty vehicle	Gasoline with three-way catalyst		17					
Light-duty vehicle	Diesel			49				
Heavy-duty vehicle	Diesel, before 1990			170				
Heavy-duty vehicle	Diesel, after 1996			135				
Moped	Two-stroke		4859					
Moped	Two-stroke, with uncontrolled catalyst		2168					
Households	Space heating	5			5		400	350
Households	Warm water	8			5			1600
Households	Cooking	25						2000
Other consumers		5			5		30	210
Industry		5				8	30	210
Power plants		0.6				7	3	65[c]

[a]Values given in kilograms of VOC per terajoule of fuel.
[b]Derived from Seier (1997) and UBA (1995).
[c]Waste incineration.

TABLE 2   Dependence of VOC Exhaust Gas Emissions of Different Vehicle Types on Traffic
Situations–Driving Modes[a,b]

		Highway mode	Rural mode	Urban mode	Traffic jam
Passenger car	Gasoline, without catalyst	0.87	0.86	2.14	4.58
Passenger car	Gasoline, with three-way catalyst, before 1991	0.14	0.17	0.37	2.80
Passenger car	Gasoline, with three-way catalyst, before EURO2	0.09	0.08	0.18	1.37
Passenger car	Gasoline, with three-way catalyst, EURO2	0.04	0.03	0.06	0.48
Passenger car	Diesel	0.05	0.07	0.15	0.27
Passenger car	Diesel, EURO2	0.04	0.05	0.12	0.23
Light-duty vehicle		0.17	0.21	0.58	1.56
Heavy-duty vehicle		0.77	1.01	3.33	9.50

[a]Values given in grams per kilometer.
[b]Derived from UBA (1995).

The values in Table 2 show that emission factors, here expressed in grams of VOC per kilometer, are exceptionally high within traffic jams. Above average emission factors also arise from driving modes in urban areas, whereas VOC emissions per kilometer on roads outside settlements and on highways with moderate speed show a minimum. However, what cannot be deduced from the values in Table 2 is that VOC emission factors also tend to increase as vehicle speed increases toward a maximum. The influence of vehicle type, engine type, and emission control technology on VOC exhaust gas emissions has been outlined see (Table 1).

Other decisive parameters of VOC emissions from motor vehicles are ambient air temperature in connection with cold starts and, furthermore, the gradient of roads. The influence of these parameters on VOC emissions is shown in Fig. 1 and 2.

Figure 1 indicates that there is an almost linear dependency of cold start emissions with ambient air temperature. At an ambient air temperature of 10° C, which represents the mean annual temperature in Central Europe, cold start emissions of VOCs are in the range of 6 to 7 grams per start. However, at a temperature of − 10° C cold start emissions are about three times higher. On the other hand, cold start emissions are rather low on warm summer days.

Exhaust gas emission factors for road traffic are commonly related to flat roads with a gradient of 0%. Figure 2 shows as an example that for highway driving modes of passenger cars, emissions are significantly higher on roads with a positive gradient, but also are higher on roads with a negative gradient.

Not only the total amount of VOCs emitted from road traffic but also the specific composition of the VOC exhaust gas emissions from motor vehicles

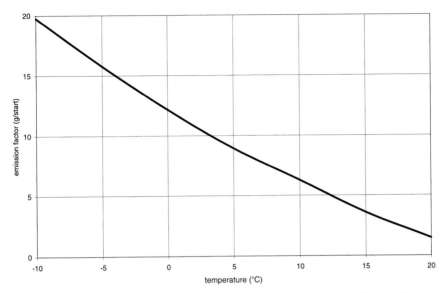

FIGURE 1 Influence of ambient air temperature on additional VOC emissions due to cold starts of passenger cars without catalysts (derived from UBA, 1995).

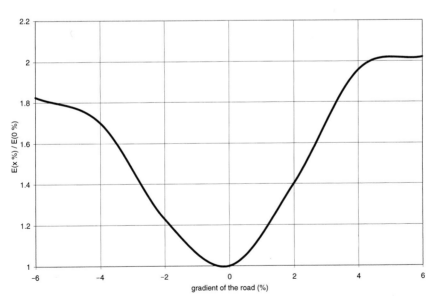

FIGURE 2 Quotient of VOC exhaust gas emissions for highway driving modes on roads with a gradient [$E(x\%)$] and on flat roads [$E(0\%)$] (derived from UBA, 1995).

is an important parameter in air quality evaluation procedures. This is at least partly influenced by the fuel composition. Table 3 gives an impression of the typical composition of different gasoline types for Germany, distinguishing also between summer grade and winter grade fuels.

The main components of paraffins are pentanes, hexanes, and octanes. The olefins are mainly composed of pentenes and hexenes. Of the aromatic hydrocarbons, toluene, xylenes, and other C_8 aromatics as well as C_9 aromatics are the major components. In Germany and the rest of the European Union (EU), the content of benzene in gasoline is limited to 5 wt%. However, typical values are in the range of 2 to 3%.

The average composition of diesel fuels is about 45 wt% paraffins, some 25 wt% naphthenes, and about 29 wt% aromatic hydrocarbons. The share of olefins is around 1 wt%. In contrast to gasoline, diesel fuel contains significant amounts of high-molecular-weight hydrocarbons ($\geq C_{10}$).

Examples of the composition of VOC tailpipe emissions are summarized in Table 4 for conventional passenger cars without catalysts, passenger cars with three-way catalysts, two-stroke engines, and diesel cars. The values represent the average composition for different driving conditions. Detailed study of the dependency of the composition of VOC tailpipe emissions on driving modes is the subject of ongoing research.

Some of the more important aspects of the VOC profiles discussed in Patyk and Höpfner (1995) are

- Of the VOC emissions from passenger cars with Otto engines, 50 to 75% consist of fuel compounds that are noncombusted
- The efficiency of three-way catalysts in passenger cars with Otto engines is around 90%, but differs from species to species
- Some hydrocarbons are produced in significant amounts by the initial combustion process or by subsequent catalytic reactions (e.g., high amounts of ethene and ethyne in conventional vehicles with Otto engines and high amounts of methane in the exhaust gas of cars with three-way catalysts).

Especially for diesel engines, the "other VOCs" indicated in Table 4 are normally high-molecular-weight paraffins (Veldt, 1992). However, Patyk and Höpfner (1995) emphasized the large uncertainty in VOC compound profiles for diesel engines. The uncertainties in VOC profiles from road traffic were also investigated by John (1997), based on measurement data from a road tunnel study (Staehelin *et al.*, 1997). Compound specific measurement data and detailed calculations are shown in Fig. 3. Figure 3 indicates a good agreement between measurements and calculations, especially for some aromatic hydrocarbons, whereas there are larger deviations in the cases of *n*-butane, *n*-pentane, and pentene.

TABLE 3 Typical Octane Numbers and Composition of Different Gasoline Types in Germany[a,b]

	Normal		Euro super		Super plus		Super leaded	
	Summer	Winter	Summer	Winter	Summer	Winter	Summer	Winter
Research octane number:	93.3	92.9	96.3	96.0	99.1	99.0	98.7	98.5
Motor octane number:	82.9	82.8	85.2	85.2	88.0	88.0	88.2	88.0
Parffins	46	54	46	50	43	44	47	52
Olefins	18	16	12	10	5	6	10	8
Aromatics	34	30	40	38	43	42	40	38
MTBE[c]	1.1	0.5	2.2	1.1	9.2	8.4	2.7	1.7
Methanol	0.1	0.2	0.2	0.2	0.1	0.1	0.1	0.1

[a]Values of chemical composition given in volume percentages.
[b]According to Patyk and Höpfner (1995).
[c]Methyl t-butyl ether.

TABLE 4   Composition of VOC Emissions from Internal Combustion Engines[a,b]

	Otto engine, four stroke without catalyst	Otto engine, four stroke three-way catalyst	Otto engine, two stroke without catalyst	Diesel engine
Paraffins				
Methane	4.0	14.0	7.0	2.4
Ethane	0.8	2.0	1.0	
Propane	0.3	0.5		
Butane	3.0	6.0	2.0	
Isobutane	2.0	3.0	0.3	
Pentane	2.0	2.5	3.0	
Isopentane	5.0	7.5	4.0	
Olefins				
Ethene	7.0	4.5	5.0	12.2
Propene	4.0	2.5	2.0	4.7
1-Butene	0.4	0.1		
Isobutene	2.5	1.5		0.8
cis-2-Butene	0.2	0.2	0.4	0.8
trans-2-Butene	0.8	0.6	0.2	0.7
1,3-Butadiene	0.6	0.4		1.1
Pentene	1.1	0.7		
Alkynes				
Ethyne	5.5	3.0	4.0	2.8
Aromatic hydrocarbons				
Benzene	5.0	6.0	5.0	1.9
Toluene	11.5	10.0	12.1	0.8
Xylene	10.0	9.0	11.0	0.8
Ethylbenzene	2.5	2.0	2.8	0.3
C_9 aromatics	7.5	6.0	8.3	
Aldehydes				
Formaldehyde	1.5	1.0	0.6	8.1
Acetaldehyde	0.7	0.6	0.2	4.2
Acrolein	0.4	0.2	0.0	2.1
Benzaldehyde	0.3	0.2	0.2	1.8
Tolualdehyde	0.5	0.4	0.2	1.0
Ketones				
Acetone	0.6	0.5	0.1	1.5
Other VOCs	20.3	15.1	30.6	52.0

[a]Values given in mass percentages.
[b]Patyk and Höpfner (1995).

At present, three-way catalysts represent the most effective measure for the control of VOC exhaust gas emissions from vehicles with Otto engines (see Tables 1 and 2). Emissions can be reduced by about 90% (average) with a three-way catalyst relative to a vehicle without a catalyst. In principle, another

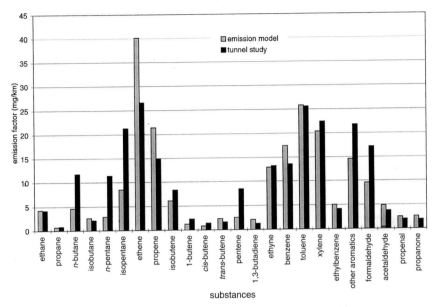

FIGURE 3   Comparison of emission factors for VOC components, derived from measurements and calculations (John, 1997).

reduction measure is to maintain the internal combustion with an excess of air ($\lambda > 1$). However, current European emission standards for exhaust gas from motor vehicles (EURO2) and further proposed restrictions (EURO3) cannot be met with this technology.

B. FLUE GAS EMISSIONS FROM COMBUSTION INSTALLATIONS

The relative importance of VOC emissions from the combustion of solid fuels has already been mentioned. The values listed for combustion installations in Table 1 represent typical values over longer time periods. However, short-term peaks with very high emission rates can be observed during the start-up and shutdown of combustion installations, during operations under partial load, or just after the feeding of solid fuel combustion installations with new material.

The relative compositions of VOC emissions from combustion installations vary considerably with different fuel types. The values in Table 5 give only rough clues about the VOC profiles from smaller furnaces. Due to their limited importance, corresponding data for large combustion installations are not yet available. However, Table 5 shows clearly that methane is the most important component in coal- and gas-fired installations. With the combustion of wood, significant amounts of alkenes, benzene, and other aromatic hydrocarbons are released into the atmosphere. The composition of oil- and gas-fired installations indicates high amounts of unburned fuel, with signifi-

TABLE 5 Estimated Composition of VOC Emissions from Combustion Installations[a,b]

	Coal	Wood	Fuel oil	Gas
Alkanes				
Methane	75	25	10	75
Ethane	3.75	7.5		12
Propane	2.5	1.13		4
Butane		0.38		4
Alkanes > C_4	1.25	2.25		
Alkanes, not specified			65	
Alkenes				
Ethene	7.5	22.5		1.8
Propene	1.25	4.5		0.6
Butene		0.38		0.6
Pentene		1.5		
Alkenes > C_3	1.25			
Alkenes, not specified			10	
Alkynes				
Ethyne	1.25	7.5		
Propyne		0.38		
Aromatic hydrocarbons				
Benzene	1.25	11.25		
Toluene	0.25	3.75		
o-Xylene	0.09	0.38		
m-Xylene	0.08	0.54		
p-Xylene	0.08	0.55		
Ethylbenzene		0.38		
Aromatics, not specified			5	
Aldehydes				
Formaldehyde		1.5		1.8
Acetaldehyde		0.75		0.12
Propionaldehyde				0.04
Acrolein		1.5		
Aldehydes ~ C_4		2.25		0.04
Aldehydes, not specified			10	
Other VOCs				
Ketones, not specified		1.13		
VOCs, not specified	4.5	3		

[a]Values given in mass percentages.
[b]Data mainly based on Veldt (1992).

cant amounts of higher molecular weight alkanes and ethane emitted, respectively.

Possibilities for reducing VOC emissions from small solid fuel furnaces come mainly from ensuring their correct construction and proper operation. Correct construction should include ensuring combustion with a stable and uniform flame even when varying amounts of coal or wood are being fed. A proper operation mode means that, for example, feeding the combustion installation with wet or damp fuel wood and completely closing ventilation flaps should be avoided.

III. MINING, TREATMENT, STORAGE, AND DISTRIBUTION OF FOSSIL FUELS

The mining, preparation, and distribution of fossil fuels represent a wide range of processes. The most important of these processes—handling of solid, liquid, and gaseous fuels and of VOCs released—will be discussed in the following subsections.

A. MINING AND PROCESSING OF SOLID FOSSIL FUELS

The primary gaseous component of coal fields is methane, but other compounds such as ethane and propane are present in minor amounts. The release of the so-called firedamp often results in emissions into the atmosphere, as it is often not economical to flare it or to use it as a fuel.

Firedamp emissions of methane from the deep mining of hard coal are of major importance among the sources discussed in this subsection. In comparison to these, methane emissions during storage and first treatment of hard coal are more than 10 times lower. Furthermore, methane emissions related to opencast mining of lignite are rather negligible, as they are more than a hundred times lower than those from the deep mining of hard coal.

B. PRODUCTION, PROCESSING, STORAGE, AND DISTRIBUTION OF LIQUID FOSSIL FUELS

Production and first treatment processing, as well as storage and distribution of liquid fossil fuels, comprise a large variety of activities that result in VOC emissions. Therefore, these sources will be discussed in three parts.

1. Production and First Treatment of Crude Oil

Crude oil production platforms on land or offshore are point sources of VOC, as a result of the following processes:

- Separation of crude oil, natural gas, and water
- Gas treatment or reinjection
- Produced water treatment
- Pressure relief and blowdown
- Drilling accompanied by the use of specially formulated drilling muds.

Examples of the composition of NMVOC emissions from crude oil production and natural gas venting are given in McInnes (1996) and shown here in Table 6. The contribution of methane to total VOC emissions primarily depends on the distribution or simultaneous occurrence of crude oil and natural gas in each deposit.

Measures for emission control during mining, storage, and distribution of fossil fuels focus on the replacement of gas venting by flaring and, furthermore, on the prevention of fugitive emissions from seals, valves, and flanges by improved monitoring and maintenance practices.

2. Petroleum Products Processing

The petroleum refining industry converts crude oil into a large variety of subproducts. Principal product categories include:

- Liquid fuels (e.g., motor gasoline, aviation gasoline and turbine fuel, diesel fuel, light fuel oil, and residual oil)

TABLE 6   NMVOC Species Profiles for Crude Oil Production and Natural Gas Venting[a,b]

	Crude oil production	Natural gas ventine
Ethane	6	72
Propane	19	14
Butanes	30	7
Pentanes	17	2
Hexanes	8	4
Heptanes	10	0
Octanes	7	0
Cycloparaffins	2	0
Benzene	0	0

[a]Values given in mass percentages.
[b]McInnes (1996).

- By-product fuels and feedstocks (e.g., naphtha, lubricants, asphalt, and liquified petroleum products)
- Primary petrochemicals (e.g., ethylene, propylene, butadiene, benzene, toluene, and xylene).

The first of these groups is of major interest here, while the latter are associated more with the chemical processes discussed later. Petroleum refining also includes a wide variety of processes. The four categories of general refinery processes are:

- Separation (atmospheric distillation, vacuum distillation, and light end recovery)
- Conversion (thermal cracking, catalytic cracking, coking, visbreaking, catalytic reforming, isomerization, alkylation, and polymerization)
- Treatment (hydrodesulfurization, hydrotreating, chemical sweetening, acid gas removal, and deasphalting)
- Blending (combining streams of various process units to produce specific products, such as motor gasoline and light fuel oil).

Depending on the type and quality of the crude oil and the intermediate products being processed as well as on the refined product demand, the types of processes employed in one refinery may be very different from the processes operated at another. Of the direct process emissions of VOCs without emission control, vacuum distillation (about 0.05 kg of VOC emissions per cubic meter of refinery feed), catalytic cracking (0.25 to 0.63 kg of VOC per cubic meter of feed), coking (about 0.4 kg of VOC per cubic meter of feed), chemical sweetening and asphalt blowing (about 27 kg of VOC per metric ton of asphalt) are the major sources. Furthermore, so-called fugitive emissions occur from leaks from all types of equipment and installations, such as static and dynamic sealings, valves, flanges, pumps, and compressors. Emission factors vary from 0.00023 kg/h for valves in streams of heavy liquids to 0.16 kg/h for pressure vessel relief valves for gases. Due to the variability of the processes mentioned, emissions released from two petroleum refineries might be very different in their specific amounts (e.g., emissions per metric ton of crude oil processed) and in their compositions. Further information on this topic can be found in, for example, McInnes (1996).

Measures to control VOC emissions from special process units include vapor recovery systems and flaring installations. Control of fugitive emissions involves minimizing leaks and spills through equipment changes, procedure changes, and improved monitoring and maintenance practices.

3. Storage and Distribution of Crude Oil and Petroleum Products

The storage and distribution of liquid fossil fuels are potential sources of VOC emissions. Gasoline is of major importance, as it is handled in large amounts and represents a product with a very high volatility (especially winter grade gasoline with a vapor pressure even higher than that of summer grade gasoline). The handling of crude oil is also an important source of VOC emissions, whereas the volatility of diesel fuel, light fuel oil, and other petroleum products is by far lower than that of gasoline or even crude oil.

The distribution chain for gasoline comprises product storage at petroleum refineries; loading of tanker ships, railroad cars and tanker trucks; filling of marketing depots outside the refinery; loading of tanker vehicles at a dispatch station; filling of underground storage tanks of gasoline service stations; refueling of vehicles at gasoline service stations; and finally storage of gasoline in vehicle tanks and its transfer through the fuel system to the engine. The occurrence of VOC emissions is caused by partial evaporation of the liquids and the displacement of vapor-loaded air above the liquid surface in tanks and vessels, induced by variations of (ambient) temperature and barometric pressure, or changes in the level of liquid in the tanks due to filling or bleeding procedures. Thus, the emissions of stationary sources are determined predominantly by the following parameters:

- Vapor pressure of the liquids
- Construction of the storage tanks (fixed roof tank, internal floating roof tank, external floating roof tank, and other tank types such as variable vapor space tank)
- Condition of the storage tanks (e.g., age, color, and condition of the tank coating)
- Construction of filling installations (splash loading, submerged fill pipe loading, bottom loading, and construction and condition of blocking and control valves)
- Meteorological conditions (variations of ambient air temperature and barometric pressure)
- Amount of liquid being handled.

VOC emission factors are given in McInnes (1996). For uncontrolled techniques, overall VOC emission factors for refinery dispatch stations are in the range of 0.3 kg/t of total gasoline handled, and transport and depots account for some 0.7 kg/t, whereas emission factors for service stations (refilling service station tanks and vehicle refueling) total 2.9 kg/t.

Breathing losses of gasoline vapor can also be significant from vehicle tanks. These are called diurnal losses. Additional emissions are caused by so-called hot-soak losses. These occur after parking a car with a warm or hot engine. Just after parking, the temperature in the engine space increases rapidly and so does that of the gasoline remaining in the fuel system within the engine space. Leakages and ventilation installations in the fuel system are then sources of gasoline vapor emissions. Depending on ambient air temperature and its diurnal variation and on the number of hot-soak loss events during a day, uncontrolled gasoline losses per vehicle can be more than 20 g per day.

The composition of VOC emissions from storage and distribution processes, of course, primarily depends on the type of liquid being handled and its temperature. At ambient temperatures low-molecular-weight and volatile product components are released predominantly from multicomponent mixtures. Therefore, for example, recovered gasoline vapor is characterized by a lower liquid density than that of the original gasoline. This fact can be illustrated by comparison of the data given in Table 7 with the data in Table 3. Gasoline vapor contains mainly paraffins, whereas the relative amount of paraffins in gasoline is in the range of some 45 to 55 vol%.

Measures to control evaporation of crude oil or gasoline components from storage tanks in refineries or interim reservoirs include the installation of internal floating decks on existing fixed roof tanks and an improved sealing of external floating roof tanks. Emissions at dispatch stations can be reduced by the installation of vapor balance lines or vapor recovery systems. Vapor balance lines are also an appropriate measure used to control emissions that

TABLE 7   Estimated Composition of
VOC Emissions from Gasoline
Evaporation at Ambient Air
Temperature[a,b]

Alkanes	
Propane	1.0
n-Butane	17.0
Isobutane	7.0
n-Pentane	8.0
Isopentane	28.0
Hexane	20.0
Other alkanes $> C_6$	5.0
Alkenes	
Butene	3.0
Pentene	5.0
Other alkenes $> C_5$	3.0
Aromatic hydrocarbons	
Benzene	1.0
Toluene	1.5
Xylene	0.5

[a]Values given in mass percentages.
[b]Veldt (1992).

occur during the filling of service station storage tanks and during car refueling. Diurnal losses and hot-soak losses from vehicles are controlled by onboard activated carbon canisters.

C. Production, Processing, and Distribution of Gaseous Products

During the production and first treatment of natural gas, the principal sources of VOC emissions are quite similar to those for crude oil (see Section III.B). However, the distribution of natural gas occurs via networks of pipelines and mains, including storage facilities, compressor stations, and pressure reduction stations. The pipeline and main networks range from national transmission systems with large-diameter pipelines and operated under high pressure to narrow low-pressure service mains that lead directly to the consumers. From the point of view of VOC emissions, low-pressure service main networks are of primary importance, with specific emissions above 0.4 vol% of natural gas handled (Schneider-Fresenius *et al.*, 1989). Methane is of course the predominant component of natural gas. However, some 10 wt% of total of emissions from this source are ethane and propane.

IV. SOLVENT USE

In industrialized countries, the application of solvents or solvent-containing products is one of the most important source categories. In this context the term solvent means organic solvents, but not other types of solvents such as water. Furthermore, the term solvent comprises not only substances that are able to dissolve other substances in a purely physical sense but also many other organic substances, for example, propellants for sprays or for the processing of synthetic foam, extraction agents, or refrigerants.

Large amounts of organic solvents are used within the chemical industry as reaction media. However, this type of application is usually not considered one of the more important source categories within the sector of solvent use emissions. The reason is that such reactions are carried out in self-contained installations or plants, leading to relatively low emissions compared to the high amounts of input. Furthermore, it is often not possible to distinguish clearly between solvent use emissions and other VOC emissions in this branch of the chemical industry. Some additional remarks on processes in the chemical industry therefore can be found in the next section. For the same reasons, the production of solvent-containing products is also attributed to the industrial processes section.

In contrast, the evaporation of solvents is deliberately intended from other solvent use processes, for example, in the context of application of paint, printing inks, adhesives, or cleansing agents. For these types of solvent use, evaporation losses are enhanced by the use of organic solvents with high vapor pressures and treatment or drying of materials at increased temperature. Where no extraction process is implemented, the solvents used will be released into the atmosphere more or less completely. Therefore, the solvent content of such products is the most important parameter in determining the amount of VOC emissions. Table 8 gives some examples of important types of products containing organic solvents.

Several aspects concerning Table 8 should be mentioned. One point is the incompleteness of the list. Due to the broad fields of application of VOCs, it is more or less impossible to list all products that might in principle contain VOCs. Therefore, Table 8 can give only the most relevant examples. Furthermore, for detailed estimations of solvent emissions it is inevitable that these product types be further subdivided; for example, the solvent content of solvent-based paints varies between some 30% (alkyd resin lacquer) to 80% and more (cellulose nitrate lacquer). Some product types (e.g., water-based paints and printing inks) are listed here to illustrate that so-called water-based products also contain some organic solvents. On the other hand, not all the products or auxiliary chemicals within a group listed in Table 8 may actually contain VOCs in significant amounts. Cleaning and maintenance agents, as an example of consumer goods used in households, may contain individual floor cleaning products with a mean solvent content of only 3 wt%, in contrast to chemicals for vehicle maintenance with an overall solvent content of some 50% listed in the same group of products.

The source category of solvent use is characterized not only by a large variety of solvent-containing products but also by broad fields of application in industrial production as well as in commercial and private use. The best example of a broad use of solvent-containing products is paint application that occurs, to a larger or smaller extent, in almost all branches of industry, as well as in some specific commercial branches and in the do-it-yourself sector. Large amounts of paints and thinners are applied in the following industrial branches:

- Shipbuilding
- Manufacturing of nonferrous metals (coil coating)
- Manufacturing of wood products
- Manufacturing of automobiles
- Mechanical engineering

TABLE 8   Selected Types of Products Containing Organic Solvents

Paints and thinners	Solvent-based paints Water-based paints[a] Wood stain Paint thinner
Printing inks and auxiliary agents	Solvent-based inks Water-based inks[a] Printing ink thinner Offset moistening agents Cleansing agents
Adhesives	Solvent-based adhesives Dispersion adhesives[a]
Surface treatment agents	Metal degreasing agents Other degreasing and technical cleansing agents Paint remover Chemicals for transport preservation Dewaxing agents Chemicals for dry cleaning of textiles
Consumer goods for private use	Personal hygiene products Products for cleaning and maintenance in house-holds Soaps and washing agents
Building protective agents	Wood protective agents Impregnant matter Other building protective agents
Refrigerants and cold protectives	Refrigerant for cooling devices and air conditioners Freezing preventives for windshield washing water Agents for aircraft defrosting
Other auxiliary agents	Plastics manufacturing Rubber manufacturing Textile finishing Production of foundry goods
Other chemicals and products	Extraction agents for production of edible fat and oil Fire extinguishing substances

[a]Important products with a low solvent content.

- Manufacturing of ironwork, sheet metals, and hardware
- Manufacturing of railroad cars
- Manufacturing of synthetic material.

Important paint consumers are also those in the car repair and building trades (painting and lacquering). The application of other types of solvent-containing products is limited to a smaller number of industrial and commercial branches. For example, chemicals for degreasing are mainly applied in some branches of metalworking industries. Furthermore, the application of printing inks and auxiliary chemicals occurs in the printing industry and, to a minor extent, in the manufacturing of paper goods. Finally, some types of solvent use operations can be assigned to only one specific branch, for example, the use of special cleaning agents in commercial dry cleaning.

As shown in the previous subsections, the handling and combustion of fossil fuels predominantly cause emissions of saturated, unsaturated, and aromatic hydrocarbons, while the relative amount of other classes of organic substances (e.g., aldehydes and ketones) is rather small. However, the VOC emissions from solvent use operations are composed of a variety of species, which are larger by far than those from the use of fossil fuels. In addition to paraffins and aromatic hydrocarbons, other groups of substances, such as alcohols, ketones, esters, and halogenated hydrocarbons, are important solvents. Examples of mean compositions of some solvent use emissions are given in Table 9.

According to Table 9, the highest contribution of pure hydrocarbons to total solvent emissions comes from adhesives, with a share around 50%. In printing inks and auxiliary chemicals, as well as in consumer goods for cleaning and maintenance in households, the contribution of pure hydrocarbons totals about 40%. Finally, industrial paints and varnishes as well as consumer goods for personal hygiene predominantly contain solvents other than pure hydrocarbons. In other words, solvents are important sources of organic compounds other than pure hydrocarbons.

It has been outlined previously that solvent emissions are attributable to a very large number and variety of sources, whereby a single source often contributes only a negligible amount to overall solvent emissions. Therefore, it is obvious that solvent emission control measures must focus on primary measures instead of waste gas treatment. Appropriate primary measures include:

- An intensified use of solvent-reduced or solvent-free products (e.g., water-based varnishes)
- An intensified use of organic solvents with high boiling point and elevated vapor pressure (e.g., high-boiling-point oils used for cleaning operations in the printing industry)
- Process modifications leading to reduced solvent use (e.g., eliminating interim cleaning and degreasing steps in metalworking industries and increasing the efficiency of paint applications by using improved application technologies).

The implementation of waste gas treatment procedures based on techniques such as absorption, adsorption, condensation, thermal incineration,

TABLE 9   Examples of the Composition of Solvent Emissions for Various Types of Products[a]

Product type	Industrial varnishes[b]	Printing ink and auxiliary agents[b]	Adhesives[b]	Cleaning and maintenance agents[b,c]	Personal hygiene agents[b,d]
Propane					2.4
Butane					20.0
Pentane					2.0
Special benzine	2.7	3.1	27.5	8.7	
White spirit	3.2	23.6	2.1	17.0	
Terpenes	0.1			6.0	
High-boiling hydrocarbons		0.2		3.2	
Toluene	2.5	7.9	20.1	1.3	
Xylene	5.7	4.1	2.0	0.8	
Solvent naphtha (light)	6.9	0.9	0.2	0.4	
Solvent naphtha (heavy)	2.1	0.2		0.2	
Methanol	1.0	1.9		2.4	
Ethanol	0.4		0.5	8.7	44.4
Isopropanol	4.2	22.3	1.7	40.8	4.9
Butanol	17.9				
Pentanol	0.1				
Benzyl alcohol	0.3				
Tetrahydrofuran			0.3		
Ethylene glycol				5.2	
Propylene glycol					5.9
Ethyl glycol	10.3			0.2	
Dimethyl ether					18.2
Methoxypropanol	6.0				
Acetone	4.9	1.2	5.2		2.2
Methyl ethyl ketone	3.1	13.1	9.7		
Methyl isobutyl ketone	1.2		1.7	0.2	
Cyclohexanone	0.4				
Diacetone alcohol	1.1				
Trimethylcyclohexanone	1.2				
Methyl acetate	1.5	4.0	3.5		
Ethyl acetate	6.5	16.6	17.5	0.3	
Propyl acetate	0.3	0.9	0.9		
Butyl acetate	15.1		0.2		
1,1,1-Trichloroethane	0.5		5.2	4.6	
Trichloroethene	0.3		1.7		
Tetrachloroethene	0.5				

[a]Values given in mass percentages.
[b]Bräutigam and Kruse (1992).
[c]IPP (1992).
[d]IKW (1992).

catalytic incineration, and biofiltering—under usual technical and economic conditions—is restricted to large plants with high solvent emissions. This includes plants that have high solvent concentrations in the waste gas and many hours of operation per year or even continuous operation.

V. INDUSTRIAL PROCESSES

A number of important processes and source types, which could be allocated also to industrial processes in principle, have been mentioned in the previous sections. This is the case for processes in mineral oil refineries as well as for many solvent use operations. However, a variety of other industrial processes cause greater or lesser emissions of VOCs. The most important processes are those related to the production of:

- Organic chemicals and solvent-containing products (e.g., ethylene, pro-pylene, vinyl chloride, and styrene, corresponding polymer products, and paints and adhesives)
- Chipboard
- Food (e.g., bread) and beverages (e.g., wine, beer, and spirits).

The production of organic chemicals causes emissions of primarily organic solvents, monomers of synthetic material, softening agents for synthetic materials, and foaming agents for the production of synthetic foams. Solvent losses connected with the production of solvent-containing paints and adhesives are usually in the range of up to 5% of the solvent input. Thus, the emissions during production are far lower than the emissions caused by the application of such products. An example of the composition of VOC emissions from a group of plants in the chemical industry located in the German Federal State of Baden–Württemberg is given in Table 10. Among the different subbranches of the chemical industry in Baden–Württemberg, the highest share of emissions comes from the production of paints and adhesives.

VOC emissions from the production of chipboard are released during drying, bonding, and pressing of the wood shavings. The process emissions consist mainly of terpenes and aldehydes (especially formaldehyde). Fermentation processes are mainly responsible for VOC emissions in the production of bread (based on yeast dough) and alcoholic beverages (beer, wine, and spirits). Ethanol is the major component released.

Measures for emission control primarily focus on production processes in the chemical industry and include process integrated measures, process modifications, substitution of input substances, and waste gas treatment.

TABLE 10   Composition of VOC Emissions from the Chemical Industry in the German
Federal State of Baden–Württemberg[a,b]

Pentane	3.0	Other ethers	0.6
Special benzine	20.4	Acetone	10.7
White spirit	8.3	Methyl ethyl ketone	0.7
Ethyne	1.3	Methyl isobutyl ketone	1.5
Toluene	4.2	Ethyl acetate	3.2
Xylene	3.1	Butyl acetate	6.2
Solvent naphtha (light)	0.6	Trichloromethane	2.4
Other hydrocarbons	1.5	Trichloroethene	0.5
Methanol	5.8	Fluorotrichloromethane	2.5
Ethanol	4.2	Other VOC compounds	0.6
Isopropanol	2.0	Identified VOCs	14.8
Tetrahydrofuran	1.9		

[a]Values given in mass percentages.
[b]Obermeier (1995).

VI. BIOLOGICAL PROCESSES

Biological processes discussed here do not include those responsible for biogenic emissions of VOCs from plants, which are the topics of Chapters 2 and 3, but are bacterial processes that synthesize methane, where the extent of these processes and of the amount of methane and other VOCs emitted is determined by human action. Large amounts of methane are released from:

* Flooded rice fields
* Livestock husbandry (including enteric fermentation processes, especially from ruminants, and anaerobic decomposition of animal wastes)
* Disposal of domestic and municipal wastes.

Methane emissions from enteric fermentation processes can be controlled to a certain extent by high-quality feed. Details are described in Leng (1991). Emissions due to anaerobic decomposition can be controlled by frequently removing animal wastes from manure storages and thus avoiding anaerobic conditions. On the other hand, it may also be the aim to create favorable conditions for methane production (e.g., in a biogas facility) and to collect and use the biogas as an energy source.

The decomposition of organic materials at waste disposal sites occurs partly under aerobic and partly under anaerobic conditions. Therefore, the gases formed during decomposition are mainly carbon dioxide and methane. The decomposition process occurs over 10 to 20 years. Measures to control methane emissions focus on good coverage of the waste disposal sites combined with the implementation of gas sampling systems, followed by collection and use of the gas as an energy source or by gas flaring.

VII. MODELING ANTHROPOGENIC VOLATILE ORGANIC COMPOUND EMISSIONS

Due to the large variety of quite different source types and categories of anthropogenic VOC emissions, it is not possible to describe here all aspects and possibilities of emission modeling in detail. However, some of the principal methods and terms will be outlined.

A. REQUIREMENTS FOR EMISSION DATA

The results of emission-inventorying activities vary according to the requirements that arise from the further use of the emission data. To get a rough idea of the more important source categories of VOCs or other air pollutants within a selected region it might be sufficient to calculate annual emission data per source category (traffic, industry, trade, households, etc.). The more accurate the emission data should be, the more it is necessary to distinguish between different source types and single processes (e.g., by subdividing the road traffic sector into the operation of various vehicle types on different road types under specific driving modes). In the case of such emission data being used for regional air quality planning purposes, it might be helpful to subdivide the inventory area into smaller territorial units or grids. This allows local differences in emission patterns to be taken into account as well. In addition, a detailed analysis of possibilities for further air pollution control requires not only source specific emission data itself but also additional information on process technologies, operation modes, and other data for a complete characterization of the sources. Finally, an important field of application of emission data for VOCs and other ozone precursors (namely, NO_x) is to use the emission data as inputs for modeling ambient air concentrations of ozone.

Due to the nonlinear nature of ozone formation processes, it is essential to disaggregate annual emission data into values with small time steps. Usually, such atmospheric models require hourly gridded emission data, including additional information such as source heights above the ground surface.

B. SOURCE CLASSIFICATION

During preparation of an emission inventory for a selected region with a huge number of single sources, it is obviously not possible to base emission calculations on measurements at each single source. Therefore, emission measurements are concentrated on a representative selection of single sources for each source type. The outcome of these measurements is then used to provide so-called emission factors, which give the amount of emissions per unit of emission relevant processes or activities. This means that the unit of an emission factor is given, for example, in grams of emission per kilometer driven by a vehicle or per terajoule of fuel consumption in a combustion installation. Thus, the method for calculating emissions from a source is to multiply the total activity rate of the source type (e.g., annual mileage of a fleet of passenger cars) with the corresponding emission factor.

An important point in such calculations is the degree of detail to which sources (with larger or smaller differences in emission characteristics) can be allocated to separate source categories. The more a source type can be subdivided into different categories, the higher the quality or accuracy that might be expected for the calculated emission data. However, it is obvious that this degree of detail is restricted by the structure and degree of detail of available input data for inventorying purposes, including emission factors and activity data. To be flexible concerning different structures of input data for emission modeling it is advantageous to define a hierarchical structure of source categories. Such a hierarchical structure has been defined within the CORINAIR (COoRdination d'Information Environmentale AIR emissions inventory) project and has been used for inventorying air pollutant emissions in 29 European countries (McInnes, 1996). This structure is defined in three levels of emission categories, the so-called Selected Nomenclature for Air Pollution (SNAP) levels. Table 11 gives an overview of SNAP level 1.

The source categories of SNAP level 1 are further subdivided within levels 2 and 3 into more specific source types. Examples of this further subdivision are the distinction between:

- Different types of industrial combustion (boilers, process furnaces with and without contact) and between different ranges of thermal power

TABLE 11   Major Source Categories of the CORINAIR 1990
Inventory (SNAP Level 1)

SNAP	Description
1	Public power, cogeneration, and district heating
2	Commercial, institutional, and residential combustion
3	Industrial combustion
4	Industrial processes
5	Extraction and distribution of fossil fuels
6	Solvent use
7	Road transport
8	Other mobile sources and machinery
9	Waste treatment and disposal activities
10	Agricultural activities

capacities in the categories of power plants, district heating plants, and industrial combustion installations

- Production of different metals, glasses, and other materials in the category of industrial combustion processes with contact
- Different processes in petroleum, iron and steel, nonferrous metal, chemical (e.g., production of a variety of inorganic and organic chemicals, and synthetic materials), wood, paper, and some other industries
- Major types of solvent use and related activities, such as paint application, metal degreasing, and chemical products manufacturing (including a variety of different chemical products)
- Different types of vehicles in the road transport sector, distinguishing between highway, rural, and urban driving modes.

With respect to the aim of calculating very detailed and accurate emission data, it is advisable to strive for an even more detailed structure of input data than is given by the source categorization of SNAP level 3. This means, for example, that, the calculation of road traffic emissions should also comprise a detailed distinction of engine types (e.g., gasoline or diesel engine and cylinder capacity classes) and emissions control technologies (e.g., without catalysts, uncontrolled catalysts, and controlled catalysts).

C. REGIONALIZATION

Another important factor in emission inventorying procedures is to achieve a high degree of accuracy in spatially resolved emission data. There are three distinct ways to do this. First, stationary sources with comparably high emission rates are treated preferably as so-called point sources. This term does not inevitably mean that emissions are released only from one single point (e.g., from one stack). It means instead that emissions are determined specifically for a selected plant or facility by using plant specific activity data and emissions factors or directly by emission measurements. Emission data can then be allocated exactly to the location of the plant or facility.

Furthermore, some source types can be treated as so-called line sources. The most important example of such line sources is road traffic on highways and other relevant road types. Based on knowledge of the location of each road section, the average traffic density on each road section, representative vehicle speed distributions for each road type, and additional information on the composition of the vehicle fleet, detailed emission calculations can be performed for each road section.

Finally, the third category comprises area sources. An area source represents a collection of single sources of the same kind within a specified area (e.g., within a municipality). This means that the location of each single source of such collections cannot be determined exactly, as this would require disproportionate effort. Usually, road traffic inside settlements, smaller combustion installations, and all other source types (except point and line sources) are regarded as area sources. Anthropogenic emissions of area sources are calculated at the level of territorial units (e.g., communities, towns, or municipalities) by using statistical data such as the number of employees or inhabitants, on the one hand, and data such as solvent use per employee or inhabitant, on the other hand. An even more detailed spatial allocation of area source emissions within a territorial unit can be reached with the help of gridded land use data, which give additional information on the location of, for example, settled areas.

D. Temporal Resolution

It is obvious that annual average emissions are not distributed equally over the year. On the one hand, there are often seasonal variations in emission rates. For example, gasoline evaporation from vehicles occurs primarily during the summer due to high ambient air temperatures, while emissions from combustion installations are significantly higher during winter than during summer. On the other hand, there are also variations in daily emissions over a week. Industrial production is normally concentrated on the working days from Monday to Friday. Thus, emissions related to industrial production occur mainly during the working week. Finally, there are large differences in the amount of emissions during daytime and during nighttime. In most cases emission-relevant activities are significantly lower during the night, with some residual road traffic activity and some large industrial plant activity occurring at night.

With respect to these facts, the generation of hourly emission data derived from annual values is based on temporal allocation factors such as:

- The annual pattern of activities (monthly share of annual totals)
- The weekly pattern of activities (deviation between different day types from average daily values)
- The daily pattern of activities (hourly share of daily totals, distinguishing different day types).

Examples of these temporal patterns are monthly statistics on industrial production rates, daily variations of vehicle mileages derived from continuous traffic countings, or commercial working time regulations. Another important parameter is the pattern of ambient air temperature (hourly values), not only influencing activity rates such as the operation of space heating systems but also determining emissions factors (e.g., for vehicle cold starts or gasoline evaporation from vehicles). Further details on the temporal disaggregation of emission data are discussed by Ebel *et al.* (1997).

VIII. EMISSION DATA

The last section of this chapter focuses on examples of available data on anthropogenic emissions of methane and NMVOCs on the global, European, and national–regional scales. In addition to data on annual emissions per source category, the spatial distribution, the temporal resolution, and the composition of NMVOC emissions will be addressed.

Estimations of total anthropogenic NMVOC emissions on the global scale vary over a wide range. FCI (1987) suggests annual emissions of 100 million tons, whereas Cullis and Hirschler (1989) estimate VOC emissions around 235 million tons per year. Concerning the contribution of important source categories, the former apportions 25 to 40% to road traffic and 8 to 16% to solvent use operations. Global emissions of methane from man-made sources are estimated to be even higher than the emissions of NMVOCs, as shown in Table 12.

Although uncertainties in the preceding emission data are significant, rice fields are considered the most important source of anthropogenic methane emissions on the global scale. The second largest source category is also closely attributed to food production, namely, the keeping of ruminants. However, the production, treatment, and distribution of fossil fuels must not be neglected in calculating total emissions of methane.

More detailed emission data for NMVOC and methane are available on the European scale (McInnes, 1996). The CORINAIR database comprises emission data for 29 European countries.[1] Table 13 clearly indicates that road transport and solvent use are by far the most important source categories of anthropogenic NMVOC emissions in Europe. About 67% of total emissions are released from these sectors. Emissions from road transport are dominated by exhaust gases and gasoline evaporation from passenger cars. Furthermore, Table 13 shows that paints and varnishes are the most important type of products whose application is connected with the release of VOCs into the atmosphere.

When the methane emissions in Europe are examined, there are only three sectors with significant emissions at all. Agricultural activities—mainly animal breeding—lead the ranking list of these three sectors with almost 15

[1]Austria, Belgium, Bulgaria, Croatia, Czech Republic, Denmark, Estonia, Finland, France, Germany, Greece, Hungary, Ireland, Italy, Latvia, Lithuania, Luxembourg, Malta, Netherlands, Norway, Poland, Portugal, Romania, Slovakia, Slovenia, Spain, Sweden, Switzerland, United Kingdom.

TABLE 12   Global Anthropogenic Emissions of Methane According to Enquete-Kommission in 1990[a]

	Best estimate	Uncertainty range
Rice fields (wet rice)	130	70–170
Enteric fermentation of ruminants	75	70–80
Biomass burning	40	20–80
Waste disposal	40	20–60
Natural gas production and distribution	30	10–50
Coal mining	35	10–80
Total anthropogenic emissions	350	200–520

[a]Values given in million tons per year.

TABLE 13   Emissions of NMVOC and Methane in 29 European Countries according to the CORINAIR 1990 Inventory[a,b]

SNAP		NMVOC	CH$_4$
1	**Public power, cogeneration, and district heating**	55	43
101	Public power and cogeneration plants	45	36
102	District heating plants	8	7
2	**Commercial, institutional, and residential combustion**	989	619
3	**Industrial combustion**	154	92
301	Industrial combustion in boilers, gas turbines, and stationary engines	52	39
302	Process furnaces without contact	46	10
303	Industrial combustion—processes with contact	54	42
4	**Industrial processes**	1220	76
401	Petroleum industries	249	9
402	Iron and steel industries and collieries	58	50
403	Nonferrous metal industry	3	1
404	Inorganic chemical industry	117	7
405	Organic chemical industry	401	4
406	Wood, paper, pulp, food, and drink industries	393	2
407	Cooling plants	0	0
5	**Extraction and distribution of fossil fuels**	1376	10,408
501	Extraction and first treatment of solid fuels	3	7,505
502	Extraction, first treatment, and loading of liquid fuels	348	133
503	Extraction, first treatment, and loading of gaseous fuels	136	71
504	Liquid fuel distribution (except gasoline)	171	1
505	Gasoline distribution	597	0
506	Gas distribution networks	121	2,698
6	**Solvent use**	4920	0
601	Paint application	1924	0
602	Degreasing and dry cleaning	525	0
603	Chemical products manufacturing	673	0
604	Other solvent use and related activities	1755	0

(continues)

TABLE 13 *(continued)*

7	**Road transport**	6756	200
701	Passenger cars	3457	150
702	Light-duty vehicles < 3.5 t	383	12
703	Heavy-duty vehicles > 3.5 t and buses	665	21
704	Mopeds and motorcycles < 50 cm³	318	6
705	Mopeds and motorcycles > 50 cm³	383	13
706	Gasoline evaporation from vehicles	1550	0
8	**Other mobile sources and machinery**	677	25
801	Off-road vehicles	417	14
802	Railways	33	1
803	Inland waterways	29	0
804	Marine activities	125	7
805	Airports [landing and takeoff (LTO) cycles]	71	4
9	**Waste treatment and disposal activities**	507	8,752
901	Wastewater treatment	32	211
902	Waste incineration	10	10
903	Sludge spreading	16	155
904	Landfills	45	7,932
905	Compost production from waste	0	27
906	Biogas production	0	40
907	Open burning of agricultural wastes (except stubble burning)	401	358
908	Latrines	0	18
10	**Agricultural activities**	759	14,793
1001	Cultures with fertilizers (except animal manure)	267	1,017
1002	Cultures without fertilizers	77	134
1003	Stubble burning	34	35
1004	Animal breeding (enteric fermentation)	0	9,385
1005	Animal breeding (excretions)	380	4,220
	Total anthropogenic emissions	17414	35,009

[a]Values given in 1000 tons per year.
[b]McInnes (1996).

million tons of methane annually. More than 10 million tons of methane are released during the extraction and distribution of fossil fuels. In this context it is remarkable that the highest share of methane emissions within this sector is not attributed to natural gas distribution networks but to the extraction and first treatment of coal. The third sector to be mentioned here is waste treatment and disposal, with methane emissions of almost 9 million tons per year. The major cause of emissions in this sector is the anaerobic decomposition of organic wastes in landfills.

Due to the unequal distribution of emissions of NMVOC and methane among the source categories shown in Table 13, it is obvious that the overall emissions of these substances are not distributed equally over Europe. Figures 4 and 5 illustrate emission rates of NMVOC and methane for the different countries, distinguishing between major source categories as outlined earlier. Figure 4 indicates that more than 50% of the total NMVOC emissions included in the CORINAIR database are allocated to only four industrialized countries in Europe, namely, Germany, United Kingdom, France, and Italy. In many countries road transport and solvent use must to be considered the major source categories. This is the case especially in Germany, where all other sectors besides road traffic and solvent use are of minor importance.

With respect to methane, the list of countries with high emissions is again led by Germany, closely followed by Poland, where the contribution from coal mining is by far above average (see Fig. 5). Again Italy, United Kingdom, and France are located in the upper part of this list, even if the contributions of single source categories to national totals show different patterns.

Examples of the spatial distribution of NMVOC emissions in Europe and parts of Europe are shown in Fig. 6. Such gridded emission data are required especially for air quality modeling purposes. The first map at the top left-hand

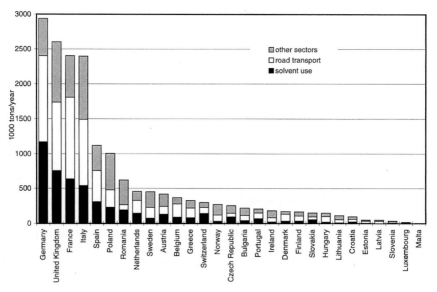

FIGURE 4   Annual NMVOC emissions in European countries according to CORINAIR 1990 (McInnes, 1996).

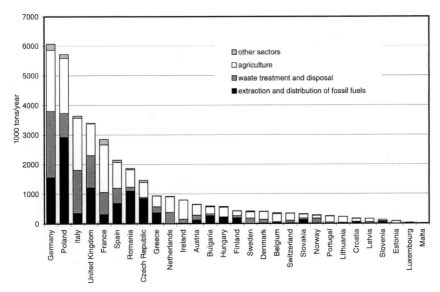

FIGURE 5   Annual CH$_4$ emissions in European countries according to CORINAIR 1990 (Mc-Innes, 1996).

side illustrates the distribution of annual NMVOC emissions—given in tons per square kilometer—for grids of 54 × 54 km. Areas with high emission rates can be found predominantly in middle-European countries. A higher spatial resolution of 18 × 18 km is given at the top right-hand side of Fig. 6, focusing on Germany and some parts of its neighboring countries. This diagram shows quite large regional differences in the spatial resolution of the emission data. Whereas emission data for Switzerland are known in this input database (derived from the CORINAIR 1990 inventory) only as a national total, the structure or pattern of emission data for the New Federal States of Germany (derived from BMBF, 1997) is much more detailed. The third map at the bottom right-hand side provides a grid resolution of 6 × 6 km for the area of the new Federal States of Germany. Large urban areas such as the city of Berlin can be easily detected at this level of spatial resolution, but also smaller cities (e.g., Dresden and Leipzig with about 0.5 million inhabitants in the southern parts of the inventory area) are locatable. Finally, a very fine grid resolution of 2 × 2 km even gives information on the emissions along single sections of highways and other roads with high traffic density, as shown for the surroundings of Berlin at the bottom left-hand side of Fig. 6. This last diagram clearly indicates that NMVOC emissions are released mainly within densely populated and industrialized areas, whereas, for example, road traffic

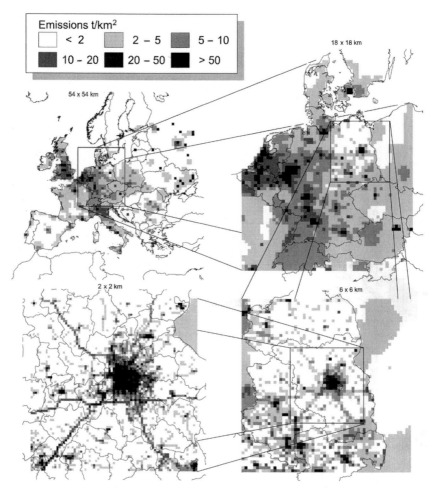

FIGURE 6   Examples of NMVOC emission inventories for different regional scales and spatial resolutions.

outside of settlements does not play the same important role as it does in the case of NO$_x$ emissions.

The emission data discussed earlier represent annual values only. Figure 7 illustrates the variations in hourly NMVOC emissions from road traffic and combustion installations for the German Federal State of Baden—Württemberg during a one-week period in September 1992, distinguishing between selected source types. Emissions from combustion installations do not show large differences between working days and weekend days. Furthermore,

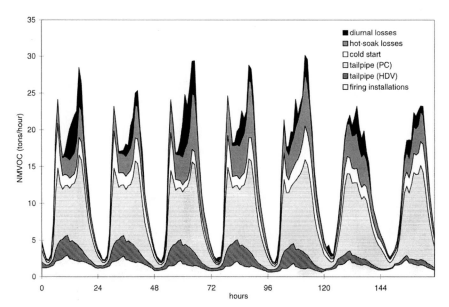

FIGURE 7   Hourly emissions of NMVOC from combustion processes and gasoline evaporation from vehicles in the German Federal State of Baden–Württemberg during a week from Monday to Sunday; values given in tons per hour.

daytime and nighttime emissions are at least of the same order of magnitude. Exhaust gas emissions from heavy-duty vehicles occur mainly during the daytime of working days, whereas emissions during the night and on Saturdays and Sundays are significantly lower. Exhaust gas emissions from passenger cars, cold start emissions, and hot-soak losses of gasoline each indicate maximum values in the morning hours and during the afternoon of working days (rush hour) as well as on Saturday morning and Sunday afternoon. Diurnal losses of gasoline from vehicles are rather different from day to day, due to a strong dependency on diurnal variations in ambient air temperature. Looking at the overall emissions, the maximum values during the afternoons of working days are about 10 times higher than the minimum values at night. Furthermore, daily emissions on weekend days are some 30% lower than on working days.

Changes in the regional distribution patterns of NMVOC emissions during selected hours of a day are shown in Fig. 8 for the area of the five New Federal States of Germany. In most parts of the region, emissions are rather low at 3 AM. Emission rates above average can be found in Berlin and some cities in the southern part of the area. Also, the locations of some large plants are indicated. These plants are operated day and night. At 6 AM, the emission

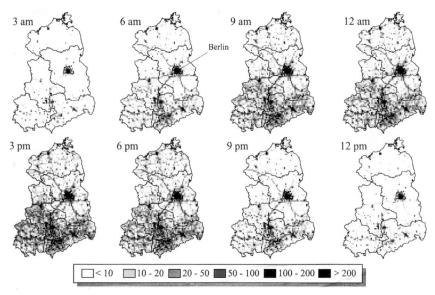

FIGURE 8   Emissions of NMVOC in the New Federal States of Germany during selected hours of a day; grid size 5 × 5 km, values given in metric tons per hour and per grid cell.

situation is determined by road traffic, leading to high emission rates within densely populated areas and also at highway intersections. Then at 9 AM road traffic emissions are still rather high, but now additional emissions from solvent use and other processes arise. These sources lead to increased emissions also in more rural area. Until 6 PM, these regional emission patterns remain more or less unchanged. Finally, until the end of the day, emissions decrease in a similar way as they increased during the early morning.

In the preceding sections some aspects of the composition of NMVOC emissions have been discussed. As a result it has been stated that processes of handling and usage of fossil fuels primarily lead to emissions of pure hydrocarbons, whereas emissions due to solvent use and other production processes mainly consist of oxygenated hydrocarbons. Figure 9 gives an example of the composition of anthropogenic NMVOC emissions (annual average) in the German Federal State of Baden–Württemberg in 1990.

Nonmethane alkanes contribute 37% to total NMVOC emissions, road traffic being the major source type. However, significant amounts of nonmethane alkanes are released due to the use of special benzene and white spirit. Aromatic hydrocarbons are second in importance with a share of more than 14% of total emissions. Again, these compounds are emitted by both main sectors, fuel consumption and solvent use. Emissions of unsaturated hydro-

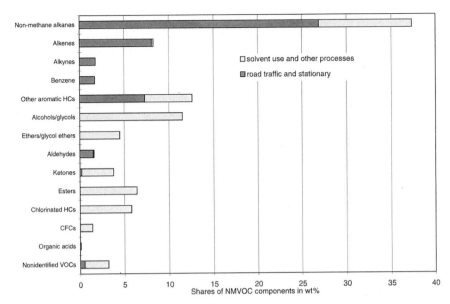

FIGURE 9   Mean composition of anthropogenic NMVOC emissions in the German Federal State of Baden–Württemberg in 1990.

carbons as well as emissions of aldehydes are attributed almost completely to combustion processes. On the other hand, all oxygenated and halogenated hydrocarbons, except aldehydes, are emitted predominantly by solvent use operations and other production processes. The ranking list is led by alcohols–glycols with a share of more than 11% of anthropogenic NMVOC emissions, followed by esters and chlorinated hydrocarbons with shares of about 6% each. In the end, about 3% of the anthropogenic NMVOC emissions remain unidentified in detail.

IX. CONCLUSIONS

The preparation of detailed databases and inventories of anthropogenic VOC emissions constitutes a rather complex topic of environmental research activities. This is due to a large variety of different source categories, source-specific parameters controlling the emissions, and source-specific profiles describing the composition of VOC emissions. All these aspects must be taken into account within emission-inventorying procedures. When such emission data are used, for identification of major emission sources and detection of

desirable measures of air pollution control, and also are applied as input data
for regional air quality models, the emission data must meet additional requirements. These requirements arise from the dynamic processes of atmospheric transport and chemical transformation, and they comprise a high
temporal and spatial resolution of emission data.

The preceding expositions indicate that on the global scale wet rice fields
and the enteric fermentation of ruminants are the most important sources of
anthropogenic emissions of methane. With respect to the European scale,
road transport and use of organic solvents are major sources of anthropogenic
NMVOC emissions, especially in highly industrialized and densely populated
regions. Whereas NMVOC emissions from combustion processes contain predominantly pure hydrocarbons (alkanes, alkenes, and aromatics), organic
solvents and their vapors also consist of oxygenated hydrocarbons such as
alcohols, ketones, and esters. Technical measures for the abatement of VOC
emissions are, in principle, available for a wide range of relevant sources,
although the implementation of such measures (e.g., three-way catalysts,
active carbon canisters for gasoline-powered vehicles, and extended use of
products with a reduced solvent content) is still not complete. Finally, it can
be stated that VOC emissions show large temporal and regional variations.
This underlines the necessity of preparing high-resolution emission inventories.

REFERENCES

BMBF (1997). Federal Ministry of Education, Science, Research and Technology, (ed.), Wissen-
    schaftliches Begleitprogramm zur Sanierung der Atmosphäre über den neuen Bundeslän-
    dern—Abschlussbericht Band 1. Bonn, Germany
Bräutigam, M., and Kruse, D. (1992). Ermittlung der Emissionen organischer Lösemittel in der
    Bundesrepublik Deutschland. Federal Environmental Agency of Germany (ed.), Berlin, Re-
    search Report No. 104 04 116/01.
Cullis, C. F., and Hirschler, M. M. (1989). Man's emission of carbon monoxide and hydrocarbons
    into the atmosphere. Atmos. Environ. 23, 1195–1203.
Ebel et al. (1997). Ebel, A., Friedrich, R., and Rodhe, H. (eds.) "Tropospheric modelling and
    emission estimation," Springer-Berlin, Heidelberg.
ENQUETE (1990). Enquete-Kommission 'Vorsorge zum Schutz der Erdatmosphäre' des 11.
    Deutschen Bundestages (ed.), Dritter Bericht der Enquete-Kommission 'Vorsorge zum Schutz
    der Erdatmosphäre'. Deutscher Bundestag, 11. Wahlperiode, Drucksache 11/8030, Bonn, Ger-
    many.
FCI (1987). Fonds der Chemischen Industrie (ed.), Folienserie 22—Umweltbereich Luft. Frank-
    furt/Main, Germany.
IKW (1992). Industrieverband Körperpflege-und Waschmittel e.V., Frankfurt/Main, Germany
    (private communications).
IPP (1992). Industrieverband Putz- und Pflegemittel e.V., Frankfurt/Main, Germany (private
    communications).
John, C. (1997). Emissionen von Luftverunreinigungen aus dem Strassenverkehr in hoher zeit-

licher und räumlicher Auflösung. Institute of Energy Economics and the Rational Use of Energy, University of Stuttgart, Germany (to be published).

McInnes, G. (ed.) (1996). Joint EMEP/CORINAIR atmospheric emission inventory guidebook, 1st Ed. European Environmental Agency, Copenhagen, Denmark.

Leng, R. A. (1991). Improving ruminant production and reducing methane emissions from ruminants by strategic supplementation. U.S. Environmental Protection Agency, Office of Air and Radiation, EPA/400/1-91/004.

Obermeier, A. (1995). Ermittlung und Analyse von Emissionen flüchtiger organischer Verbindungen in Baden–Württemberg. Institute of Energy Economics and the Rational Use of Energy, University of Stuttgart, Germany, Research Report No. 19.

Patyk, A., and Höpfner, U. (1995). Komponentendifferenzierung der Kohlenwasserstoff-Emissionen von Kfz. ifeu Institut für Energie und Umweltforschung, Heidelberg, Germany.

Schneider-Fresenius et al. (1989). Schneider-Fresenius, W., Hintz, R.-A., Klöpfer, W., and Wittekind, J. Ermittlung der Methan-Freisetzung durch Stoffverluste bei der Erdgasversorgung der Bundesrepublik Deutschland. Batelle-Institut e.V., Frankfurt/Main, Germany.

Seier, J. (1997). Luftschadstoffemissionen aus Feuerungsanlagen. Institute of Energy Economics and the Rational Use of Energy, University of Stuttgart, Germany (to be published).

Staehelin et al. (1997). Staehelin, J., Keller, C., Stahel, W., Schläpfer, K., Bürgin, T., and Schneider, S. Modelling emission factors of road traffic from a tunnel study. Environmetrics (submitted).

UBA (1995). Umweltbundesamt (UBA), Berlin, Bundesamt für Umwelt, Wald und Landschaft (BUWAL), Bern, INFRAS AG, Bern (ed.), Handbuch für Emissionsfaktoren des Strassenverkehrs—Version 1.1. Published as software on CD-ROM, Berlin, Germany.

Veldt, C. (1992). Updating and upgrading the PHOXA emission data base to 1990. TNO Institute of Environmental and Energy Technology, Apeldoorn, Netherlands.

CHAPTER 2

Biogenic Emissions of Volatile Organic Compounds from Higher Plants

RAY FALL

Department of Chemistry and Biochemistry, and Cooperative Institute for Research in Environmental Sciences, University of Colorado, Boulder, Colorado

---

The biosphere, especially vegetation, releases a complex mixture of volatile organic compounds (VOCs) into the atmosphere.

Some of these biogenic VOCs are emitted in surprisingly large amounts and have high enough chemical reactivity to significantly affect the chemistry of the atmosphere. In this chapter the major biogenic VOCs emitted by higher plants and the mechanisms of their formation are described. In addition, the regulation and roles of formation of VOCs and their emission from plant surfaces are described. The details of these biochemical and biological processes are complex, comparable in complexity to the photochemistry of VOCs in the atmosphere, but need to be appreciated to accurately model regional and global biogenic VOC emissions.

I. INTRODUCTION

A. THE BIOSPHERE EXCHANGES CHEMICALS WITH THE ATMOSPHERE AND IMPACTS ATMOSPHERIC CHEMISTRY

Most living systems directly or indirectly exchange volatile chemical compounds with the atmosphere. Over biological time the major gases exchanged with living organisms have included carbon dioxide, with production by respiring organisms and uptake by photosynthetic and chemoautotrophic organisms; oxygen (O_2), with production by photosynthetic organisms and uptake by aerobic respiration; and nitrogen (N_2), which is produced by denitrifying bacteria and taken up by nitrogen-fixing bacteria. These exchanges have altered the chemical composition and reactivity of the atmosphere (Warneck, 1993). Especially significant is the rise of photosynthetic organisms that have produced an oxygen-rich atmosphere. This in turn has resulted in the formation of ozone and OH radical, reactive species that have great consequence for the chemistry of the atmosphere (Chapter 7).

Living organisms also exchange *trace* gases with the atmosphere. Interest in these compounds stems from the discovery in the 1950s that photochemical smog in the Los Angeles area was formed from the reactions of hydrocarbons and nitrogen oxides (Haagen-Smit, 1952). Subsequently, Went (1960) suggested that the "blue hazes" that form in the summertime in forested areas, such as the Blue Mountains (Australia) or Smoky Mountains (United States), might similarly result from photochemical reactions of volatile organic substances and isoprenoids emitted by vegetation. An important outcome of these investigations was the idea that emissions of trace amounts of biogenic volatile organic compounds (VOCs) might contribute to ozone formation in the atmosphere. The photochemistry of tropospheric ozone formation is discussed in detail in Chapter 7. Ozone is a reactive oxidant that is damaging to plants and animals, and its measurement and control are major goals in the United States and in the European Union. Despite continuing concern about anthropogenic emissions of VOCs, it is now also clear that in some cases trace emissions of biogenic VOCs are so large that they contribute to ozone episodes. For example, the oak forests surrounding metropolitan areas of the southeastern United States, such as Atlanta, contribute the major fraction of reactive hydrocarbons responsible for photochemical formation of ozone (Chameides *et al.*, 1988).

The findings that trace VOC emissions are mediators of photochemical oxidant formation and may control the oxidative capacity of the troposphere, have led to an explosion of articles on the role of these biogenics in tropospheric and stratospheric chemistry and on modeling these emissions (reviewed elsewhere in this book). Much less is known about the underlying biochemistry and physiology of biogenic VOC formation, especially in plants.

B. WHAT ARE THE MAJOR BIOGENIC VOLATILE ORGANIC COMPOUNDS EMITTED TO THE GLOBAL ATMOSPHERE?

Early estimates of the types of organic compounds released from vegetation resulted in a list of over 300 different compounds, including hydrocarbons and isoprenoids, plus some alcohols and esters known to occur in floral fragrances (Graedel, 1979). Isidorov *et al.* (1985) expanded this list by detecting additional VOCs in emissions from cut branches of a variety of plants; in addition to hydrocarbons and isoprenoids, various alcohols, aldehydes, ketones, esters, and furans were detected. The list of natural VOCs known to enter the atmosphere now extends to over 1000 compounds with the publication of lists of VOCs in floral scents (Knudsen *et al.*, 1993), and with surveys of VOCs from additional plants and other natural sources (Puxbaum, 1997). In addition, there are potentially thousands of volatile or semivolatile isoprenoids known to exist in plants, but not yet reported in atmospheric samples or floral scents. Just the monoterpene family alone includes some 1000 structures, many of which are volatile, and semivolatile sequiterpenes include more than 6500 structures (Gershenzon and Croteau, 1991). If these natural compounds are considered, the possible number of minor VOCs emitted from plants numbers in the several thousands.

Given the diversity of biogenic VOCs, it is useful to ask: What are the major biogenic VOCs released to the atmosphere? Table 1 summarizes our current knowledge of the major biogenic VOCs or VOC classes; and indicates their primary sources, estimated global emissions, chemical lifetimes in the atmosphere, and whether the mechanisms for their formation are known. For completeness Table 1 also includes methane, the single most abundantly emitted biogenic VOC, produced by bacteria in soils, sediments, and other anaerobic environments (reviewed in Ferry, 1993); and the organosulfur compound dimethylsulfide (DMS), a VOC primarily produced in marine environments as the result of processes associated with phytoplankton (reviewed in Andreae and Crutzen, 1997). There is a fair degree of certainty concerning the global emissions of the top four VOCs or VOC classes listed, methane, isoprene, monoterpenes, and DMS. Knowledge of the ambient levels of these

TABLE 1   The Major Biogenic VOCs or Families of Biogenic VOCs

Species	Primary natural sources	Estimated annual global emission (Tg C; $Tg = 10^{12}$ g)	Reactivity (atmospheric lifetime in days)	Mechanism of formation known?
Methane	Wetlands, rice paddies	319–412	4000	Yes
Isoprene	Plants	175–503	0.2	Yes
Monoterpene family	Plants	127–480	0.1–0.2	Yes
Dimethylsulfide	Marine phytoplankton	15–30	<0.9	Yes
Ethylene	Plants, soils, oceans	8–25	1.9	Yes
Other reactive VOCs (e.g., acetaldehyde, 2-methyl-3-buten-2-ol (MBO), hexenal family)	Plants	~260	<1	Yes, except MBO
Other less reactive VOCs (e.g., methanol, ethanol, formic acid, acetic acid, acetone)	Plants, soils	~260	>1	Uncertain, except ethanol

Data presented are derived from Singh and Zimmerman (1992), Conrad (1995), Guenther *et al.* (1995), Andreae and Crutzen (1997), and Rudolph (1997).

compounds, their chemistries and biogenesis, and their global emission rates has resulted from intense interest and numerous field and laboratory investigations. For the other biogenic VOCs in Table 1, there is much less certainty of their global emissions, in part because (1) many of these other VOC emissions were identified only recently; (2) there are analytical difficulties associated with sampling and analysis of these compounds; and (3) many of these VOCs were previously thought to have a limited role in atmospheric chemistry, resulting in a lack of intensive field studies that focus on these compounds. Following the recommendation of an international group of investigators, these additional VOCs can be grouped into two categories: other reactive VOCs with estimated atmospheric lifetimes of less than one day, and other less reactive VOCs with lifetimes of more than one day (Guenther *et al.,* 1995). Examples of other reactive VOCs include acetaldehyde, 2-methyl-3-buten-2-ol (MBO), and the hexenal family of aldehydes and alcohols. Other VOCs in the lower reactivity category include methanol, ethanol, formic acid, acetic acid, and acetone.

It is noteworthy that except for methane and DMS the major biogenic VOCs listed in Table 1 are primarily emitted from terrestrial plants. Soils, sediments, freshwater aquatic systems, oceans, and animals are much weaker sources of VOCs, other than methane and DMS (Duce *et al.,* 1983; Singh and Zimmerman, 1992). The lack of significant emissions of other VOCs from soils and sediments resides in part from their large capacity for microbial oxidation of hydrocarbons and other organic species, which largely prevents release of VOCs to the atmosphere (Conrad, 1995). For example, only about 10% of methane produced in soils and sediments escapes microbial oxidation. Presumably other VOCs produced in soils are similarly oxidized. Nevertheless, small amounts of hydrocarbons, organic acids, and other VOCs are released from soils, sediments, and marine systems.

C. Plant Volatile Organic Compounds Are Formed in Many Different Tissues and Cellular Compartments

Figure 1 illustrates a central theme of this chapter: plants have the metabolic potential to produce and emit a large variety of VOCs. The figure illustrates a hypothetical "VOC tree" that emits all the major plant-derived VOCs plus floral scent VOCs. In reality, different plant families emit different subsets of these VOCs. The VOC tree also emphasizes the fact that biogenic VOCs are (1) produced in many different plant tissues and compartments, and (2) are the products of diverse physiological processes. The following sections con-

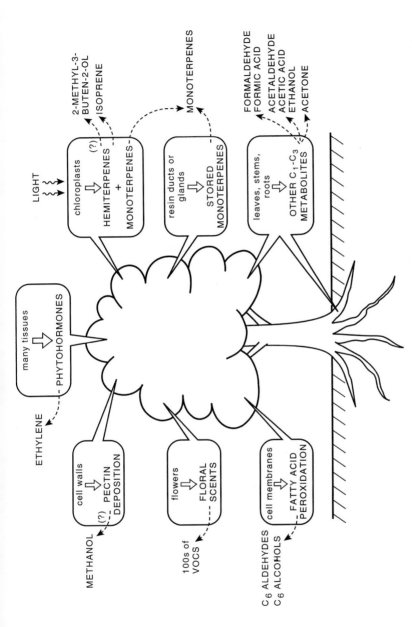

FIGURE 1 A "VOC tree" illustrates that plants have the metabolic potential to produce and emit a variety of VOCs. In this scheme a hypothetical tree emits all known major plant VOCs plus floral scents. The probable plant tissues and compartments for VOC formation are indicated. Major uncertainties, indicated by question marks, and the pathways of VOC exit from plants are discussed in the text.

sider each of these VOCs or VOC families, except for floral scents, and the underlying mechanisms for their production.

Many aspects of VOC emissions from vegetation, including impacts on atmospheric chemistry, global and regional emission inventories, ecology, and physiology, have been reviewed elsewhere (e.g., Sharkey *et al.*, 1991; Fehsenfeld *et al.*, 1992; Singh and Zimmerman, 1992; Guenther *et al.*, 1994, 1995; Monson *et al.*, 1995; Sharkey, 1996; Harley *et al.*, 1998b; Helas *et al.*, 1997; Lerdau *et al.*, 1997). Rather than repeat this information, some of which is also reviewed elsewhere in this book, this chapter focuses on the following important questions. What are the underlying mechanisms of biogenic VOC formation in plants? How is plant VOC formation regulated? Why are these VOCs produced? How are VOCs released from plants to the atmosphere? The answers to these questions, when known in detail, should allow biologists and atmospheric scientists to develop realistic models of regional and global VOC emissions. An understanding of these biological mechanisms may also provide a basis for predicting regional and global responses of plant VOC emissions in various ecosystems to changing global climate and anthropogenic impacts (Monson *et al.*, 1995), and guide thinking about strategies to preserve or improve air quality.

II. ETHYLENE: A VOLATILE PLANT HORMONE

One of the first biogenic VOCs identified was the plant hormone ethylene ($CH_2{=}CH_2$), which has the systematic name of ethene. In 1901, Neljubov demonstrated that the active constituent in coal gas that produced shedding of leaves from plants near leaking gas mains was ethylene, and that this gas could induce a variety of growth responses in treated plants (reviewed in Abeles *et al.*, 1992). It was later suggested that plants themselves, especially ripening fruit, produced volatile ethylene, and this was confirmed in the 1930s.

As the result of a remarkable number of investigations, we now know that ethylene plays a key role in controlling plant growth and development, including enhancement of fruit ripening, seed germination and flowering, inhibition of stem and root elongation; in control of senescence of leaves and flowers; and in sex determination. In addition, ethylene production is greatly enhanced in response to stresses, such as wounding, chemical exposure, and pathogen infection, triggering protective responses. Details of ethylene's role in plant biology have been described in a large number of books and review articles (e.g., Abeles *et al.*, 1992; McKeon *et al.*, 1995).

These discoveries led to interest in determining if air polluted with ethylene might alter plant growth properties, and evidence of crop damage due to anthropogenic ethylene was obtained (Abeles *et al.*, 1992). As a consequence of this work, attempts were made to estimate the magnitude of natural and anthropogenic sources of atmospheric ethylene. As summarized by Rudolph (1997), the natural source estimates are complicated by the issue of whether plants, soils, and oceans are both sources and sinks for ethylene, and the difficulty of accounting for higher stress ethylene emission rates in terrestrial vegetation (see later). Global values for biogenic ethylene emission have ranged widely. By leaving stress ethylene out of the estimates, Rudolph (1997) concluded that plants are the largest natural source of ethylene (several terragrams (Tg) per year) with smaller amounts from soils (3 Tg per year) and oceanic sources (1 Tg per year). Soil bacteria and fungi are known to produce this gas, explaining why soils are an ethylene source. Oceanic ethylene may be formed primarily from the photochemical degradation of dissolved organic carbon released by phytoplankton.

In addition to its role as a volatile plant hormone in controlling normal plant development, the phenomenon of "stress ethylene" production is relevant to atmospheric chemistry, since emissions of ethylene to the atmosphere increase dramatically in stressed plants. Typical stresses that increase ethylene formation include physical (mechanical wounding, drought, high temperature, chilling, freezing, water-logging, and gamma irradiation), chemical (phytotoxic chemicals, ozone, and sulfur dioxide), and biotic factors (pathogen infection, and insect or nematode damage) (Abeles *et al.*, 1992). Increases in ethylene formation following such stresses can last for hours or days, and it is not unusual for the maximal stress ethylene emission rate to increase 10- to 400-fold over the basal emission rate. The current view of stress ethylene is that it serves to induce protective responses in affected plants.

A. Mechanism of Formation

In plants ethylene is produced in an interesting enzymatic pathway that is summarized in Fig. 2. The two carbons of ethylene are derived from the common amino acid, L-methionine. Methionine can be activated to the biological methyl donor S-adenosylmethionine, which is then transformed into a cyclopropane-containing amino acid derivative, 1-aminocyclopropane-1-carboxylate (ACC) by action of the enzyme ACC synthase. ACC synthase has been found associated with a variety of cell membranes, especially vacuole membranes, and it is thought that some fraction of ACC produced in plant cells is sequestered or stored in the vacuole for later conversion to ethylene. ACC can also be translocated between plant tissues.

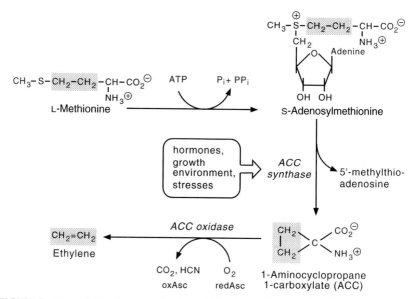

FIGURE 2 Biosynthesis of ethylene in plants. This scheme illustrates that the carbon backbone of ethylene is derived from the amino acid methionine by the reactions shown. Abbreviations: ATP, adenosine triphosphate; oxAsc, oxidized ascorbic acid; P_i, inorganic phosphate; PP_i, pyrophosphate; and redAsc, reduced ascorbic acid.

ACC is converted to ethylene by the enzyme ACC oxidase, which catalyzes a remarkable reaction in which C-3 and C-4 of ACC are released as ethylene, C-1 is released as CO_2, and C-2 is converted to HCN (Kende, 1993). This type of biochemical reaction is described as a mixed function oxidation, since both a reductant (ascorbic acid) and an oxidant (oxygen) are required. The enzyme also requires iron as a cofactor and bicarbonate or CO_2 as an activator.

B. REGULATION OF FORMATION

Given the multiple roles of ethylene in plant growth, development, dormancy, and fruit ripening as well as in stress responses, it is not surprising that regulation of ethylene formation is complex; this topic has been reviewed by Abeles *et al.* (1992) and McKeon *et al.* (1995). Although these details will not be recounted here, it is noteworthy that the ACC synthase reaction, instead of ACC oxidase, is the major site of ethylene biosynthesis control. For example, in the case of cucumbers a chilling stress induces a large increase (e.g., > 50-fold) in ethylene formation several hours later. This increase is preceded first by a large increase in ACC synthase activity, and second by a large rise in ACC concentration. Thus, it appears that the rate of ethylene synthesis is controlled primarily by the level of ACC, the substrate of the ethylene-forming enzyme, ACC oxidase. However, under some circumstances, ethylene production rates are also controlled by the levels of ACC oxidase.

As indicated on Fig. 2, the levels of ACC synthase are governed by plant hormone levels, by changes in the growth environment, and by stresses. The picture is complicated by the findings that plants contain multiple ACC synthase genes, and that different members of this gene family are expressed (transcribed) preferentially in response to physiological and developmental events, such as fruit ripening, changes in other hormone levels (including abscisic acid, auxin, or gibberellins), or in response to various stresses. The exact mechanisms by which different signals induce transcription of ACC synthase genes are uncertain. In the case of stress ethylene formation, there appear to be both abiotic and biotic signals that control genes involved in ethylene biosynthesis (Abeles et al., 1992).

It has been pointed out that in ozone-polluted environments stress ethylene formation may actually result in additional plant stress and damage (reviewed in Wellburn and Wellburn, 1996). This can occur because tropospheric ozone levels in the Northern Hemisphere have been rising in the last several decades, and ozone-exposed plants generally produce and emit much more ethylene (Tingey et al., 1976). Ozone–ethylene reactions in plant tissues can lead to the formation of toxic organic hydroperoxides, which themselves might trigger even more stress ethylene formation. When ethylene formation in plants is blocked by addition of specific inhibitors, ozone exposure results in less plant damage, supporting the idea of a role of ethylene in this damage. In addition, comparison of six ozone-sensitive and ozone-tolerant pairs of cultivars from different plant species, including woody species (pine and poplar), crop species (radish and tobacco), and clover and plantain, showed that in response to ozone fumigation the ozone-tolerant cultivars did not produce stress ethylene or produced significantly less than comparable ozone-sensitive cultivars (Wellburn and Wellburn, 1996).

Based on these results, one could argue that an important control of ethylene emissions from vegetation is related to the inherent ability of most ozone-sensitive plants to respond to air pollution episodes by overproducing ethylene, and that global ethylene emissions may be rising in response to increasing global air pollution.

C. WHY PRODUCE A VOLATILE HORMONE?

Ethylene is a volatile gas (boiling point $-103.7°C$) that readily escapes plant tissues. This leads to the question of why plants produce such a volatile hormone. This can be best understood by considering ethylene's role in response to a stress such as wounding. At the site of wounding stress ethylene formation is induced. Ethylene's volatility and ability to readily diffuse across cellular membranes allow the molecule to penetrate through plant tissues, aided in leaves and stems by the extensive air spaces, from sites of synthesis to sites of ethylene signal reception where a defense or response to the wounding can be mounted. The receptors for ethylene are under current investigation and are not fully understood, but it is clear ethylene induces factors, such as enzymes and phytoalexins (plant antibiotics), that inhibit invading microorganisms that are present on plant surfaces and can enter plant tissue at wound sites (O'Donnell et al., 1996). Stress ethylene also triggers complex events, such as abscission of diseased or damaged leaves and flowers.

Since ethylene is a signaling molecule, a method of attenuating the signal is needed. Ethylene has a small solubility in water, about 6.5 nM at 25°C, and it is thought that cellular receptors that bind ethylene operate in this concentration range (Abeles et al., 1992). However, because of its high volatility, ethylene will rapidly partition into the air spaces in plants and exit to the atmosphere, effectively terminating its binding to the receptors. This sequence of events attenuates signal transduction by the receptor, and has the consequence for atmospheric chemistry that a volatile hydrocarbon is released from vegetation to the atmosphere. It is noteworthy that some plant ethylene can be metabolized by oxidation in a few plant species (Abeles et al. 1992). However, it appears that most of the ethylene produced in plants is eventually released to the atmosphere.

III. ISOPRENE AND METHYLBUTENOL: LIGHT-DEPENDENT HEMITERPENES OF UNCERTAIN FUNCTION

One of the most studied biogenic VOCs is the hydrocarbon isoprene, which has the structure 2-methyl-1,3-butadiene (Fig. 3). Interest in this molecule has been fueled by the findings that its global emission from vegetation is large, 175 to 503 Tg of carbon per year (see Table 1), and that it is very reactive in the troposphere, contributing to the formation of ozone and other oxidants (Chapter 7). Emission of isoprene from woody plants, such as locust, poplar, and willow, was discovered by Sanadze around 1957 (reviewed in Sanadze, 1991) and later confirmed by Rasmussen (1970). Sanadze and co-workers established that the hydrocarbon was emitted by these plants in response to light, that the light response curves of isoprene emission and photosynthetic CO_2 assimilation were very similar, that isoprene was readily labeled by administration of $^{13}CO_2$ to leaves, and that isolated chloroplasts from poplar leaf protoplasts produced some isoprene (Sanadze, 1991). All these results indicated a linkage of isoprene production to photosynthetic processes, and it is ironic that decades after isoprene's discovery we are still searching for the details of this linkage (see later).

Rasmussen (1978) initiated a screening of plants to determine the phylo-genetic basis of isoprene emission; although the results have not been pub-lished in detail, it appeared that of the hundreds of plant species tested about 30% were isoprene emitters. A compilation of published and unpublished screenings of plants for isoprene emissions (Harley et al., 1998b) also supports the view that while most plants are not isoprene emitters, still a significant fraction of species has this capability. In general, most isoprene emitters are woody species, although some ferns, vines, and other herbaceous species are emitters. Few crop species emit significant amounts of isoprene.

There is no clear phylogenetic basis for isoprene emission in plants. Iso-prene is produced in many plant families, and in plant families that contain isoprene-emitting species not all species are emitters. A good illustration of this latter point is that isoprene emission potential is variable in oak trees (genus *Quercus*). All the North American oaks are high isoprene emitters, but many European oak species are nonisoprene emitters; and, curiously, many of the latter emit monoterpenes in a light-dependent manner (Seufert et al., 1997). Light-dependent monoterpene emission is discussed in Section IV. While these findings raise many unanswered questions about isoprene's func-tion, they clearly indicate that in many species of oak trees isoprene produc-tion is not essential.

Quantitative isoprene emission rates from a large number of North Ameri-can and European plants have now been measured in field and greenhouse experiments with isoprene fluxes usually reported as $\mu gC/gdw/h$, where gdw is gram dry weight. Guenther et al. (1994) and Steinbrecher (1997) have summarized these data, and isoprene emission rates at representative light [photosynthetically active radiation (PAR) = 1000 $\mu mol/m^2/sec$] and leaf temperature (30°C) can be categorized in four ranges ($\mu gC/gdw/h$): negligible, <0.1; low, 14 ± 7; moderate, 35 ± 17.5; and high, 70 ± 35. High isoprene emitters include species of *Casuarina* (ironwood), *Eucalyptus, Populus* (pop-lars and aspen), and *Quercus* (oaks). The basis for the variability in isoprene emission rate is not known with certainty, but, as described in more detail later, isoprene emission rate depends not only on instantaneous light and

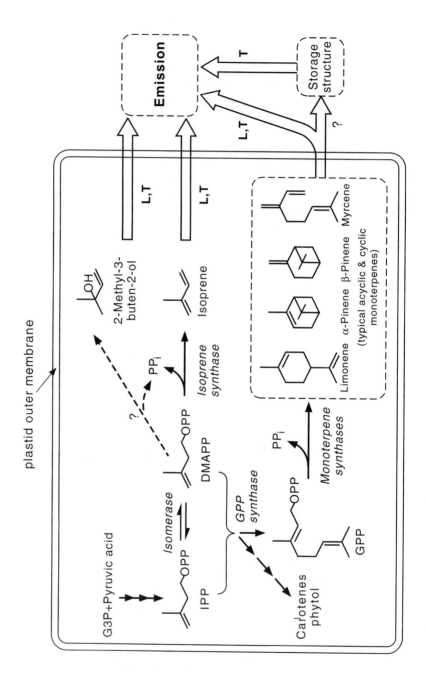

temperature, but also on the growth environment of plants and leaf development. Variations in the latter might account in part for finding isoprene emitters in the low to high range. Where sensitive isoprene detection methods have been used, plants considered nonemitters have been found to emit small amounts of isoprene (<0.1 μC/gdw/h) (Harley et al., 1998b). Thus, it may be true in the strictest sense that most plants are isoprene emitters.

Isoprene production and emission are not limited to higher plants. Isoprene is also emitted in small amounts from organisms as diverse as bacteria, fungi, marine phytoplankton, and humans and other mammals (cited in Kuzma et al., 1995). The reasons for these emissions are as puzzling as isoprene emission from plants (see later).

Isoprene was the only significant hemiterpene (C_5 isoprenoid) found in the atmosphere until studies conducted in a Colorado pine forest by Goldan et al. (1993) revealed that the C_5 alcohol, MBO, (see Fig. 3), was a very abundant VOC. MBO levels in the forest atmosphere showed a diurnal cycle, peaking at midday (around 2 to 3 ppbv) and declining to near zero at night. This pattern paralleled that for isoprene, which was also detected, but at five to eight times lower levels. Since isoprene is known to be produced by vegetation, it was proposed that MBO was also derived from light-dependent processes in the pine forest vegetation. This proved to be the case because subsequently MBO was found be emitted from lodgepole pine (*Pinus contorta*) branches at this site (B. Baker and R. Fall, unpublished), and lodgepole pine needles in a gas exchange cuvette were shown to be significant source of light-dependent MBO emission (i.e., 20 μgC/gdw/h at 30°C and 1000 μmol/m²/sec; Harley et al., 1998a). Ponderosa pine (*P. ponderosa*) was also shown to emit MBO

FIGURE 3 Biosynthesis of hemiterpenes (C_5) and monoterpenes (C_{10}) occurs in plastids. This diagram illustrates the view that various isoprenoids are synthesized in plant organelles called plastids; plastids may be photosynthetic (chloroplasts) or nonphotosynthetic (chromoplasts and leucoplasts). For simplicity only the outer plastid membrane is shown. The synthesis of isoprene in chloroplasts has been established, but the formation of 2-methyl-3-buten-2-ol (MBO) in this compartment in pine needles is conjecture. Monoterpene synthesis occurs in nonphotosynthetic plastids in plants that store monoterpenes and are likely to be formed in chloroplasts in plants that emit light-dependent monoterpenes. For completeness the figure indicates that higher isoprenoids, such as carotenes (C_{40}) and the phytol (C_{20}) side chain of chlorophyll, are synthesized in chloroplasts [McGarvey and Croteau (1995)]. As described in the text, release of plastidic VOCs to the atmosphere involves light- or temperature-dependent processes; and some monoterpenes are transferred to, and emitted from, storage structures. Abbreviations: DMAPP, dimethylallyl diphosphate; G3P, glyceraldehyde-3-phosphate; GPP, geranyl diphosphate; IPP, isopentenyl diphosphate; L, light-dependent emission; PP$_i$, pyrophosphate; T, temperature-dependent emission.

(Harley *et al.*, 1998a); and because lodgepole and ponderosa pines are two of the most common and widely distributed pine species in North America, the emissions of MBO may be a significant contributor to photochemistry above North American pine forests. MBO is almost as reactive with the OH radical as isoprene (Rudich *et al.*, 1995), giving rise to acetone and other oxygenated products (Fantechi *et al.*, 1996).

A. MECHANISMS OF FORMATION

The isoprenoids, also called terpenoids, are assembled from five-carbon "isoprene" units (but not isoprene itself), and comprise over 20,000 compounds, ranging from hemiterpenes (C_5), like isoprene, to monoterpenes (C_{10}), sesquiterpenes (C_{15}), diterpenes (C_{20}), triterpenes (C_{30}), and tetraterpenes (C_{40}), to name a few (McGarvey and Croteau, 1995). The five-carbon building blocks of all isoprenoids include isopentenyl diphosphate (IPP) and dimethylallyl diphosphate (DMAPP) shown in Fig. 3. IPP can be isomerized to DMAPP (see Fig. 3), and these two isoprenyl diphosphates may then undergo condensation, initiating a series of reactions leading to higher level isoprenoids. It was assumed that IPP was derived exclusively from three molecules of acetyl-coenzyme A (CoA) by the well-known mevalonic acid pathway (see details in McGarvey and Croteau, 1995). Instead, it has now become clear that the biosynthesis of IPP in plants can occur by two different pathways depending on the cellular compartment. The sesquiterpenes and triterpenes are synthesized from mevalonic acid in nonplastid compartments, including the cytosol and endoplasmic reticulum (McGarvey and Croteau, 1995), while plastidic isoprenoids, such as isoprene, monoterpenes, phytol (C_{20}), and carotenoids (C_{40}), are formed from IPP generated from a nonmevalonate pathway; this pathway was termed the glyceraldehyde-3-phosphate–pyruvic acid pathway by Lichtenthaler *et al.* (1997) and Zeidler *et al.* (1997).

Various lines of evidence support the view that isoprene is formed enzymatically from DMAPP in chloroplasts (see Fig. 3). Leaf extracts from the isoprene emitter, aspen (*Populus tremuloides*), were found to contain a soluble enzyme, isoprene synthase, which was purified 4000-fold and found to catalyze isoprene and pyrophosphate formation from DMAPP (Silver and Fall, 1995). Like other prenyl diphosphate-dependent enzymes (McGarvey and Croteau, 1995), isoprene synthase requires a divalent cation, such as Mg^{2+} or Mn^{2+}, which is likely to assist in the ionization of the pyrophosphate (PP_i) moiety to generate a carbocation intermediate as shown in the following scheme:

DMAPP   Mg²⁺   OPPᵢ   H⁺ PPᵢ   ISOPRENE

Abstraction of a proton from the carbocation and elimination of PP_i would yield isoprene. The enzyme has also been detected in, and partially characterized from, leaves of other isoprene emitters, including velvet bean (*Mucuna* sp.; Kuzma and Fall, 1993), willow (*Salix discolor;* Wildermuth and Fall, 1996), and oaks (*Quercus;* Schnitzler *et al.,* 1997), but not in leaf extracts and organelles of a nonemitting species like spinach (Wildermuth and Fall, 1996). Wildermuth and Fall (1996, 1998) fractionated willow leaves, and found that isoprene synthase is located primarily in chloroplasts. Fractionation of chloroplasts revealed that two forms of isoprene synthase were present: a soluble stromal form and a form tightly bound to thylakoid membranes. The bound form can be solublized at alkaline pH, suggesting it is anchored to the membranes by a cleavable lipid anchor (Wildermuth and Fall, 1998). The significance of the two forms of isoprene synthase is unknown, but as discussed later, observed light and temperature dependencies of isoprene emission in leaves are interpretable in the context of isoprene synthase behavior.

Curiously, the isolated isoprene synthases are relatively poor catalysts. The catalytic efficiencies of enzymes are usually expressed in terms of the Michaelis constant (K_M), which is the substrate concentration that produces one-half maximal reaction rate, and the turnover number (k_{cat}), which is the number of substrate molecules that each enzyme processes per unit time. Aspen isoprene synthase exhibits a very high K_M for its substrate, DMAPP, indicative of the enzyme's low affinity, and a low turnover number of about 1.7/sec (Silver and Fall, 1995). Most enzymes have turnover numbers ranging from 1 to 10^4/sec (Stryer, 1995). In the willow system, the membrane-bound and stromal forms of isoprene synthase have similar, sluggish catalytic properties (Wildermuth and Fall, 1998). It is possible that during purification of the enzymes, a cofactor, an activator, or activated forms of the enzymes have been lost. Since isoprene formation is light dependent, it is possible that isoprene synthase is a light-activated enzyme and that only low activity forms of the enzyme have been isolated so far. Many light-activated enzymes are known to exist in plants (Dey and Harborne, 1997).

The low activity of isoprene synthase *in vitro* also raises the issue of the relevance of the enzyme *in vivo*. Can this enzyme account for the relatively high rates of isoprene emission seen in some plants, or is isoprene derived from nonenzymatic processes? Several experimental findings indicate that the activity of isoprene synthase in leaf extracts closely parallels the capacity of those leaves to emit isoprene. First, in velvet bean leaves the youngest leaves do not emit isoprene, and maximal isoprene emission is seen only in fully expanded leaves (Kuzma and Fall, 1993). Extracts of developing and fully expanded velvet bean leaves contained soluble (i.e., stromal) isoprene synthase activity that correlated with emission potential. Similar results have been found in developing willow leaves (Wildermuth and Fall, 1998). Second, the activity of extractable isoprene synthase in oak leaves has been shown to parallel simulated isoprene emissions from oak trees over the course of a growing season (Schnitzler *et al.*, 1997). Finally, when the temperature dependence of isoprene synthase activity has been compared to that for intact leaves from which the enzyme was derived, very similar, although not identical, profiles have been obtained, these include experiments with aspen (Monson *et al.*, 1992), velvet bean (Kuzma *et al.* 1995), and willow (see Fig. 4C later). The idea that some leaf isoprene production might arise by a nonenzymatic pathway has not been ruled out. For example, low levels of isoprene emitted from most "nonemitters," described earlier, could be due to small amounts of nonenzymatic hydrolysis of cellular DMAPP.

The mechanism of MBO synthesis in pine needles is unknown, so the biochemical pathway for its formation shown in Fig. 3 is speculative. The light-dependence of MBO emission is suggestive of a chloroplast location for its synthesis. There is limited evidence to support the idea that MBO is derived from DMAPP. MBO was found to be the major product of (1) the nonenzymatic solvolysis of DMAPP in studies with rat liver extracts (Deneris *et al.*, 1985), and (2) the enzymatic hydrolysis of DMAPP in aspen leaf extracts (G. Silver and R. Fall, unpublished). It should also be noted that MBO is known to be a pheromone for bark beetles, and that injection of beetles with [^{14}C] mevalonic acid gave rise to small, but significant, incorporation of radioactivity into MBO (Lanne *et al.*, 1989); this result is also consistent with the formation of MBO as the result of an isoprenoid pathway, although not by the non-mevalonic acid pathway proposed in Fig. 3.

B. REGULATION OF FORMATION

Studies of the regulation of isoprene formation at the chloroplast level, the site of leaf isoprene synthesis, would be very desirable, but so far it has been difficult to isolate intact, functional chloroplasts from isoprene emitters (Wildermuth and Fall, 1996). The regulation of isoprene emission has been inferred mainly from laboratory gas exchange experiments with modern leaf cuvette technology. With these devices, gas exchanges—including isoprene emission, CO_2 uptake, and water vapor release (transpiration)—have been measured simultaneously under controlled conditions of light intensity (PAR), temperature, and gas flow and gas composition (Monson and Fall, 1989). As reviewed in Lerdau et al. (1997), leaf isoprene emission occurs essentially without a leaf reservoir; and although isoprene is released from stomatal pores, its emission is not effectively controlled by stomatal conductance. As a result, isoprene emission is tightly linked to its instantaneous production in the isoprene synthase reaction.

One of the most striking features of isoprene emission is its dependency on light. Figure 4A shows the response of isoprene emissions by an aspen leaf to light–dark transitions. This experiment shows the following: (1) the emissions are almost completely light dependent, taking about 30 min to reach a steady state; (2) in light-to-dark transitions, isoprene emission rate falls rapidly in a few minutes to near zero; and (3) there is a postillumination burst of isoprene, which is unexplained (Monson et al., 1991). When light levels are varied, as shown in Fig. 4B, isoprene emission rate responds similarly to leaf photosynthetic CO_2 assimilation or photosynthetic electron transport, showing evidence of the well-known phenomenon of light saturation of photosynthetic processes. Closer inspection of Fig. 4B reveals that in these experiments isoprene emission rate did not completely level off at the highest PAR levels; in experiments with sun leaves in temperate and tropical forest canopies, it has also been found that isoprene emission rate does not fully saturate at high light (reviewed in Harley et al., 1998b), an indication of a difference between isoprene formation and photosynthesis. Similarly, while the light dependence of isoprene emission generally correlates with photosynthetic CO_2 assimilation, withdrawal of CO_2 does not lead to decreased isoprene emissions (Monson and Fall, 1989), suggesting that the effect of light on emissions may not directly relate to photosynthesis. It is possible that there may be light activation of isoprene synthase, as mentioned earlier, or light-dependent formation of DMAPP, or both; and these processes do not fully saturate at high light. The levels of DMAPP in chloroplasts are unknown, but the finding of two forms of isoprene synthase in the chloroplast stroma and thylakoid membranes (Wildermuth and Fall, 1998) places the enzymes in proximity to light-activation signals for photosynthetic processes, and would allow rapid responses to changing light conditions.

Leaf temperature is also a major controller of isoprene emissions. For example, Fig. 4C shows the effect of temperature on isoprene emission rate from a willow leaf, and on two forms of isoprene synthase obtained from the same leaves. In each case maximal isoprene production occurs in the range of 40 to 45°C, followed by a rapid decline in production at higher temperatures.

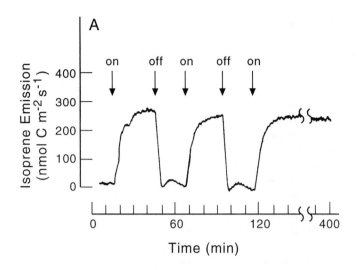

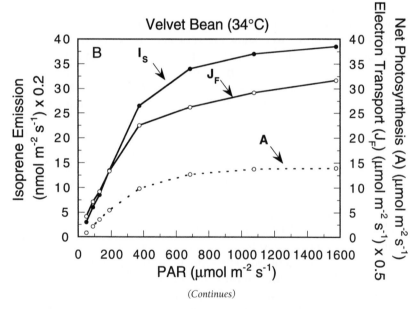

(Continues)

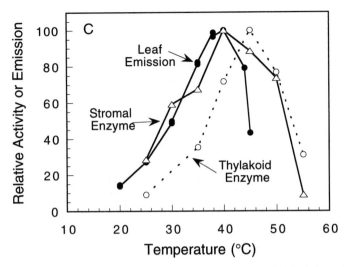

FIGURE 4 Short-term regulation of leaf isoprene emission is controlled by light and temperature. (A) Aspen leaf isoprene emission occurs as soon as light is applied [photosynthetically active radiation (PAR) = 1000 μmol/m^2/sec], reaching steady state in about 30 min, and declines to near zero in 2 to 3 min after a light–dark transition, often with a small second peak of isoprene released in the dark [redrawn from Monson *et al.* (1991); used with permission]. (B) Velvet bean leaf isoprene emission (I_s), photosynthetic electron transport (J_F), and photosynthetic CO_2 assimilation (A) rates are dependent on PAR and show light saturation [redrawn from Monson *et al.* (1992); used by permission of the publisher. © American Society of Plant Physiologists.]. (C) The temperature response of willow leaf isoprene emission rate is similar but not identical to that for isolated stromal and thylakoid-bound forms of isoprene synthase (from Fall and Wildermuth (1998); used with permission).

The exact temperature optima of these reactions are hard to establish due to irreversible effects at high temperatures. However, results of this type are notable in several respects. First, they suggest that leaf isoprene emission response to temperature could be due to effects on isoprene synthase itself. Second, the decline in leaf emissions at high temperature is indicative of denaturation of an enzyme or irreversible membrane damage; a nonenzymatic path for production of isoprene would have a very different temperature response curve (Kuzma *et al.*, 1995). Third, this temperature-dependent behavior has helped in the formulation of an algorithm for isoprene emission (Chapter 3; Fall and Wildermuth, 1998).

Major long-term regulators of isoprene emissions are leaf development, as mentioned earlier, and the seasonal dependence of leaf maturation. Ohta

(1986) demonstrated that oak trees show a seasonal dependence of isoprene emission, with peak emission rates occurring at midsummer on the hottest days. Seasonal effects on isoprene emission potential have also been studied in aspen clones growing at three different altitudes in the Rocky Mountains of Colorado (Monson et al., 1994). Emission occurred later in higher altitude trees, which leafed out later and were exposed to lower ambient temperature than lower altitude trees. These results are consistent with the idea that exposure of leaves to some critical high temperature or cumulative daily temperatures is necessary for the appearance of isoprene synthase and isoprene emission. Subsequent studies of these same clonal plants revealed that the patterns of leaf isoprene emission and extractable isoprene synthase activity paralleled each other during the first three months after bud break, both peaking in August (M. Wildermuth and R. Fall, unpublished). Experiments with oak trees also demonstrated that soluble isoprene synthase shows a seasonal pattern of activity (Schnitzler et al., 1997). However, in the aspen experiments, while leaf isoprene emission increased 46-fold, extractable isoprene synthase increased only 10- to 15-fold. The reason for this difference may be the discovery that a significant fraction of isoprene synthase is bound to thylakoid membranes, in aspen as well as in willow (Wildermuth and Fall, 1996) and in oak leaves (J.-P. Schnitzler, 1997), and these experiments underestimated total leaf isoprene synthase activity. Taken together it seems likely that maximal isoprene emission potential occurs when leaves are fully expanded, a time when isoprene synthase is at its maximum.

Another important factor controlling isoprene emission rate is the growth environment. This was revealed dramatically in greenhouse experiments with kudzu (Sharkey and Loreto, 1993) and aspen (Monson et al., 1994) grown at low ambient temperatures. Fully expanded leaves could be found that emitted no detectable isoprene. However, exposure of these leaves to a higher temperature (e.g., 30°C) in leaf cuvettes led to an appearance of isoprene emission in the next few hours, an emission that persisted in heat-treated leaves but not other leaves on the same plants. These experiments demonstrated that growth temperature environment exerts controls on isoprene emission rate, consistent with the results obtained from studies of the seasonal dependence of the onset of isoprene emission. The intensity of light in the growth environment also controls the isoprene emission potential of a leaf. For example, shade-adapted leaves of oak and sweetgum have a two- to fourfold lower isoprene emission potential than leaves at the top of a plant canopy when measured at identical light and temperature (reviewed in Harley et al., 1998b).

The controls of MBO emission from pine needles have not yet been investigated in detail, but it is noteworthy that, as for isoprene, light, temperature, and leaf development are major regulators of the emission (Harley et al., 1998a).

C. ROLES

Since its discovery in 1957 the role of leaf isoprene emission has been an enigma, especially since high emitters typically convert 1 to 2% of CO_2 fixed by photosynthesis into the hydrocarbon (Harley et al., 1998b). The biochemical energy required to reduce CO_2 to isoprene is considerable, and it is puzzling why plants would expend so much energy for isoprene formation. It is known that plants make impressive arrays of secondary compounds, many of which have no known function (Harborne, 1993). Where functions have been identified, they are mainly protective: antibacterial, antiviral, antifungal, or herbivore repellant. In some cases such compounds serve as attractants to pollinators or predators of herbivores, or act as allelochemical agents (i.e., chemicals that interfere with interactions between species). Two reports hint that isoprene might have such a role. Terry et al. (1995) showed that administration of isoprene gas to barley and mustard family plants caused a significant early flowering. Conceivably, this might disrupt the timing of plant pollination in competing plants. This model does not explain why plant isoprene emissions would continue through most of the growing season, long past when competitor plants might be normally flowering. Michelozzi et al. (1997) found that the feeding preference of the insect Sinella coeca was altered by isoprene; the insects avoided food supplemented with isoprene at levels as low as 100 nM isoprene. Although the relevance of this finding to living leaves is unclear—springtail insects like Sinella feed primarily on decaying leaves that would not be expected to produce isoprene—it is possible that other insects may also be repelled from leaves emitting isoprene.

A more compelling model for isoprene's role—as a thermal protectant for the photosynthetic apparatus—has been presented (Singsaas et al., 1997). Isoprene fumigation increases the temperature at which thermal damage occurs in the photosynthetic apparatus of two isoprene emitters, kudzu (Pueraria lobata) and white oak (Quercus alba), but not in bean (Phaseolus vulgaris), a species that does not emit isoprene. In this model, leaf temperature can rise rapidly in sun leaves, leading to an increase in the fluidity of cellular membranes, and especially thylakoid membranes. Because of its nonpolar nature, isoprene production would allow the hydrocarbon to partition into these membranes, reducing the fluidity of membrane bilayers. As leaves cool, isoprene production rate decreases, and because of its relatively low boiling point, levels of isoprene in membranes and leaves would decline. Thus, isoprene is proposed to serve as a volatile mediator of membrane fluidity. In support of this model is the finding that leaves of plants grown at low temperature make little or no isoprene, and then rapidly induce isoprene synthesis when exposed to high temperature (discussed earlier). In addition, the location of forms of isoprene synthase in the chloroplast stroma and

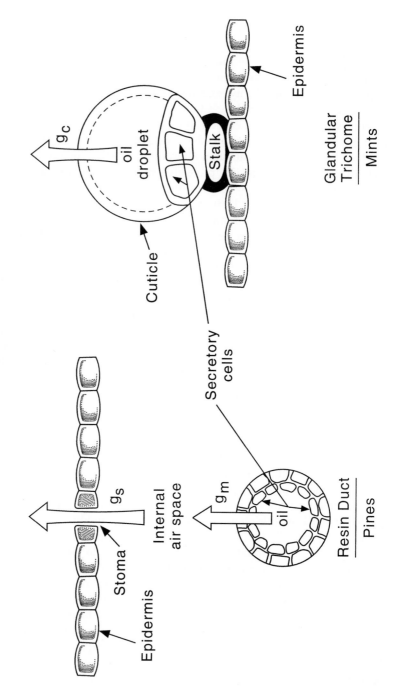

MONOTERPENE EMISSION

thylakoid membrane (Wildermuth and Fall, 1998) would allow the product, isoprene, to be directly inserted into chloroplast membranes. The model does not explain why some plants in the same genus (e.g., oak trees) contain isoprene-emitting and nonemitting species, or why many species common in temperate climatic areas are isoprene emitters while most species in hot, desert environments are not.

It should be noted that some groups of non-isoprene-emitting plants emit light-dependent MBO or monoterpenes, again at a significant fraction of CO_2 fixation rates. Examples are lodgepole pine, which emits MBO (Harley et al., 1998a), and Holm oak, which emits monoterpenes (Loreto et al., 1996). The light and temperature dependencies of these emissions are remarkably similar to those for leaf isoprene emission. However, at this time much less is known about the roles of light-dependent MBO or monoterpene emissions than is known for isoprene.

IV. MONOTERPENES: STRUCTURALLY DIVERSE, MULTIFUNCTIONAL ISOPRENOIDS

Certain plants, especially those in the conifer, mint, composite and citrus families, are capable of the synthesis of monoterpenes. The structural diversity of these C_{10} isoprenoids, assembled from just two "isoprene" units, is remarkable. Over 1000 monoterpene structures are known (Gershenzon and Croteau, 1991), and include an array of acyclic and cyclic structures. Figure 3 shows typical examples of acyclic, monocyclic, and bicyclic monoterpenes found in mint leaves; mint leaves actually contain a variety of additional monoterpenes as described later. In every case that has been examined, monoterpene-producing plants synthesize an array of monoterpenes instead of just one compound. Some of this structural diversity is due to the unusual finding that monoterpene synthases often make multiple products.

FIGURE 5   Monoterpene formation and storage in pine needles and mint leaves occur in specialized structures. The figure illustrates (left) that pine monoterpenes are primarily produced in resin ducts. Specialized secretory cells lining resin ducts produce and secrete monoterpenes into the central cavity of the ducts. Monoterpene emission to the atmosphere is governed by mesophyll conductance (g_m) and stomatal conductance (g_s) [redrawn from Tingey et al. (1991); used with permission]. Plants in the mint family contain specialized structures, glandular trichomes, on their surfaces, and the trichome is the site in which monoterpenes are synthesized in secretory cells and stored as an oil droplet [redrawn from Gershenzon and Croteau (1991); used with permission]. Monoterpene emission in undisturbed mint plants is probably controlled by cuticular conductance (g_c); if the cuticle is ruptured, the terpenoid contents can exude onto and coat the plant's surfaces, facilitating volatilization.

Monoterpene-producing plants usually accumulate pools of monoterpenes, and store them in specialized structures, such as resin ducts (pines), resin blisters (firs), glandular trichomes (mints), or leaf storage cavities (eucalypts) (Gershenzon and Croteau, 1991). Figure 5 illustrates two of these storage structures: a pine resin duct, and a mint leaf glandular trichome. In each case specialized secretory cells lining a cavity synthesize monoterpenes and secrete them into an oil storage cavity. Stored forms of monoterpenes act as a deterrent to feeding insects and other herbivores, and also as solvent for diterpene acids that are also contained in these structures. Monoterpenes have finite vapor pressures, as anyone who has smelled a Christmas tree or picked mint can attest, and thus are released from plants to the atmosphere.

Additional sources of volatile plant monoterpenes are now being characterized, including wound-induced monoterpene synthesis in conifers, pollinator attractant synthesis in flowers, and light-dependent monoterpene emission from certain tree species. These developments are discussed later.

A. Mechanisms of Formation

The biosynthesis of several monoterpenes has been worked out in detail (see references in McGarvey and Croteau, 1995). In each case the acyclic C_{10}-diphosphate, geranyl diphosphate (GPP, see Fig. 3), serves as the precursor. GPP itself is derived from condensation of IPP and DMAPP. Most monoterpenes are cyclic structures, and monoterpene cyclases (sometimes called monoterpene synthases) catalyze their formation from GPP; the cyclases that have been characterized are unusual enzymes in that they form multiple products. For example, the four monoterpene structures shown in Fig. 3 (inset) are all products of the enzyme limonene synthase. The mechanism of this enzyme has been studied in detail (Rajaonarivony et al., 1992), and is shown in Fig. 6. Purified limonene synthase from the glandular trichomes of peppermint converts GPP to limonene as a result of (1) isomerization of GPP to a tertiary diphosphate, linalyl diphosphate (LPP), and then (2) cyclization of LPP to limonene via a carbocation intermediate (α-terpinyl cation). A divalent metal ion such as Mg^{2+} or Mn^{2+} assists in the ionization of the pyrophosphate group of GPP or LPP. In the active site of the enzyme, various other carbocation intermediates can form, and proton abstraction from each leads to release of small amounts of myrcene, α-pinene, and β-pinene. Thus, one monoterpene synthase can form acyclic, monocyclic, and bicyclic structures; this finding was verified by the cloning and expression of the spearmint limonene synthase gene in bacteria, and demonstrated that the recombinant enzyme formed the same distribution of four monoterpene products (Colby et al., 1993).

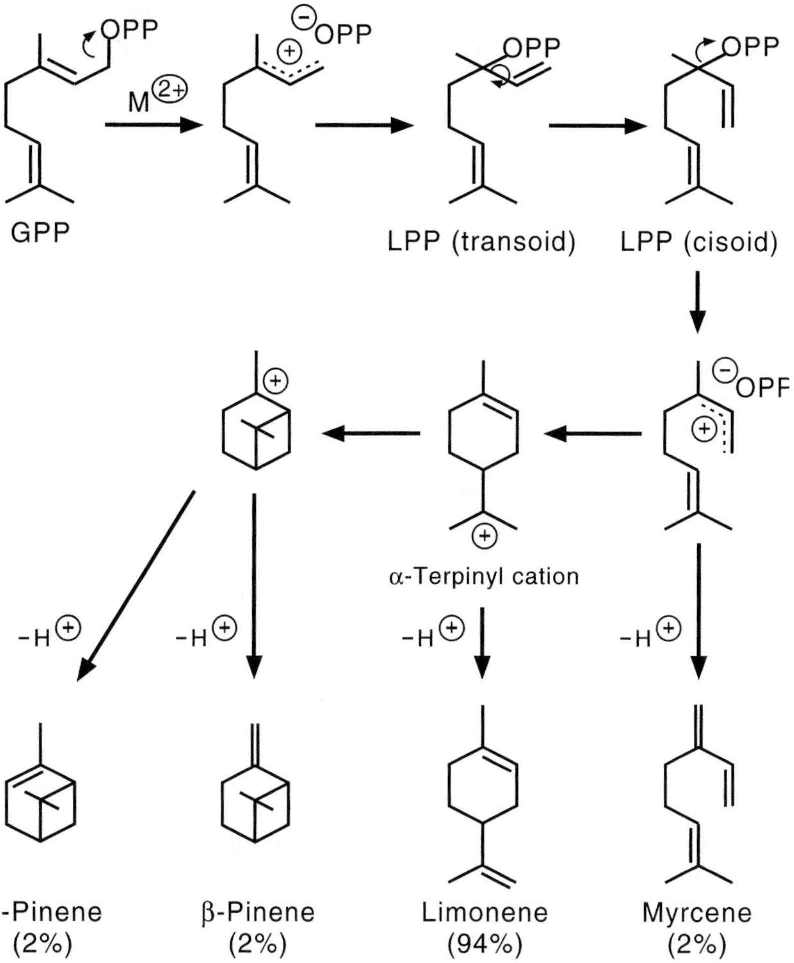

FIGURE 6   The proposed catalytic mechanism for limonene synthase illustrates how one enzyme can produce four monoterpene products. Purified limonene synthase from the glandular trichomes of peppermint converts geranyl diphosphate (GPP) to limonene and as a result (1) isomerization of GPP to linalyl diphosphate (LPP), and then (2) cyclization to limonene via a carbocation intermediate (α-terpinyl cation). In the active site of the enzyme, various other carbocation intermediates can form, and proton abstraction from each leads to release of small amounts of myrcene, α-pinene, and β-pinene. Other abbreviations: M^{2+}, divalent metal ion; OPP., inorganic pyrophosphate. [from McGarvey and Croteau (1995); used by permission of the publisher. © American Society of Plant Physiologists.]

Multiple monoterpene synthases (cyclases) have now been characterized from conifers and sage. Each requires a divalent metal ion and produces a mixture of products (see Steele *et al.*, 1995).

It is noteworthy that in many plants the initial monoterpene synthase products are converted to oxygenated monoterpenes. A good example is the mint family, in which limonene produced in glandular trichomes is converted to a complex mixture of oxygenated monoterpenes, such as the ketones menthone and isomenthone, and the alcohols, menthol and isomenthol. Complex oxygenation pathways of this type, described in Croteau and Gershenzon (1994), explain in part why so many monoterpenoid structures are known.

Little is known about the enzymatic mechanism of monoterpene formation in Mediterranean oaks that produce light-dependent monoterpenes (see Loreto *et al.*, 1996). Since these plants (e.g., *Quercus ilex*) also make a mixture of monoterpene products, primarily α-pinene and β-pinene plus smaller amounts of at least 12 other monoterpenes, it is likely that monoterpene synthases that form multiple products are present in their leaves. Labeling of several of these VOCs with $^{13}CO_2$ in the light suggests that these synthases are primarily located in leaf chloroplasts (Loreto *et al.*, 1996), but they have not yet been characterized.

B. REGULATION OF FORMATION

The regulation of monoterpene biosynthesis in plants that store these compounds has been reviewed by Croteau and colleague (McGarvey and Croteau, 1995), who point out that the regulation is complex because of (1) the multiple functions of monoterpenes and (2) the existence of both spatial and temporal control of monoterpene formation during plant development. For example, monoterpene synthesis is highest in tissues with specific functions, such as young leaves that form resin ducts, blisters, or glandular trichomes to deter feeding insects and other herbivores. These specialized structures are enriched in monoterpene synthases that are most highly expressed in these expanding leaves, and then decline later in leaf development. In conifers, such as grand fir (*Abies grandis*), the biosynthesis of wound-inducible monoterpenes also occurs in woody tissues (see Steele *et al.*, 1995). After wounding by insects or from physical injury, formation of monoterpene cyclases is induced in tissues in proximity to the wound site. This aids in oleoresin formation and sealing of the wound. In the flowering plant, *Clarkia brewerii*, the monoterpene linalool is synthesized mainly in the flowers as a pollinator attractant; linalool emission correlates with the appearance of the enzyme, linalool synthase (McGarvey and Croteau, 1995). Thus, at one level the control of monoterpene formation is correlated with the expression of genes for the monoterpene synthases over the course of plant development or in response to wounding signals.

The subcellular compartments where stored monoterpenes are synthesized are fairly well established. It appears that GPP synthase and monoterpene cyclases are located exclusively in nonphotosynthetic plastids as indicated in Fig. 3 (McGarvey and Croteau, 1995). A topic of current research is the regulation of formation of substrate GPP for monoterpene synthesis. It was previously thought that plastidic GPP and monoterpenes were derived from the mevalonic acid pathway, but as discussed earlier in Section III, it appears that plastidic isoprenoids are derived instead from the glyceraldehyde-3-phosphate–pyruvic acid pathway. The controls over carbon flow in this pathway and the levels of GPP in plastids are unknown.

The control of light-dependent monoterpene emissions in oaks is similarly of interest (Loreto *et al.*, 1996), but largely unknown. As mentioned previously, these monoterpenes are probably formed in chloroplasts in the light, but whether light activates the cyclase(s), production of substrate (GPP), or both is not known. It has been shown that monoterpene emission in Norway spruce (*Picea abies*), a widespread tree in European forests, includes both light-dependent and light-independent components (Schürmann *et al.*, 1993). This suggests that controls of monoterpene formation in some plants, such as Norway spruce, may involve both the short-term and long-term mechanisms suggested earlier, and that regulation occurs in a variety of tissues and cellular compartments.

C. ROLES OF MONOTERPENE FORMATION

As mentioned in the preceding discussion, monoterpenes play a variety of roles in plants. This topic has been reviewed in detail elsewhere (Harborne, 1993; Langenheim, 1994). The ecological roles of monoterpenes include direct defense against herbivores and pathogens, indirect defense by attraction of enemies of herbivores, attraction of pollinators, and allelopathic effects on competing plants. Here we will discuss roles of monoterpenes, including a role as a solvent, that are closely related to their emission to the atmosphere.

The finding that many monoterpene-producing plants develop highly specialized structures for the storage of monoterpenes (e.g., see Fig. 5) indicates an important biological role for these compounds. In conifers, monoterpenes serve a protective role in defense against invading insects and their associated fungal pathogens, as well as other herbivores (Gershenzon and Croteau, 1991). After wounding, conifers release oleoresin, a viscous, odoriferous material that is composed of about equal amounts of monoterpenes and diterpene resin acids, both of which are toxic to insects and fungi. Oleoresin monoterpenes, collectively called turpentine, serve as a solvent for the diterpenoid resin acids, and after wounding the oleoresin serves to seal the wound by evaporation of the turpentine, leaving a hardened rosin barrier. In some conifers (e.g., pines) large amounts of oleoresin are stored in extensive resin duct systems, and in others (e.g., true firs) only small amounts of oleoresin are stored in resin blisters, but oleoresin production is induced upon wounding (Steele *et al.*, 1995). Clearly, evaporation of monoterpene solvents from oleoresins released to plant surfaces contributes to atmospheric VOCs.

In the mints (Lamiaceae) the formation of glandular trichomes (see Fig. 5) also serves a defensive role against herbivores. For example, feeding insects on leaf surfaces cause rupture of the oil glands, releasing the toxic monoterpene and diterpene acid contents. Again, evaporation of the monoterpene solvent leads to VOC emissions.

Monoterpenes are also directly emitted from plants into the atmosphere to attract pollinators and enemies of herbivores. Flowering plants give off complex mixtures of VOCs, usually isoprenoids, benzenoids, and fatty acid derivatives (Knudsen *et al.*, 1993), that provide a chemical concentration gradient that can guide pollinators to the flowers. The mixture of VOCs in the flower scent is important, as often individual components are not usually sufficient as attractants (Langenheim, 1994). For example, alfalfa flowers contain a relatively simple mixture of honeybee attractants, only five major VOCs; when tested individually as attractants, only one, the monoterpene linalool, was an attractant and the other four were neutral or repellent (Dobson, 1993). The mixture was a more effective attractant, illustrating the complexity of insect olfactory systems. Some plants defend themselves by attracting enemies of herbivores. For example, VOC mixtures containing the monoterpenes linalool and ocimene are emitted by apple leaves; both of these monoterpenes are known to attract a predatory spider mite that feeds on spider mites that infest apple foliage (Takabayashi *et al.*, 1994). Similar results have been found in attraction of parasitic wasps to pines infected with pine bark beetles. In these pines enhanced monoterpene emissions due to wounding by beetles attract the wasps; this and other examples are cited in Langenheim (1994).

Stored monoterpenes or those induced by wounding play the roles mentioned earlier. What is the role of the production and emission of light-dependent monoterpenes from oaks? At this time there is no clear answer, although a thermal protective role like that for isoprene has been suggested (Loreto *et al.*, 1996). In this regard it is interesting to note that monoterpene-storing plants usually make complex mixtures of monoterpenes, and it has been argued that the production of such mixtures of toxic compounds reduces the probability of the evolution of insects that can detoxify all of them (Langenheim, 1994). As with stored monoterpenes, light-dependent monoterpenes from oaks are a complex mixture (discussed earlier), and it will be interesting to see if production of this mixture has a functional significance.

V. C_6 ALDEHYDES AND ALCOHOLS: PRODUCTS OF LEAF DAMAGE AND MEMBRANE LIPID PEROXIDATION

It is now known that leaves of most plants have the potential to produce and emit a family of C_6 aldehydes and C_6 alcohols, referred to here as the hexenal family. This topic has been reviewed in detail by Hatanaka (1993). These volatile compounds are readily sensed in the odor of new mown grass or of brewed green tea. Their production is linked to physical wounding of leaves or to signals produced by herbivores and plant pathogens as they attack plants. Since their formation is a result of the breakdown of C_{18} fatty acids, or octadecanoids, it is pertinent to discuss first the octadecanoid pathway in plants, then the regulation and role of hexenal production, and finally the relation of herbivory and grazing on emission of these VOCs on larger ecosystem scales.

A. MECHANISMS OF FORMATION

The hexenal family is produced in leaves by the octadecanoid pathway, which begins with the release of free fatty acids from cellular lipids (reviewed in Wasternack and Parthier, 1997). As shown in Fig. 7, when leaves are physically wounded or invaded by pathogenic microorganisms, signals are received in leaf membranes activating lipases that act on neutral lipids and phospholipids, releasing fatty acids. The lipases may include both neutral lipid lipases or phospholipases (Hatanaka, 1993), but surprisingly little information exists on the nature of these enzymes, or on the mechanisms of signal reception and activation of these enzymes. As a result of increased lipase activity, fatty acids are released, and of these products the polyunsaturated C_{18} species, α-linolenic acid and linoleic acid, are especially important for C_6 aldehyde and alcohol production. α-Linolenic acid and linoleic acid are thought to be the natural substrates for the enzyme lipoxygenase (LOX), even though the enzyme will act on a broad range of fatty acids. The structure of α-linolenic acid and its reaction with LOX are illustrated in Fig. 7. LOX catalyzes oxygen insertion at C-13 of α-linolenic acid resulting in the production of the hydroperoxide, 13-(S)-hydroperoxylinolenic acid, which can then be cleaved by the enzyme hydroperoxide lyase, a membrane-bound enzyme found in chloroplast membranes. In the case of 13-(S)-hydroperoxylinolenic acid, (3Z)-hexenal and a C-12 oxoacid are produced. (3Z)-Hexenal is the precursor of the rest of the hexenal family. The C-12 oxoacid is rearranged by isomerization to produce a wound hormone, which is reported to assist in wound healing (Hatanaka, 1993).

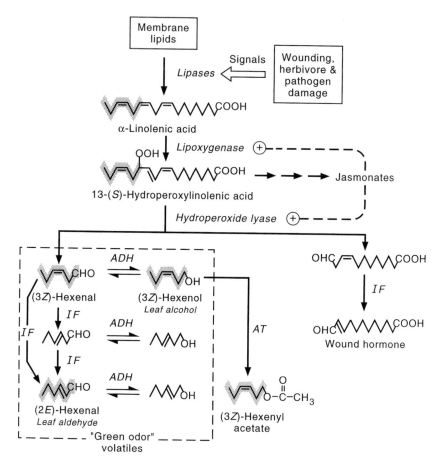

FIGURE 7 Production of the hexenal family of VOCs and other signaling factors after leaf damage and peroxidation of C_{18} fatty acids (octadecanoids). In this scheme, also known as the octadecanoid pathway, the peroxidation of the polyunsaturated C_{18} fatty acid, α-linolenic acid, gives rise to a mixture of C_6 aldehydes, C_6 alcohols, and the acetate ester of (3Z)-hexenol. This mixture produces the characteristic odor of green tea [Redrawn with permission from Phytochemistry, 34, A. Hatanaka, "The biogeneration of of green odour by green leaves," pp. 1201–1218, 1993, with kind permission from Elsevier Science Ltd., The Boulevard, Langford Lane, Kidlington OX5 1GB, UK]. Peroxidation of another leaf polyunsaturated C_{18} fatty acid, linoleic acid, gives rise to the VOCs n-hexanal and n-hexanol (not shown). The octadecanoid pathway also produces the signaling factors, jasmonates and wound hormones, discussed in the text. Abbrevations: ADH, alcohol dehydrogenase; AT, acetyltransferase; and IF, isomerization factor.

As reviewed by Hatanaka (1993), the "green odor" of damaged leaves is a combination of several C_6 compounds derived from (3Z)-hexenal. As summarized in Fig. 7, (3Z)-hexenal can by converted by enzymatic and nonenzymatic paths to a variety of products, the most important of which are (2E)-hexenal, often referred to as "leaf aldehyde," and (3Z)-hexenol, often referred to as "leaf alcohol." For reasons that are not clear, some of this latter compound can be acetylated, presumably enzymatically, to give (3Z)-hexenyl acetate (Fig. 7). Thus, wounded leaves release up to six volatile C_6 aldehydes and alcohols plus (3Z)-hexenyl acetate, all derived from α-linolenic breakdown. By similar pathways, breakdown of linoleic acid in wounded leaves gives rise to two additional volatile C_6 products, n-hexanal and n-hexanol. These pathways explain the complex mixtures of C_6 aldehydes and alcohols emitted from wounded plants.

B. Regulation of Formation

What signals control the octadecanoid pathway and the formation of the hexenal family of VOCs? Wounding and pathogen infection trigger a variety of responses in plant tissues, and much current work is focusing on identifying the signals that elicit these responses. A current view of control of the octadecanoid pathway is that either physical damage to cells or entry of pathogen cell wall components triggers release of fatty acids from cellular membranes, providing α-linolenic acid for the LOX reaction. As shown in Fig. 7, the octadecanoid pathway provides not only the precursors of the hexenal family and wound hormone, but also the precursor of lipid-signaling molecules known as jasmonates (Wasternack and Parthier, 1997). Water-soluble jasmonic acid and its volatile methylated derivative, methyl jasmonate, are known to be effective activators of plant defensive gene expression. An additional effect of the production of jasmonates after wounding is that these signaling molecules can diffuse to unwounded sites in affected plants, and elicit preemptive defense responses away from the wound site. One of these responses is the induction of LOX and hydroperoxide lyase (see Fig. 7), which can result in the formation of more products of the LOX pathway, including the hexenal family. An example of this effect has been noted in cotton plants, where damage induced by caterpillars on lower leaves elicits the production of volatiles such as (3Z)-hexenyl acetate, monoterpenes, and sesquiterpenes in undamaged upper leaves (Röse et al., 1996). These responses in unwounded leaves are probably due to detection of water-soluble and volatile jasmonates released from the wounded leaves.

Another potential regulator of the octadecanoid pathway in plants may be ultraviolet (UV) radiation. Conconi et al. (1996) reported that exposure of tomato leaves to UV radiation (254 nm) triggered the induction of numerous plant defensive genes, just as occurred when tomato leaves were physically wounded. It was proposed that these responses occur as a result of UV light damage, with release of α-linolenic acid from membrane lipids and operation of the octadecanoid pathway, and that increases in UV light exposure, as a result of stratospheric ozone depletion, might stimulate such reactions and the formation of hexenal family VOCs in the world's vegetation.

C. ROLES

It has been known for many years that various C_6 alcohols and aldehydes have antibiotic properties, and this provides a physiological rationale for the formation and emission of these VOCs. For example, the presence of (2E)-hexenal in *Ginkgo* leaves was suggested to be the agent responsible for the ginkgo's noted resistance to diseases and pests. Subsequently, (2E)-hexenal was found to be inhibitory to a variety of protozoa, fungi, insects, and mites (reviewed in Croft *et al.*, 1993). In studies by Croft *et al.* (1993) inoculation of bean leaves with a pathogenic bacterium led to the production of the hexenal family, including (3Z)-hexenol and (2E)-hexenal as the major volatiles. Control treatment of leaves with a buffer or a non-pathogen resulted in very low levels of these products. Tests of the antibacterial effect of two of these bean leaf products revealed that (2E)-hexenal was an effective bactericide, causing cell lysis of the bacterium used to infect the leaves; (3Z)-hexenol was approximately 20 times less effective. These results lead to the conclusion that one reason for the production of hexenals in leaves is to inhibit invasion and infection by opportunistic bacteria residing on leaf surfaces.

By similar reasoning it is possible to explain the large increase in production of the hexenal family in leaves after physical wounding or maceration. It is known that these compounds form within less than a minute of leaf damage (Hatanaka, 1993), and it is likely that this response reflects the plant's attempt to inhibit pathogen invasion of leaf tissue. However, it is not clear why so many different C_6 products are formed in wounded leaves, or how C_6 product distribution is regulated. For example, (3Z)-hexenol and its acetylated product (3Z)-hexenyl acetate are often the most abundant VOCs detected after wounding, yet these are the least bactericidal. The controls on the partitioning of the parent compound, (3Z)-hexenal, into either (2E)-hexenal (more potent bactericide) or (3Z)-hexenol (less potent bactericide) (see Fig. 7) are unknown. Perhaps an unsaturated aldehyde like (3Z)-hexenal is also toxic to the host plant, and the formation of (3Z)-hexenol and its acetate ester is a detoxification mechanism.

Whether or not stored pools of the hexenal family exist in undamaged leaves is not clear. Kitamura *et al.* (1992) studied the patterns of C_6 aldehydes that accumulated in tea leaves during the growing season, and whether the accumulated aldehydes paralleled the emission of these compounds from leaves. Growing leaves accumulated much more (3Z)-hexenal than (2E)-hexanal or n-hexanal, but the pattern of emission of aldehydes from detached leaves did not reflect this leaf composition. However, the methods used for leaf extraction may not have prevented continual formation of products during handling, homogenization, and solvent extraction. It is known that "rough handling" or intentional damaging of foliage leads to greatly enhanced hexenal family emissions, probably as a result of their production as described earlier (Arey et al., 1993; Turlings et al., 1995). However, even when precautions are taken during emission sampling to prevent disturbance of foliage, significant emissions of hexanal family compounds have been detected (see later) suggesting that pools of these VOCs may exist in apparently undamaged leaves. Even this conclusion is subject to the consideration that the leaves of plants tested might have barely visible microbial, insect, or mite damage; or have small surface damage due to, for example, wind or hail. The existence of leaf pools of hexenal family VOCs needs to be carefully studied with rapid freezing and extraction methods that will block octadecanoid pathway enzymes during analysis.

D. How Widespread Is the Hexenal Pathway in Plants?

Hatanaka et al. (1987) surveyed 37 different plant species to determine the distribution of enzymes for C_6 aldehyde production in leaf and fruit extracts. The survey included many dicotyledonous plants, several monocotyledonous plants, and some other plant groups. They found that virtually all plant extracts produced detectable (3Z)-hexenal, (2E)-hexenal, and n-hexanal. Although rates of production of the hexenals varied substantially in different plants—leafy vegetables, fruits, and monocotyledonous plants exhibited lower rates—these results are generally in accord with the view that enzymatic production of C_6 aldehydes and derivatives is a common feature of the green leaves of higher plants. Only a few surveys of hexenal family emissions from intact plants have been reported; and as summarized by Arey et al. (1993), these measurements are complicated by (1) the problem, mentioned earlier, of enhanced emissions due to rough handing associated with chambers used to enclose foliage, and (2) analytical difficulties in the recovery of these compounds from sampling devices. With these caveats in mind, Arey et al. (1993) reported that all 18 agricultural plants tested emitted 3-hexenyl acetate, or 3-hexenol or both as the major VOCs. Although the effects of temperature and light on the emissions of these two compounds were not studied, emission rates were significant, in a range similar to that seen for typical monoterpene emissions from plants (e.g., 0.1 to 3 μgC/gdw/h; Guenther *et al.*, 1994). *n*-Hexanal and (2*E*)-hexenal emissions were also detected from several crops, but emission rates were not determined.

The preceding findings, coupled with the knowledge that herbivory in forest ecosystems and cropland is extensive, suggest that there is continual input of hexenal family VOCs into the atmosphere, but it is not yet known how large these inputs are. For example, traditional estimates of annual rates of herbivory (percentage of leaf area damaged per year) for forest communities range from 7.5% (temperate forests) to 10.9% (tropical forests) (summarized in Coley and Aide, 1991). When consumption of young expanding leaves, those most attractive to herbivores, is factored in, herbivory rates are substantially higher. Average annual canopy losses of 16 to 42% have been reported for Australian rain forest trees (Lowman, 1992). Such continual leaf damage during the growing season is likely to cause significant releases of hexenal family VOCs in the world's forests, especially tropical forests where leaf productivity and populations of herbivores are high. Grasslands may also be a significant source of hexenal family emissions. Galbally and co-workers have shown that wounding of pasture grass and clover leads to large increases in the emissions of biogenic VOCs, including hexenal family VOCs (Kirstine *et al.*, 1998). These workers conclude that since large areas of the earth's surface are covered by grasslands, subjected to seasonal grazing by animals, the VOCs released could be quite significant.

It is of interest that Arey *et al.* (1993) estimated the reactions of 3-hexenyl acetate and 3-hexenol with OH and NO_3 radicals and O_3, and calculated tropospheric lifetimes of about 1 to 3 h. From this we can conclude that volatile hexenal family compounds may contribute significantly to tropospheric photochemistry in forest and grassland systems.

VI. C_1 to C_3 OXYGENATED VOLATILE ORGANIC COMPOUNDS: BY-PRODUCTS OF CENTRAL PLANT METABOLISM

There are a variety of oxygenated VOCs (oxVOCs) detectable in plant emissions, but it has been thought that they occurred in only minor amounts. For example, in Zimmerman's survey of VOC emissions from vegetation in the United States (Zimmerman, 1979) a large number of unknown VOCs (referred to as "other VOCs"), which may have included various oxygenates, were detected. While the individual other VOC emission rates were relatively low, typically 0.5 to 5 μg/gdw/h, when totaled they represented on average 30% of the total VOC emissions. As mentioned in Section I, Isidorov *et al.*

(1985) qualitatively surveyed various plants, representing mainly species characteristic of forests of northern Europe and Asia, for biogenic VOC emissions, and detected a diverse range of oxVOCs, including alcohols, aldehydes, ketones, esters, and ethers. Measurements, including studies with agricultural plants, have also shown low-level emissions of complex mixtures of oxVOCs (reviewed in Puxbaum, 1997).

Interest in biogenic oxVOC emissions, especially those in the C_1 to C_3 range, have increased with the discoveries that methanol and acetone are major components of forest air, and are also very abundant in the free troposphere. Methanol and acetone were two of the major VOCs detected, day and night, at U.S. pine forest sites in rural Alabama and Colorado (reported in Fehsenfeld et al., 1992; Goldan et al., 1995). For example, at a loblolly pine plantation site (Alabama) over a six-week period in the summer, of the total VOCs detected hourly, oxVOCs accounted for 46% (day) and 40% (night) of volatile carbon, and the major oxVOCs both day and night were methanol and acetone (Goldan et al., 1995). Methanol levels in the air at the top of the pine canopy showed a diurnal variation similar to that for isoprene (arising from nearby oaks), with peak levels of both VOCs in the afternoon, suggestive of vegetation sources for both. While acetone levels did not show such a clear diurnal cycle, it was suggested that this ketone might be derived from direct biogenic emission plus photochemical breakdown of other biogenic VOCs. In studies of VOCs in the upper troposphere Singh et al. (1995) found surprisingly high concentrations of methanol and acetone. In the free troposphere (5 to 10-km height) at northern midlatitudes, methanol and acetone were present at about 700 and 500 pptv, respectively, declining to about 400 and 200 pptv, respectively, at southern latitudes. These levels of acetone may contribute to the formation of odd hydrogen radicals (OH plus HO_2) and to the sequestering of NO_x in the form of peroxyacetylnitrate (Singh et al., 1995; McKeen et al., 1997).

Are there biogenic sources of methanol and acetone detected in the atmosphere? There are relatively few reports on the emission of oxVOCs from plants. However, studies with enclosed leaves demonstrated that significant emissions of methanol occur from a variety of plants (reviewed in Fall and Benson, 1996). The magnitude of these emissions, typically 0.3 to 17 μgC/gdw/h, lies between those of other major biogenic VOCs, isoprene and monoterpenes. Higher emissions of methanol occurred in young leaves, sometimes more than 40 μgC/gdw/h, than occurred in fully expanded leaves, leading to the conclusion that methanol emissions from crops and forests may be higher during the early part of the growing season. The finding that methanol emissions are correlated with light-dependent stomatal conductance is consistent with the observation that there are diurnal profiles of methanol in the air of forest canopies. Small emissions of acetone have been detected from the buds of all conifers tested (MacDonald and Fall, 1993), suggesting

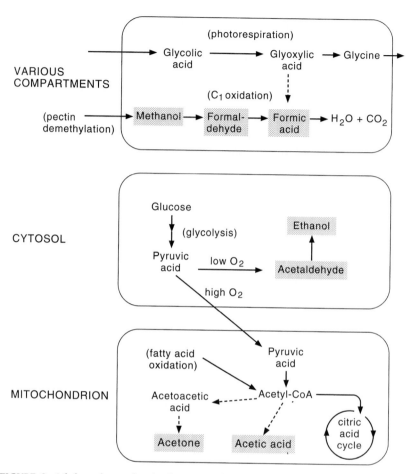

FIGURE 8 Likely pathways for the formation of C_1 to C_3 oxVOCs in plants. This scheme summarizes metabolic pathways in various cellular compartments that are likely to give rise to C_1, C_2, and C_3 oxVOCs, which are highlighted. As discussed in the text, the pathways to acetone and acetic acid in plants are uncertain.

that some of the acetone in pine forest air may derive from direct emissions from the trees. In addition, measurements by Lindinger and co-workers have suggested that decaying vegetation is a significant source of oxHCs (Warneke *et al.*, 1998). They found that the wetting of dried leaves, grass, and conifer needles leads to the emission of many oxVOCs, especially methanol and acetone, and it is estimated that this source could account for global annual emissions of 6 to 8 Tg of acetone and 18 to 40 Tg of methanol.

Significant emissions of formic and acetic acid from the biosphere have been suspected for some time, but the measurements are complicated by methodological considerations. Nevertheless, significant emissions of both acids from tropical trees in the Amazon, various trees in Europe, and savanna soils have been described (reviewed in Bode et al., 1997). In extensive field experiments, formic and acetic acid, as well as formaldehyde and acetaldehyde, were detected as small emissions from European oaks (Quercus ilex) and pines (Pinus pinea) (Kesselmeier et al., 1997). Emissions of formaldehyde, formic acid, and acetaldehyde under standard conditions (30°C, PAR = 1000/m²/sec) were low, ranging from about 0.2 to 1 µg/gdw/h, but similar to monoterpene emission rates from pine, thus representing about 50% of the measured VOC emissions from pine branches. Emissions of acetic acid from oaks and pines were lower, but significant. Formic and acetic acid fluxes generally correlated with light-dependent leaf transpiration, suggesting that they are formed within these plants and emitted via leaf stomata. The correlation of formaldehyde and acetaldehyde emissions with plant physiological parameters was difficult to establish because of complications due to relatively high ambient levels and atmospheric deposition of these aldehydes during the emission measurements. In subsequent laboratory experiments, using more controlled conditions, formic and acetic acid were found to be emitted from a variety of tree species, again in a light-dependent manner, however, interestingly, four crop species showed no emission, only uptake of these acids (Kesselmeier et al., 1998).

These scattered observations of C_1 to C_3 oxVOC emissions from foliage suggest that the biosphere may be a significant source of many low-molecular-weight oxVOCs. As summarized in Table 1, a recent global estimate of total other biogenic VOC emissions, which primarily include the C_1 to C_3 oxVOCs, was approximately 260 Tg per year (Guenther et al., 1995). This estimate was made with the knowledge that large uncertainties exist in emission estimates for these VOCs. It is clear that many more emission measurements are needed to establish more accurate regional and global fluxes of these compounds to the atmosphere.

A. Mechanisms of Formation

Except for acetaldehyde and ethanol formation, our understanding of the underlying mechanisms of formation of C_1 to C_3 oxVOCs in plants is rudimentary. For this reason it is not useful to speculate in detail on these mechanisms. Instead, the following information is presented as a brief summary of likely biochemical pathways for formation of these VOCs (additional discussion is presented in Fall and Benson, 1996; Sharkey, 1996; Bode et al., 1997). Figure 8 summarizes these pathways.

C_1 oxVOC formation may be related to operation of the C_1 oxidation pathway. Methanol is known to be oxidized in a variety of plant tissues in a stepwise fashion by the C_1 oxidation pathway to formaldehyde, to formic acid, and then ultimately to CO_2 and water (see Fig. 8). The subcellular location of this oxidative pathway is uncertain. Carbon is likely to enter the C_1 pathway from two sources: (1) as methanol from pectin demethylation during cell wall growth and expansion, a process which produces substantial methanol (Fall and Benson, 1996); and (2) as formic acid from a side reaction of photorespiratory metabolism that converts glyoxylic acid to formic acid (rather than to glycine) (Bode et al., 1997). It seems possible that if an enzyme of this pathway becomes limiting, methanol or the intermediates, formaldehyde and formic acid, may diffuse out of cells, and exit plants with the transpiration stream. Thus, higher rates of cell wall synthesis in growing plants, or elevated rates of photorespiration under conditions of high light, may increase pools of C_1 intermediates and result in the release of C_1 oxVOCs. It has been noted that like methanol, formic acid emissions from foliage are light dependent, consistent with linkages to photorespiration and light-dependent stomatal conductance (Kesselmeier et al., 1997). Bode et al. (1997) suggest that formate (and acetate) may be released from cells to the apoplast, the extracellular aqueous compartment of a leaf cell. The apoplast may be acidic enough, pH 5 to 6.5, to allow for protonation to the corresponding acids, which could then be emitted by the transpiration stream.

C_2 oxVOCs, like the methanol–formaldehyde–formic acid family, includes three different oxidation states: ethanol, acetaldehyde, and acetic acid. However, unlike the C_1 family, they are not related metabolically by a stepwise oxidation series. Instead, acetaldehyde and ethanol are related by redox reactions that are controlled by cellular anaerobiosis in the cytosol, and acetic acid is a side product of mitochondrial metabolism. As shown in Fig. 8, glucose breakdown by glycolysis produces pyruvic acid, which under aerobic conditions is transported to the mitochondrion for further oxidation. When cellular oxygen levels fall, to maintain the energy production of glycolysis, pyruvic acid is irreversibly converted to acetaldehyde, and acetaldehyde is reduced to ethanol. This process in plants is essentially identical to alcoholic fermentation in anaerobic yeast, a well-known process at the heart of the brewing industry (Stryer, 1995). In plants, alcoholic fermentation occurs to the greatest extent in roots, where, under conditions of flooding, soil oxygen levels are depleted. Ethanol and some acetaldehyde are then produced and secreted by roots; in addition, some fraction of these C_2 oxVOCs also enters the transpiration stream and is released when stomata open (reviewed in Roshchina and Roshchina, 1993). Surprisingly, even highly aerobic plant tissues (i.e., leaves) have the potential to produce ethanol and acetaldehyde if allowed to become anaerobic (Kimmerer and MacDonald, 1987). This leaf process has been proposed to reflect a mechanism to avoid transient acidosis, since pyruvic acid decarboxylation results in formation of neutral species (acetaldehyde plus CO_2). Leaf ethanol and acetaldehyde production is also enhanced by a variety of stresses, including air pollution stress (MacDonald *et al.*, 1989). One or more of these processes might account for some of the acetaldehyde and ethanol seen in experiments with pine forest canopies (Goldan *et al.*, 1995). The control of the proportion of acetaldehyde to ethanol released has not been established.

The biochemistry of acetic acid formation in plants has not been investigated. It is known that acetic acid can be incorporated into plant products, via activation to acetyl-CoA, but whether its formation might occur by oxidation of acetaldehyde or by hydrolysis of acetyl-CoA (processes that occur in microorganisms) is unknown (Bode *et al.*, 1997). In Fig. 8 it is speculated that under metabolic conditions where high levels of acetyl-CoA are generated, such as during fatty acid oxidation (Stryer, 1995), some hydrolysis of acetyl-CoA might occur, leading to formation and release of acetic acid.

Few C_3 oxVOC emissions have been detected in plants; acetone is the exception. Murphy (1985) noted that acetone emission from germinating seeds was higher in those with a high lipid (triglyceride) content. It is likely that as these oil seeds germinate acetoacetic acid pools rise as a result of high rates of fatty acid oxidation. Acetoacetic acid could then spontaneously decarboxylate to produce acetone. This sequence of events is identical to ketone body formation in humans, which lead to acetone in the breath (Stryer, 1995). A similar metabolic sequence might explain acetone emission from conifer buds, where acetone emission correlated with acetoacetic acid levels (MacDonald and Fall, 1993); some plants are known to store triglyceride in their stems and to metabolize these energy sources prior to bud break.

B. REGULATION OF FORMATION

There is a general lack of understanding of oxVOC formation in plants, and as a result it is not feasible to discuss detailed controls of these processes. However, if the pathways described in Fig. 8 are correct, the main controls on C_1 to C_3 oxVOC biosynthesis can be predicted: rates of cell wall growth and photorespiration would control formation of C_1 oxVOCs; root flooding and the resulting anaerobiosis or leaf stress would control production of C_2 oxVOCs; and triglyceride mobilization would control acetone levels. An extension of these ideas is that plant development and seasonal factors may be important regulators of some oxVOC emissions from forests (Sharkey, 1996). For example, methanol emissions that are highest in expanding leaves would be predicted to be highest during bud break and leaf expansion that occurs at the beginning of the growing season. Likewise, acetone from conifers might peak just at bud break. Since some fast-growing conifers planted extensively for paper pulp production (e.g., loblolly pine) "flush" needles several times during the growing season, this might lead to repeated releases of acetone during the growing season. Formic acid emissions might peak during mid-summer when the highest levels of photosynthesis and photorespiration occur. Plants may emit the most acetaldehyde and ethanol after extensive periods of rain or seasonal flooding. It is clear that much remains to be learned about plant processes controlling C_1 to C_3 oxVOC formation and emission.

VII. EMISSION FROM THE PLANT TO THE ATMOSPHERE

Much of the preceding discussion concerns the mechanisms and regulation of formation of biogenic VOCs. How do these compounds exit plants, and what factors regulate their emissions from plant surfaces? The following discussion will focus primarily on isoprene and monoterpenes, since varieties of experiments have been conducted on the physical controls of isoprene and monoterpene emissions from foliage. By using these VOCs as examples, the following will be described: diffusion processes governing VOC emissions, the role of stored pools, and how wounding and physical damage expose VOC pools to volatilization.

The main paths available for VOC exit from a leaf into the atmosphere are summarized in Fig. 9. For simplicity we can consider four paths: (1) diffusion across the cuticle of the leaf epidermis; (2) conductance into the atmosphere through stomatal pores; (3) release from the leaf air spaces as a result of wounding; and (4) emission or evaporation of VOCs from material released to plant surfaces after wounding or physical damage. This scheme will be referred to in the following discussion.

A. SIMPLE DIFFUSION GOVERNS VOLATILE ORGANIC COMPOUND EMISSIONS

The fundamental processes governing emissions of VOCs like isoprene and monoterpenes from leaves have been summarized elsewhere (Tingey et al., 1991; Monson et al., 1995; Lerdau et al., 1997). Essentially, these emissions are a result of simple diffusion of the VOCs along a vapor pressure gradient

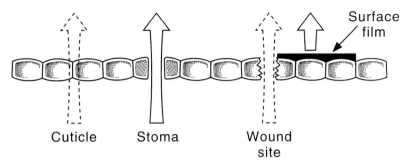

FIGURE 9   Summary of the pathways of emission of VOCs from leaf surfaces to the atmosphere. This scheme illustrates that the four major exit pathways for leaf VOCs are cuticular and stomatal conductance, emission from wounded surfaces, and emission (or evaporation) from surface films that form around wound sites. For simplicity, potential emission sites from stems and bark are not shown.

from cellular sites of relatively high VOC concentration and corresponding high VOC vapor pressure to the air surrounding the leaf, where there are relatively low VOC concentrations and correspondingly low effective vapor pressures. According to basic laws of diffusion, the emission flux of a VOC is related to the diffusion coefficient (k) of the compound, its vapor pressure in the leaf (VP_{leaf}) and the atmosphere ($VP_{atmosphere}$), and the resistance to diffusion:

$$\text{emission flux} = k\,(VP_{leaf} - VP_{atmosphere})/\text{resistance} \qquad (1)$$

A simplification of Eq. 1 is that the relatively low concentrations of VOCs in the atmosphere means the $VP_{atmosphere}$ term can essentially be neglected. However, the diffusion coefficient and resistance terms are complex, and unknown to some extent. As pointed out by Tingey *et al.* (1991), the diffusion coefficient of a VOC is determined by its size and shape, the viscosity of the medium through which it diffuses, and the temperature. For most VOCs the diffusion coefficients through cells, membranes, cell walls, and leaf air spaces are unknown. The rate of diffusion is also governed by path length, which is also uncertain, given the complexity of the architecture of a leaf. Often, VOC diffusion is related to the diffusion of water vapor or CO_2, whose diffusion properties in leaves are better understood, but which have very different polarities and molecular sizes than most VOCs.

The resistance term in Eq. 1 is actually composed of many resistances, whose effects are additive. For example, in considering that isoprene is produced in chloroplasts (Wildermuth and Fall, 1996) its passage out of the leaf is governed by the resistances it would encounter crossing, in turn, the chloroplast inner and outer membranes, the cytosol, the plasmalemma (cell membrane), the cell wall, the intercellular air space, and finally the stomatal pores. For monoterpene-emitting plants, at least three different cases are now evident, each with different resistance terms. First, for pines with stored internal pools of monoterpenes (see Fig. 5, left), the major resistances to diffusion to the atmosphere are passage across the secretory plus mesophyll cells forming the resin ducts, the leaf air space, and the stomatal pores. Second, for the case of pools of monoterpenes stored in external structures like glandular trichomes in mints (Fig. 5, right), the main resistance term is cuticular resistance. Third, in the case of plants that exhibit light-dependent monoterpene emissions (e.g., Mediterranean oaks), it is probable that the resistance pathways are similar to those for isoprene-emitting leaves.

It is now clear that isoprene is emitted from the surfaces of leaves that have stomata, since in the cases of hypostomatous leaves (stomata only on the lower surface) of oak, eucalyptus, and aspen (reviewed in Monson *et al.*, 1995), isoprene is essentially only emitted from that surface. For amphistomatous leaves (stomata on both surfaces) of cottonwood, isoprene is emitted from both surfaces. Similarly, for light-dependent α-pinene emission from *Quercus ilex* leaves, α-pinene was emitted only from the lower leaf surface that contains the stomata (Loreto *et al.*, 1996). Given these observations, it is somewhat ironic that isoprene emission rate is relatively unaffected by artificial closure of stomata induced by administration of the hormone abscisic acid or by interrupting the transpiration stream (Fall and Monson, 1992). The explanation for these results came from quick-freeze analysis of the isoprene content of treated leaves, which demonstrated that as stomata were induced to close in the light (when isoprene still continues to be formed in chloroplasts) the internal vapor pressure of isoprene, VP_{leaf} term, rose linearly, maintaining the emission flux (Eq. 1).

B. Emission from Stored Pools

Unlike isoprene, which is emitted from a small transient pool almost as soon as it is made, monoterpene emissions from most conifer needles result from volatilization from a large leaf pool or reservoir (reviewed in Tingey *et al.*, 1991; Monson *et al.*, 1995). As discussed in Section IV, conifer monoterpenes are stored in the leaf in resin ducts (see Fig. 5) or related structures. Daily emissions of monoterpene from these storage structures usually represent only a small fraction of the total monoterpene pool. The basal monoterpene emission rate is then primarily governed by the pool size that can vary as a function of season, and the instantaneous emission rates for monoterpenes from this pool are governed by leaf temperature and the effect of temperature on the vapor pressures of individual monoterpenes. Thus, the emissions of monoterpenes from conifer needles generally increase exponentially with temperature and show no dependence on light (light-dependent monoterpene emissions from spruce needles are exceptions; Schürmann *et al.*, 1993). Accordingly, monoterpene emission rates from pine needles do not change after a transition from light to dark, a treatment that normally produces stomatal closure (reviewed in Monson *et al.*, 1995). These results, as with isoprene emissions, can also be explained using Eq. 1. If the internal monoterpene vapor pressure (VP_{leaf} term) increases when stomata close, it can compensate for an increase in diffusional resistance, and the instantaneous rate of emission will be largely unchanged. The VP_{leaf} term can increase under these conditions because the large internal pool of conifer needle monoterpenes continues in the dark to be the source of volatile monoterpenes.

Similar thermal control of monoterpene emission rate is seen in the mint family (see Tingey *et al.*, 1991), which has external stores of these VOCs in the glandular trichomes described in Fig. 5. Exit of monoterpene from these structures occurs by simple diffusion across the epidermis, and is controlled by the waxy cuticle surface of epidermal cells (Fig. 9). In this case, relative humidity is also an important mediator of the emission rate, with higher emissions rates occurring at higher relative humidity. The explanation for these increased emissions is that hydration of the oil gland cuticle results in an increase in cuticular permeability to these compounds.

C. Wounding and Physical Damage Exposes Leaf Volatile Organic Compound Pools to Volatilization

The classic disturbance response of monoterpenes illustrates the importance of wounding and physical damage on VOC emissions. Early studies on monoterpene emissions from conifers, sages, and mints revealed that even minor physical disturbance led to long-lasting emissions of this class of VOCs (Tingey *et al.*, 1991). Disruption of monoterpene storage sites in these plants will expose oil surfaces to the atmosphere, allowing direct oil-to-air volatilization. In addition, physical damage leads to the exudation of oleoresin onto the plant surface in conifers, or the rupture of surface oil glands in mints. As mentioned earlier, monoterpenes can act as solvents for resin acids contained in resin ducts, and are subject to evaporation when oleoresins are released to plant surfaces. Ruptured oil glands spill their contents onto external leaf surfaces, where subsequent evaporation can occur. Epidermal damage induced by wind and hail or herbivore feeding also leads to exposure of internal surfaces and enhanced emissions of monoterpenes by similar processes. Figure 9 illustrates that the path of exit of some VOCs, such as wound-induced monoterpenes and hexenals, occurs by such processes.

Other VOCs that have significant leaf pools might also exit leaves by this path after similar damage. For example, Kirstine *et al.* (1998) studied the effects of cutting pasture grasses on VOC emission rates. They observed that VOC emission rates from clover or grass clippings increased immediately after cutting 80 to 180 times the maximal emission rates from undamaged plants at similar temperature and PAR. The emission rates increased with temperature and degree of maceration, and decreased exponentially with time. The pattern of VOCs emitted from these damaged tissues was quite different from undamaged plants, suggesting that wounding exposes internal pool of VOCs to evaporation or triggers formation of new VOCs in response to wounding. In agreement with the latter, the major VOCs released from damaged pasture grass were (expressed as percentage of total detected VOCs): (3Z)-hexenyl acetate (40%), (3Z)-hexenal (13%), (3Z)-hexenol (9%), and (2E)-hexanal (8%), members of the wound-induced hexenal family discussed in Section V. Other significant VOCs released included methanol (9%) plus other oxVOCs. Damaged clover emitted methanol (32%), butanone (16%), and acetone (9%) plus smaller amounts of other oxygenates including the hexenal family. Why clover has such large pools of methanol, butanone, and acetone is unknown.

Finally, it was mentioned earlier that decaying vegetation is a potentially large source of atmospheric oxVOCs (Warneke *et al.*, 1998). It is interesting that many of the same oxVOCs reported by Kirstine *et al.* (1998) in wounded plants have been seen when dried vegetation is hydrated. In addition to acetone and methanol, acetaldehyde and butanone were significant volatile oxgenated hydrocarbons (oxHCs) in both studies. These results suggest that plant matter, even when dried and rehydrated, can release oxHCs contained in internal pools or surface films.

VIII. CONCLUDING REMARKS AND UNCERTAINTIES

What can we conclude about the current state of knowledge of biogenic VOC emissions? Instead of focusing on the complex details presented earlier, it may be useful to address our general progress in understanding these larger questions concerning biogenic VOCs from plant sources. What biogenic VOCs are emitted to the atmosphere? When are VOCs produced? Where are

VOCs produced? How are VOCs synthesized? What regulates VOC formation and emission? Why are VOCs produced? Hopefully, the emerging answers to these questions may benefit a wider atmospheric science audience who is addressing current problems in biogenic VOC research, such as designing better field VOC sampling strategies and devices (including choosing when to sample and how to sample VOC emissions without disturbing or wounding plant surfaces), using patterns of VOC emissions as indicators of ecosystem health or stress, modeling regional and global VOC emissions, and predicting future changes in VOC emission patterns in response to climate and land use changes.

A. What Biogenic Volatile Organic Compounds Are Emitted from Plants?

We have seen that several major VOCs are emitted from plants. These VOCs represent a diverse range of chemical structures, from relatively simple hydrocarbons, alcohols, carbonyls, and acids to increasingly complex structures in the hexenal and monoterpene families. As mentioned in Section I, if floral scents and the monoterpene and sesquiterpene families are considered, the possible number of minor VOCs emitted from plants is in the thousands. It is unlikely that we will ever fully understand the complex biological details of so many compounds. Hopefully, knowledge gained for the few major VOCs presented here will provide a framework to understanding these minor VOCs and those yet to be discovered.

It is noteworthy that not all plants emit the same set of VOCs. Indeed, in some plant families, such as the oaks, individual species exhibit puzzling patterns of emission of isoprene or monoterpenes, but usually not both. Thus, additional complexities of biogenic VOC research are to understand the diversity of VOC formation, in particular, plant families, and to apply these principles to the problem of scaling up VOC emission estimates to landscapes, ecosystems, regions, and global vegetation.

B. When Are Plant Volatile Organic Compounds Produced in Plants?

As reviewed here, plant VOCs are produced on many different timescales. Minute-to-minute changes in the formation of VOCs might occur as a result of environmental changes, such as rapid variation of light intensity in a forest canopy. For example, Fig. 4A shows that instantaneous leaf isoprene emission varies within less than a minute in response to alterations in incident light. Formation of the hexenal family VOCs can also occur within a few minutes of wounding or physical damage. Day-to-day changes in the emissions of most VOCs are controlled by both biological processes and the physical environment, so that we see diurnal cycles of emissions of isoprene, MBO, monoterpenes, and methanol from forest canopies in response to changing light or temperature. Longer term seasonal changes are also very important for most VOC emissions, as the basal rate of formation of many of these compounds is controlled by physiological factors accompanying leaf growth, development, and senescence. These longer term aspects of biogenic VOC formation and emission are the least well understood, in large part because of the tremendous human effort required to perform seasonal field experiments, even on one plant species.

C. WHERE ARE VOLATILE ORGANIC COMPOUNDS PRODUCED IN PLANTS?

As illustrated in the VOC tree (see Fig. 1), volatile compounds are produced throughout plant structures. While much attention has focused on VOC formation in leaves, as we begin to understand the origins of oxVOCs there is a need to address processes that occur in stems and roots, where much of the emitted methanol and acetone may be formed. The transport of water-soluble VOCs from these tissues to leaves by the transpiration stream of the plant may be an important aspect of oxVOC dynamics. Other processes occurring in nonleaf tissues that impact VOC emissions include ethylene formation and its controls, and formation of wound-induced monoterpenes in conifers. Even when leaf VOCs are considered, the issue of where formation occurs is complex. The major leaf VOCs discussed here are produced in an array of different cellular compartments. This is especially significant in modeling leaf VOC emissions, since as discussed in Section VII, the diffusive processes controlling VOC passage through plant compartments and exit from the leaf are not completely understood.

D. HOW ARE VOLATILE ORGANIC COMPOUNDS FORMED IN PLANTS?

The biochemical mechanisms of VOC formation are very complex, comparable in complexity to the multiple channels for VOC oxidation in the atmosphere (Chapter 7). The pathways of VOC formation include many aspects of plant cell metabolism: amino acid metabolism (ethylene), lipid metabolism (isoprene, monoterpenes, hexenals, and probably MBO, acetone, and acetic acid), carbohydrate metabolism (acetaldehyde and ethanol), and perhaps the overlap of carbohydrate and C_1 metabolism (methanol, formaldehyde, and formic acid). Why do we need to know these metabolic details? One answer is that modeling of complex chemical processes, such as predicting the ozone-formation potential of isoprene photochemistry in the atmosphere or the isoprene-emission potential of an oak forest canopy, requires a understanding of the relevant chemical or biochemical reactions underlying and controlling the processes. It is fair to say that we have a good understanding of the biochemical mechanisms for the formation of ethylene, isoprene, many monoterpenes, the hexenal family, acetaldehyde, and ethanol. The biosynthetic mechanisms of other VOCs, such as methanol, formaldehyde, formic acid, acetic acid, and MBO, are much less certain (see Table 1).

E. WHAT REGULATES PLANT VOLATILE ORGANIC COMPOUND FORMATION AND EMISSIONS?

The regulation of VOC formation in plants is relatively uncertain and is an important area of active investigation. Major issues concern (1) the light-dependent regulation of isoprene, monoterpene, and MBO formation in certain plant groups; (2) the signals that control wound responses in hexenal and monoterpene formation; and (3) the metabolic controls of oxVOC formation in leaves, stems, and roots. Much is now known about the exceedingly complex regulation of ethylene formation in plants; presumably these controls are complex because ethylene is a hormone.

F. WHAT FACTORS MEDIATE VOLATILE ORGANIC COMPOUND EXIT FROM PLANT SURFACES?

As reviewed here (see Fig. 9), some VOCs, such as isoprene, methanol, and many monoterpenes (from unwounded leaves), have orderly exits from plants, passing out through stomatal pores, subject under some circumstance to restricted flow by controls on stomatal aperture. Certain monoterpenes exit from external storage structures, and the permeability properties of plant cuticles control the emission flux. Other VOCs exit without control by the host plant, as in the cases of monoterpenes, hexenals, and many oxVOCs, which are released after wounding or physical damage. As detailed in the text, the controls on this type of emission may be exerted by the degree to which herbivorous insects and other herbivores feed on and wound a large fraction of leaves in forest canopies or grasslands. Since the major driving force for most VOC emissions from plants, regardless of path, is the leaf-to-air VOC concentration gradient, further work is needed to understand diffusive paths and diffusion coefficients of VOCs in plant compartments.

G. WHY ARE VOLATILE ORGANIC COMPOUNDS PRODUCED IN PLANTS?

The reasons for the formation of some plant VOCs are now becoming clear, although the physiological rationales for formation of many VOCs are un-known. For ethylene, the hexenal family, and stored monoterpenes, our understanding of metabolic rationales is well advanced. Major questions persist in explaining the formation of light-dependent isoprenoids (including isoprene, MBO, and monoterpenes), and volatile organic acids and acetone. It is especially puzzling why leaves of plants producing isoprene may commit so much assimilated CO_2 and energy to formation of this hydrocarbon. Solving these questions is not just an academic exercise, because it is likely that the answers will contribute significantly to future models and predictions of regional and global VOC emissions.

It is also pertinent to point out that some VOC emissions from plants occur simply because biochemical products in leaves come in contact with highly ventilated surfaces—the internal volume of a leaf may contain 50% air spaces. Thus, compounds like acetone and methanol that have a significant air–water partition coefficient may exit leaves simply by inevitable losses to the exiting transpiration stream that drives water flow from roots to leaves. The gains in understanding photosynthetic CO_2 assimilation and water transpiration obtained from leaf gas exchange technology should continue to enhance our understanding of processes controlling VOC emissions from highly aerated leaf structures.

H. WHERE DO WE GO FROM HERE?

All scientific endeavors seem to generate more questions than they answer, and the same is true for biogenic VOC research. As pointed out in this chapter, there are numerous problems remaining to be solved. Perhaps the most important future direction involves interdisciplinary research. As summarized by Guenther *et al.* (1995) and Monson *et al.* (1995), continued progress in understanding biogenic VOC processes at different scales—from genes, proteins, and metabolic pathways; to plant compartments, cells, and tissues; to forest and crop canopies; and eventually to landscapes, ecosystems, regions, and the entire terrestrial biosphere—will require the cooperative interaction of biologists, biochemists, chemists, physicists, and atmospheric scientists. Attempts to construct useful VOC emission inventories will require two-way communication between these scientists, and their appreciation of the complex chemical pathways in the atmosphere and the equally complex biochemical pathways and regulation in living systems.

ACKNOWLEDGMENTS

The author is grateful to the following individuals who shared unpublished data or proved insights for this review: Ian Galbally, Alex Guenther, Peter Harley, Jürgen Kesselmeier, Manuel Lerdau, Werner Lindinger, Russ Monson, A. (Ravi) Ravishankara, Joerg-Peter Schnitzler, Guenther Seufert, Tom Sharkey, Hanwant Singh, Rainer Steinbrecher, and Pat Zimmerman.

REFERENCES

Abeles, F. B., Morgan, P. W., and Saltveit, M. E. (1992). "Ethylene in Plant Biology," 2nd ed., Academic Press: New York.

Andreae, M. O., and Crutzen, P. J. (1997). Atmospheric aerosols: biogeochemical sources and role in atmospheric chemistry. *Science.* **276**, 1052–1058.

Arey, J., Winer, A. M., Atkinson, R., Aschmann, S. M., Long, W. D., and Morrison, C. L. (1993). The emission of (Z)-3-hexen-1-ol, (Z)-3-hexenylacetate and other oxygenated hydrocarbons from agricultural plant species. *Atmos. Environ.* **25A**, 1063–1075.

Bode, K., Helas, G., and Kesselmeier, J. (1997). Biogenic contribution to atmospheric organic acids. *In* "Biogenic Volatile Organic Compounds in the Atmosphere" (G. Helas, J. Slanina, and R. Steinbrecher, eds.), pp. 157–170. SPB Academic Publishing: Amsterdam.

Chameides, W. L., Lindsay, R. W., Richardson, J., and Kiang, C. S. (1988). The role of biogenic hydrocarbons in urban photochemical smog: Atlanta as a case study. *Science* **241**, 1473–1475.

Colby, S. M., Alonso, W. R., Katahira, E. J., McGarbey, D. J., and Croteau, R. (1993). 4S-Limonene synthase from the oil glands of spearmint (*Mentha spicata*). cDNA isolation, characterization, and bacterial expression of the catalytically active monoterpene cyclase. *J. Biol. Chem.* **268**, 23016–23024.

Coley, P. D., and Aide, T. M. (1991). Comparison of herbivory and plant defenses in temperate and tropical broad-leaved forests. *In* "Plant-Animal Interactions: Evolutionary Ecology in Tropical and Temperate Regions" (P. W. Price, T. M., Lewinsohn, G. W., Fernandes, and W. W. Benson, eds.), pp. 25–49. John Wiley & Sons, New York.

Conconi, A., Smerdon, M. J., Howe, G. A., and Ryan, C. A. (1996). The octadecanoid signaling pathway in plants mediates a response to ultraviolet radiation. *Nature* **383**, 826–829.

Conrad, R. (1995). Soil microbial processes and the cycling of atmospheric trace gases. *Phil. Trans. R. Soc. London Ser. A.* **351**, 219–230.

Croft, K. P. C., Jüttner, F., and Slusarenko, A. J. (1993). Volatile products of the lipoxygenase pathway evolved from *Phaseolus vulgaris* (L.) leaves inoculated with *Pseudomonas syringae* pv. *phaseolicola*. *Plant Physiol.* **101**, 13–24.

Croteau, R., and Gershenzon, J. (1994). Genetic control of monoterpene biosynthesis in mints (*Mentha*:Lamiaceae). *Rec. Adv. Phytochem.* **28**, 193–229.

Deneris, E. S., Stein, R. A., and Mead, J. F. (1985). Acid-catalyzed formation of isoprene from a mevalonate-derived product using a rat liver cytosolic fraction. *J. Biol. Chem.* **260**, 1382–1385.

Dey, P. M., and Harborne, J. B. (1997). "Plant Biochemistry," Academic Press: San Diego.

Dobson, H. (1993). Floral volatiles in insect biology. *In* "Insect-Plant Interactions" (E. A. Bernays, ed.), pp. 47–81. CRC Press: Boca Raton, FL.

Duce, R. A., Mohnen, V. A., Zimmerman, P. R., Grosjean, D., Cautreels, W., Chatfield, R., Jaenicke, R., Ogren, J. A., Pellizzari, E. D., and Wallace, G. T. (1983). Organic matter in the global troposphere. *Rev. Geophys. Space Phys.* **21**, 921–952.

Fall, R. and Benson, A. A. (1996). Leaf methanol—the simplest natural product from plants. *Trends Plant Sci.* **1**, 296–301.

Fall, R. and Monson, R. K. (1992). Isoprene emission rate in relation to stomatal distribution and stomatal conductance. *Plant Physiol.* **100**, 987–992.

Fall, R. and Wildermuth, M. C. (1998). Isoprene synthase: From biochemical mechanism to emission algorithm. *J. Geophys. Res.*, in press.

Fantechi, G., Jensen, N. R., Hjorth, J., and Peeters, J. (1996). Mechanistic study of the atmospheric degradation of isoprene and MBO. EUROTRAC Symposium (Transport and transformation of pollutants in the troposphere over Europe), Garmisch-Partenkirchen, Germany.

Fehsenfeld, F., Calvert, J., Fall, R., Goldan, P., Guenther, A., Hewitt, C. N., Lamb, B., Liu, S., Trainer, M., Westberg, H., and Zimmerman, P. (1992). Emissions of volatile organic compounds from vegetation and the implications for atmospheric chemistry. *Global Biogeochem. Cycles* **6**, 389–430.

Ferry, J. G. (ed.) (1993). "Methanogenesis. Ecology, Physiology, Biochemistry and Genetics," Chapman & Hall: New York.

Gershenzon, J., and Croteau, R. (1991). Terpenoids. *In* "Herbivores. Their Interactions with Secondary Plant Metabolites," 2nd ed., Vol. 1: The Chemical Participants, G. A., Rosenthal, and M. R. Berenbaum, (eds.), pp. 165–219 Academic Press: San Diego.

Goldan, P. D., Kuster, W. C., Fehsenfeld, F. C., and Montzka, S. A. (1993). The observation of a C_5 alcohol in a North American pine forest. *Geophys. Res. Lett.* **20**, 1039–1042.

Goldan, P. D., Kuster, W. C., Fehsenfeld, F. C., and Montzka, S. A. (1995). Hydrocarbon measurements in the southeastern United States: the Rural Oxidants in the Southern Environment (ROSE) Program 1990. *J. Geophys. Res.* **100**, 25,945–25,963.

Graedel, T. E. (1979). "Chemical Compounds in the Atmosphere," Academic Press: New York.

Guenther, A., Hewitt, C. N., Erickson, D., Fall, R., Geron, C., Graedel, T., Harley, P., Klinger, L., Lerdau, M., McKay, W. A., Pierce, T., Scholes, B., Steinbrecher, R., Tallamraju, R., Taylor, J., and Zimmerman, P. (1995). A global model of natural volatile organic compound emissions. *J. Geophys. Res.* **100**, 8873–8892.

Guenther, A., Zimmerman, P., and Wildermuth, M. (1994). Natural volatile organic compound emission rate estimates for U.S. woodland landscapes. *Atmos. Environ.* **28**, 1197–1210.

Haagen-Smit, A. J. (1952). Chemistry and physiology of Los Angeles smog. *Ind. Eng. Chem.* **44**, 1342–1346.

Harborne, J. B. (1993). "Introduction to Ecological Biochemistry," Academic Press: London.

Harley, P., Fridd-Stroud, V., Greenberg, J., Guenther, A., and Vasconcellos, P., (1998a). Emission of 2-methyl-3-buten-2-ol by pines: a potentially large natural source of reactive carbon to the atmosphere. *J. Geophys. Res.* in press.

Harley, P., Lerdau, M., and Monson, R. (1998b). Ecological and evolutionary aspects of isoprene emission. *Oecologia* (submitted).

Hatanaka, A. (1993). The biogeneration of green odour by green leaves. *Phytochem.* **34**, 1201–1218.

Hatanaka, A., Kajiwara, T., and Sekiya, J. (1987). Biosynthetic pathway for C_6-aldehydes formation from linolenic acid in green leaves. *Chem. Phys. Lipids.* **44**, 341–361.

Helas, G., Slanina, S., and Steinbrecher, R. (eds.) (1997). "Biogenic Volatile Organic Compounds in the Atmosphere," SPB Academic Publishing: Amsterdam.

Isidorov, V. A., Zenkevich, I. G., and Ioffe, B. V. (1985). Volatile organic compounds in the atmosphere of forests. *Atmos. Environ.* **19**, 1–8.

Kende, H. (1993). Ethylene biosynthesis. *Annu. Rev. Plant Physiol. Plant Mol. Biol.* **44**, 283–307.

Kesselmeier, J., Bode, K., Gerlach, C., and Jork, E.-M. (1998). Exchange of atmospheric formic and acetic acid with trees and crop plants under controlled chamber and purified air conditions. *Atmos. Environ.* **32**, 1765–1775.

Kesselmeier, J., Bode, K., Hofmann, U., Müller, H., Schäfer, L., Wolf, A., Ciccioli, P., Brancaleoni, E., Cecinato, A., Frattoni, M., Foster, P., Ferrari, C., Jacob, V., Fugit, J. L., Dutaur, L., Simon, V., and Torres, L. (1997). Emission of short chained organic acids, aldehydes and monoterpenes from *Quercus ilex* L. and *Pinus pinea* L. in relation to physiological activities, carbon budget and emission algorithms. *Atmos. Environ.* **31**(S1), 119–133.

Kimmerer, T. W., and MacDonald, R. C. (1987). Acetaldehyde and ethanol biosynthesis in leaves of plants. *Plant Physiol.* **84**, 1204–1209.

Kirstine, W., Galbally, I., Ye, Y., and Hooper, M. (1998). Emissions of volatile organic compounds (including oxygenated species) from pasture. *J. Geophys. Res.* **103**, 10605–10619.

Kitamura, A., Matsui, K., Kajiwara, T., and Hatanaka, A. (1992). Changes in volatile C_6 aldehydes emitted from and accumulated in tea leaves. *Plant Cell Physiol.* **33**, 493–496.

Knudsen, J. T., Tollsten, L., and Bergström, L. G. (1993). Floral scents—a checklist of volatile compounds isolated by head-space techniques. *Phytochemistry.* **33**, 253–280.

Kuzma, J. and Fall, R. (1993). Leaf isoprene emission rate is dependent on leaf development and the level of isoprene synthase. *Plant Physiol.* **101**, 435–440.

Kuzma, J., Nemecek-Marshall, M., Pollock, W. H., and Fall, R. (1995). Bacteria produce the volatile hydrocarbon isoprene. *Curr. Microbiol.* **30**, 97–103.

Langenheim, J. H. (1994). Higher plant terpenoids: a phytocentric overview of their ecological roles. *J. Chem. Ecol.* **20**, 1223–1280.

Lanne, B. S., Ivarsson, P., Johnsson, P., Bergström, G., and Wassgren, A.-B. (1989). Biosynthesis of 2-methyl-3-buten-2-ol, a pheromone component of Ips typographus (Coleoptera: Scolytidae). *Insect Biochem.* **19**, 163–167.

Lerdau, M., Guenther, A., and Monson, R. (1997). Plant production and emission of volatile organic compounds. *BioSci.* **47**, 373–383.

Lichtenthaler, H. K., Schwender, J., Disch, A., and Rohmer, M. (1997). Biosynthesis of isoprenoids in higher plant chloroplasts proceeds via a mevalonate-independent pathway. *FEBS Lett.* **400**, 271–274.

Loreto, F., Ciccioli, P., Brancaleoni, E., Cecinato, A., Frattoni, M., and Sharkey, T. D. (1996). Different sources of reduced carbon contribute to form three classes of terpenoid emitted by *Quercus ilex* L. leaves. *Proc. Natl. Acad. Sci. U.S.A.* **93**, 9966–9969.

Lowman, M. D. (1992). Leaf growth dynamics and herbivory in five species of Australian rainforest canopy trees. *J. Ecol.* **30**, 433–447.

MacDonald, R. C., and Fall, R. (1993). Acetone emission from conifer buds. *Phytochemistry.* **34**, 991–994.

MacDonald, R. C., Kimmerer, T. W., and Razzaghi, M. (1989). Aerobic ethanol production by leaves: evidence for air pollution stress in trees in the Ohio River Valley, USA. *Environ. Pollut.* **62**, 337–351.

McGarvey, D. J., and Croteau, R. (1995). Terpenoid metabolism. *Plant Cell.* **7**, 1015–1026.

McKeen, S. A., Gierczak, T., Burkholder, J. B., Wennberg, P. O., Hanisco, T. F., Keim, E. R., Gao, R.-S., Liu, S. C., Ravishankara, A. R., and Fahey, D. W. (1997). The photochemistry of acetone in the upper troposphere: a source of odd-hydrogen radicals (submitted).

McKeon, T. A., Fernández-Maculet, J. C., and Yang, S. F. (1995). Biosynthesis and metabolism of ethylene. *In* "Plant Hormones. Physiology, Biochemistry and Molecular Biology," 2nd ed., P. J. Davies, (ed.), pp. 118–139. Kluwer Academic Publishers: Dordrecht.

Michelozzi, M., Raschi, A., Tognetti, R., and Tosi, L. (1997). Eco-ethological analysis of the interaction between isoprene and the behavior of Collembola. *Pedobiologia.* **41**, 210–214.

Monson, R. K., and Fall, R. (1989). Isoprene emissions from aspen leaves. Influence of environment and relation to photosynthesis and photorespiration. *Plant Physiol.* **90**, 267–274.

Monson, R. K., Hills, A. J., Zimmerman, P. R., and Fall, R. (1991). Studies of the relationship between isoprene emission rate and CO_2 or photon-flux density using a real-time isoprene analyzer. *Plant Cell Environ.* **14**, 517–523.

Monson, R. K., Jaeger, C. H., Adams III, W. W., Driggers, E. M., Silver, G. M., and Fall, R. (1992). Relationships among isoprene emission rate, photosynthesis, and isoprene emission rate as influenced by temperature. *Plant Physiol.* **98**, 1175–1180.

Monson, R., Lerdau, M., Sharkey, T., Schimel, D., and Fall, R. (1995). Biological aspects of constructing biological hydrocarbon inventories. *Atmos. Environ.* **29**, 2989–3002.

Monson, R. K., Harley, P. C., Litvak, M. E., Wildermuth, M., Guenther, A. B., Zimmerman, P. R., and Fall, R. (1994). Environmental and developmental controls over the seasonal pattern of isoprene emission from aspen leaves. *Oecologia.* **99**, 260–270.

Murphy, J. (1985). Acetone production during germination of fatty seeds. *Physiol. Plant.* **63**, 231–234.

O'Donnell, P. J., Calvert, C., Atzorn, R., Wasternack, C., Leyser, H. M. O., and Bowles, D. J., (1996). Ethylene as a signal mediating the wound response of tomato plants. *Science.* **274**, 1914–1917.

Ohta, K. (1986). Diurnal and seasonal variations in isoprene emission from live oak. *Geochem. J.* **19**, 269–274.

Puxbaum, H. (1997). Biogenic emissions of alcohols, ester, ether and higher aldehydes. *In* "Biogenic Volatile Organic Compounds in the Atmosphere" (G., Helas, J., Slanina, and R. Steinbrecher, eds.), pp. 79–99. SPB Academic Publishing: Amsterdam.

Rajaonarivony, J. I. M., Gershenzon, J., and Croteau, R. (1992). Characterization and mechanism of (4S)-limonene synthase, a monoterpene cyclase from the glandular trichomes of peppermint (*Mentha* × *piperita*). *Arch. Biochem. Biophys.* **296**, 49–57.

Rasmussen, R. A. (1970). Isoprene: identified as a forest-type emission to the atmosphere. *Environ. Sci. Technol.* **4**, 667–671.

Rasmussen, R. A. (1978). Isoprene plant species list, Special report of the Air Pollution Section, Washington State University, Pullman, WA.

Röse, U. S. R., Manukian, A., Heath, R. R., and Tumlinson, J. H. (1996). Volatile semiochemicals released from undamaged cotton leaves. *Plant Physiol.* **111**, 487–495.

Roshchina, V. V., and Roshchina, V. D. (1993). "The Excretory Function of Higher Plants," Springer-Verlag: Berlin.

Rudich, Y., Talukdar, R., Burkholder, J. B., and Ravishankara, A. R. (1995). Reaction of methylbutenol with hydroxyl radical: mechanism and atmospheric implications. *Phys. Chem.* **99**, 12188–12194.

Rudolph, J. (1997). Biogenic sources of atmospheric alkenes and acetylene. In "Biogenic Volatile Organic Carbon Compounds in the Atmosphere" (G. Helas, J. Slanina, and R. Steinbrecher, eds.), pp. 53–65. SPB Academic Publishing: Amsterdam.

Sanadze, G. A. (1991). Isoprene effect—light-dependent emission of isoprene by green parts of plants. In "Trace Gas Emissions by Plants" (G. A. Sanadze, ed.), pp. 135–152. Academic Press: San Diego.

Schnitzler, J.-P., (1997) Fraunhofer-Institut für Atmosphärische Umweltforschung, Garmisch-Partenkirchen, Germany, personal communication.

Schnitzler, J.-P., Lehning, A., and Steinbrecher, R. (1997). Seasonal pattern of isoprene synthase activity in Quercus robur leaves and its significance for modeling isoprene emission rates. Bot. Acta 110, 240–243.

Schürmann, W., Ziegler, H., Kotzias, D., Schönwitz, R., and Steinbrecher, R. (1993). Emission of biosynthesized monoterpenes from needles of Norway spruce. Naturwissenschaften. 80, 276–278.

Seufert, G., Bartzis, J., Bomboi, T., Ciccioli, P., Cieslik, S., Dlugi, R., Foster, P., Hewitt, C. N., Kesselmeier, J., Kotzias, D., Lenz, R., Manes, F., Perez Pastor, R., Steinbrecher, R., Torres, L., Valentini, R., and Versino, B. (1997). An overview of the Casteporziano experiments. Atmos. Environ. 31(S1), 5–18.

Sharkey, T. D. (1996). Emission of low molecular mass hydrocarbons from plants. Trends Plant Sci. 1, 78–82.

Sharkey, T. D., Holland, E. A., and Mooney, H. A. (eds.) (1991). "Trace Gas Emissions by Plants," Academic Press: San Diego.

Sharkey, T. D., and Loreto, F. (1993). Water stress, temperature, and light effects on the capacity for isoprene emission and photosynthesis of kudzu leaves. Oecologia. 95, 328–333.

Silver, G. M., and Fall, R. (1995). Characterization of aspen isoprene synthase, an enzyme responsible for leaf isoprene emission to the atmosphere. J. Biol. Chem. 270, 13010–13016.

Singh, H. B., Kanakidou, M., Crutzen, P. J., and Jacob, D. J. (1995). High concentrations and photochemical fate of oxygenated hydrocarbons in the global troposphere. Nature 378, 50–54.

Singh, H. B., and Zimmerman, P. R. (1992). Atmospheric distribution and sources of nonmethane hydrocarbons. In "Gaseous Pollutants: Characterization and Cycling" (J. O. Nriagu, ed.), pp. 177–235. Wiley-Interscience: New York.

Singsaas, E. L., Lerdau, M., Winter, K., and Sharkey, T. D. (1997). Isoprene increases thermotolerance of isoprene-emitting species. Plant Physiol. 115, 1413–1420.

Steele, C. L., Lewinsohn, E., and Croteau, R. (1995). Induced oleoresin biosynthesis in grand fir as a defense against bark beetles. Proc. Natl. Acad. Sci. U.S.A. 92, 4164–4168.

Steinbrecher, R. (1997). Isoprene: production by plants and ecosystem-level estimates. In "Biogenic Volatile Organic Carbon Compounds in the Atmosphere" (G., Helas, J., Slanina, and R., Steinbrecher, eds.), pp. 101–114. SPB Academic Publishing: Amsterdam.

Stryer, L. (1995). "Biochemistry," W. H. Freeman, New York.

Takabayashi, J., Dicke, M., and Posthumus, M. A. (1994). Volatile herbivore-induced terpenoids in plant-mite interactions: variation caused by biotic and abiotic factors. J. Chem. Ecol. 20, 1329–1354.

Terry, G. M., Stokes, N. J., Hewitt, C. N., and Mansfield, T. A. (1995). Exposure to isoprene promotes flowering in plants. J. Exp. Bot. 46, 1629–1631.

Tingey, D. T., Standley, C., and Field, R. W. (1976). Stress ethylene evolution: a measure of ozone effects on plants. Atmos. Environ. 10, 969–974.

Tingey, D. T., Turner, D. P., and Weber, J. A. (1991). Factors controlling the emissions of monoterpenes and other volatile organic compounds. In "Trace Gas Emissions by Plants" (T. D. Sharkey, E. A. Holland, and H. A. Mooney, eds.), pp. 93–119 Academic Press: San Diego.

Turlings, T. C. J., Loughrin, J. H., McCall, P. J., Röse, U. S. R., Lewis, W. J., and Tumlinson, J. H. (1995). How caterpillar-damaged plants protect themselves by attracting parasitic wasps. *Proc. Natl. Acad. Sci. U.S.A.* **92**, 4169–4174.

Warneck, P. (1993). Chemical changes of the atmosphere on geological and recent time scales. *In* "Global Atmospheric Chemical Change" (C. N. Hewitt, and W. T. Sturges, eds.), pp. 1–52. Elsevier Applied Science: London.

Warneke, C., Karl, T., Judmaier, H., Hansel, A., Jordan, A., Lindinger, W., and Crutzen, P. J. (1998). Acetone, methanol and other partially oxidized volatile organic emissions from dead plant matter by abiological processes: significance for atmospheric chemistry. *Global Biogeochem. Cycles*, in press.

Wasternack, C., and Parthier, B. (1997). Jasmonate-signaled plant gene expression. *Trends Plant Sci.* **2**, 302–307.

Wellburn, A. R., and Wellburn, F. A. M. (1996). Gaseous pollutants and plant defence mechanisms. *Biochem. Soc. Trans.* **24**, 461–464.

Went, F.W. (1960). Blue hazes in the atmosphere. *Nature* **187**, 641–643.

Wildermuth, M. C., and Fall, R. (1996). Light-dependent isoprene emission. Characterization of a thylakoid-bound isoprene synthase in *Salix discolor*. *Plant Physiol.* **112**, 171–182.

Wildermuth, M. C., and Fall, R. (1998). Biochemical characterization of stromal and thylakoid-bound isoforms of isoprene synthase in willow chloroplasts. *Plant Physiol.* **116**, 1111–1123.

Zeidler, J. G., Lichtenthaler, H. K., May, H. U., and Lichtenthaler, F.W. (1997). Is isoprene emitted by plants synthesized via the novel isopentenyl pyrophosphate pathway? *Z. Naturforsch.* **52c**, 15–23.

Zimmerman, P. (1979). Determination of emission rates of hydrocarbons from indigenous species of vegetation in the Tampa/St. Petersburg, Florida area. Appendix C, Tampa Bay Area Photochemical Study, EPA 904/9-77-028 U.S. Environmental Protection Agency, Region IV, Atlanta, GA.

CHAPTER 3

Modeling Biogenic Volatile Organic Compound Emissions to the Atmosphere

ALEX GUENTHER

Atmospheric Chemistry Division, National Center for Atmospheric Research, Boulder, Colorado

This chapter describes methods of estimating the magnitude and distributions of biogenic volatile organic compound (VOC)

Reactive Hydrocarbons in the Atmosphere
Copyright © 1999 by Academic Press. All rights of reproduction in any form reserved.

emissions. Although large uncertainties are associated with biogenic VOC emission estimates, there is a need for best estimates that can be used in global and regional chemistry and transport models. Numerical methods for characterizing some of the better known sources, including vegetation foliage, are reviewed. Initial global emission estimates are derived for other sources. Some sources that have a negligible annual global flux may still be important, at least in certain regions or seasons.

I. EMISSION MODELING

Hydrocarbon emission models are necessary components of the chemistry and transport models that synthesize our understanding of atmospheric chemistry and are important tools for investigating the processes that control the composition of the atmosphere. Global models are used to predict how human perturbations might result in changes in climate and chemical composition of the atmosphere. Emissions of biogenic hydrocarbons are sensitive to global changes in climate and landcover (Turner et al., 1991). While both biogenic and anthropogenic emissions are important for global atmospheric chemistry, the total annual emission rate of reactive volatile organic compounds (VOCs) to the global atmosphere is dominated by biogenic sources. For example, Guenther et al. (1995) estimate that the total global emission rate of biogenic VOC of about 1150×10^{12} grams (Tg) per year is more than seven times greater than the total global emission rate of anthropogenic VOC.

On a regional basis, regulatory air quality models are used to identify the relative contribution of various emission sources to pollution levels so that the most effective emission reduction scenarios can be developed. This process is relatively straightforward for pollutants that are emitted directly into the atmosphere and undergo few chemical transformations. This is not the case for ozone, which has a very complex origin. Ozone reduction is difficult and has been elusive in many regions. Numerical studies have shown that an effective ozone control strategy for many regions requires an understanding of both anthropogenic and biogenic emissions of its precursor compounds (Chameides et al., 1988; Roselle et al., 1991) even though only anthropogenic emissions are considered for emission reduction. A good example is the Atlanta, Georgia urban area. Chameides et al. (1988) estimated that biogenic VOC emissions in excess of 50 kg/km²/d, in the presence of anthropogenic NO_x, could result in harmful ozone levels, above 120 ppb, even if there were no anthropogenic VOC emissions. Geron et al. (1995) estimated that biogenic VOC emissions in the Atlanta urban area are about 56 kg/km²/d, indicating that it is not possible to reduce ozone to below 120 ppb by reducing anthropogenic VOC emissions and that instead reductions in anthropogenic NO_x emissions are required.

How can we estimate biogenic emissions for the highly variable, and in many cases little understood, sources described in Chapter 2? The first step in the emission estimation process is to calculate the order of magnitude of a source. Rasmussen (1972) based the first U.S. estimate of isoprene and monoterpene emissions on an emission rate that was determined by placing a mix of oak and pine foliage in a 1 liter flask. This rate was multiplied by 1800 h (180 ten-hour days) and a canopy depth that was assumed to range from 10 to 200 cm. The resulting estimated range of 2.3 to 46.4 Tg per year demonstrated that natural emissions could be a significant component of the total U.S. VOC flux. The next step in the emission estimation process is to compile a database of emission estimates, which is known as an emission inventory. Zimmerman (1979) generated the first biogenic hydrocarbon emission inventory by dividing the United States into several biomes and latitude zones and then estimating emissions for four seasons. A decade later, there was considerable interest in developing emission inventories for Europe as well (Anastasi et al., 1991; Hewitt and Street, 1992; Molnar, 1990). Although these emission estimates were very uncertain, they were incorporated into regulatory oxidant models and demonstrated that, in some cases, ozone predictions were sensitive to estimates of biogenic VOC emissions.

Emission inventories can be used as inputs to photochemical models, but a more useful tool is a dynamic emission model that responds to changes in environment, land use, and other factors. This enables the photochemical modeler to predict emissions for the specific scenario under investigation. Emission models also benefit from biochemical and ecological studies of the processes controlling biogenic emissions (e.g., Silver and Fall, 1991; Monson et al., 1992). These studies can identify the factors controlling emissions and suggest a framework for numerical algorithms. Models predict emission rates, E, as the product of (1) an emission factor, ϵ, that represents the emission rate at a given activity level; (2) activity factor, γ, that represents the activity level, and (3) a source density factor, D, that represents the number of sources within an area. The same general approach,

$$E = \epsilon\gamma D, \tag{1}$$

can be used for both anthropogenic and natural emissions.

How are these factors determined? Figure 1 illustrates some of the techniques used to investigate biogenic VOC emissions. Various flux measurement techniques are described in detail by Guenther et al. (1996a). Flux measurement systems that operate on small scales, such as enclosures, are used to investigate how biological and environmental conditions control emission rates. Individual leaves, branches, or entire trees can be enclosed and their

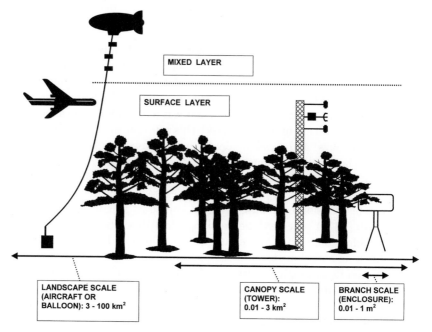

MIXED LAYER

SURFACE LAYER

LANDSCAPE SCALE (AIRCRAFT OR BALLOON): 3 - 100 km²

CANOPY SCALE (TOWER): 0.01 - 3 km²

BRANCH SCALE (ENCLOSURE): 0.01 - 1 m²

FIGURE 1   Illustration of the scales and measurement techniques used to develop and evaluate biogenic emission models.

environment carefully characterized and controlled. Micrometeorological flux systems measure above canopy fluxes and are used to evaluate the net flux predicted by emission models. These systems can be deployed on towers, tethered balloons, and aircraft. The horizontal canopy area characterized by these flux techniques, which are dependent on factors that include the height of the flux system and the meteorological conditions, is typically about 0.01 to 3 km² for tower-based systems and 3 to 100 km² for tethered balloon or aircraft systems. Tower flux systems are particularly useful for investigating how emissions vary with time on diurnal, seasonal, and year-to-year time-scales. The tethered balloon and aircraft systems are used to characterize large areas.

In this chapter, methods for estimating trace gas emissions and the factors ϵ, γ, and D are described in detail for isoprene and monoterpenes, which are the dominant reactive VOCs in the atmospheric boundary layer above many terrestrial surfaces. Methods include numerical algorithms for characterizing some sources, and simple estimates for other sources. Although these estimates are very uncertain, it is useful to calculate the order of magnitude of each emission source to determine its relative importance. This often produces a value that is within the range of estimates predicted by much more detailed methods. For example, the first global estimate of biogenic hydrocarbon emissions, 432 Tg per year (Rasmussen and Went, 1965), was based simply on estimates of the amount of terpenes in plants and the global total biomass. This estimate is within a factor of three of all estimates, based on detailed global models, demonstrating the usefulness of a first-order estimate.

II. ISOPRENE

Isoprene is only one of more than a hundred biogenic VOCs emitted into the atmosphere (see Chapter 2) but this single compound has a very important role in global atmospheric chemistry. The global annual isoprene flux is of a similar magnitude to that of methane but isoprene is more than four orders of magnitude more reactive. Table 1 indicates that together these two compounds account for over half of the total global VOC flux. Isoprene is the largest sink of the OH radical in the boundary layer over many land surfaces while methane and CO dominate in most of the rest of the atmosphere.

A. CANOPY FOLIAGE

The foliage in vegetation canopies is the largest source of isoprene emissions and considerable efforts have been directed toward the development of emission models that can simulate emissions from this source. The resulting models provide reasonable predictions of ambient isoprene concentrations over forests in at least some locations and times of year (Guenther *et al.*, 1996a; Geron *et al.*, 1997; Lamb *et al.*, 1996). The three factors (Eq. 1) used to estimate isoprene emissions are described in the following sections.

1. Source Factor, D

The source factor for vegetation foliage, foliar density, is usually expressed as grams of dry weight foliar mass per square meter of ground area (gdw/m^2). There are two components to foliar density estimates: the peak (the seasonal maximum) and the seasonal fractions. Peak foliar densities range from less than 50 gdw/m^2 for semibarren landscapes to around 1500 gdw/m^2 for some forests. Guenther (1997) compared six models that predict biogenic hydrocarbon emissions from vegetation foliage in the United States. Some models assigned a constant foliar density to broad vegetation types while others used more complex algorithms that predict CO_2 uptake, carbon partitioning into

TABLE 1 Annual Global VOC Emission Estimates[a]

Source	Isoprene	Monoterpenes (Tg of carbon)	Methane	Other VOCs[b]
Canopy foliage	460	115	<1	500
Terrestrial ground cover and soils	40	13	175	50
Flowers	0	2	0	2
Ocean and freshwater	1	<0.001	15	10
Animals, humans, and insects	0.003	<0.001	100	0.003
Anthropogenic (including biomass burning)	0.01	1	220	93
Total	~500	~130	~510	~650

[a]Based on Watson et al., 1992; Guenther et al., 1995; Singh and Zimmerman, 1992; and this study.
[b]Other VOCs include all volatile organic compounds other than methane, isoprene, and monoterpenes.

leaves, and leaf longevity. The average U.S. peak foliar density estimate ranged from 360 to 705 gdw/m^2 while estimates for specific biomes varied by as much as a factor of three. The foliar density estimates predicted by the global hydrocarbon emission model of Guenther et al. (1995) have been compared with estimates of species-specific foliar density data for these regions. The global model overpredicted foliar densities by about 60% in Canada and by a factor of two to three in southern Africa (Guenther et al., 1996b). The uncertainty associated with peak foliar density estimates is a large part of the total uncertainty in biogenic VOC emission model estimates and efforts to improve these estimates are needed.

There are two approaches being developed that should result in significantly improved estimates of peak foliar density. Geron et al. (1994) describe a method for estimating peak foliar densities based on vegetation coverage (the fraction of a landscape covered by vegetation) and species composition. Land surfaces are considered mosaics of areas covered by individual plant species and barren areas. Peak foliar density tends to be fairly uniform for a plant species and this method can provide accurate estimates of foliar density. The limitation of this approach is the availability of suitable databases. The data have been compiled in the United States and some other regions, in an effort to inventory natural resources. In other regions, these data may be obtained from ground measurement studies that have been conducted to calibrate satellite-derived landcover classifications.

Foliar density has been directly estimated from optical satellite measurements for over a decade (Running *et al.*, 1989), but these data are of limited application because they saturate at high foliar densities and are seriously degraded by cloud cover. Foliage databases are available on global scales at high resolution (EDC-NESDIS, 1992) but the estimates are uncertain. New techniques, particularly those that combine optical measurements with microwave, lidar, and other sensors, may improve remote-sensing estimates of foliage on regional and global scales.

Foliar density varies with season in landscapes with deciduous plants but is relatively constant for evergreen vegetation. Models using climatological data (Lamb *et al.*, 1987) or satellite observations (Guenther *et al.*, 1995) have estimated seasonal variations in foliar density. The climatological data that can be used to predict seasonal variations in foliage include temperature and precipitation. It is likely that reasonable relationships can be defined for specific locations but it is difficult to apply these relationships globally. Satellite observations are more suitable for estimating foliar variations on global scales although clouds and sensor saturation at high foliar densities limit the accuracy of the optical techniques that are in use.

2. Emission Factor, ϵ

Landscape average isoprene emission factors can be estimated using either "bottom-up" or "top-down" approaches. The bottom-up approach integrates estimates of species composition with emission factors for each plant species. The emission factor for each plant species is estimated using the enclosure measurement methods described later. The landscape average emission factor is the weighted average of the emission factors associated with each plant species. The top-down approach uses the micrometeorological or inverse modeling techniques (see Guenther *et al.*, 1996a) to directly estimate emission rates from which landscape average emission factors can be estimated using foliar density and emission activity factors.

Emission factors for individual plant species can be estimated using enclosure measurement techniques (Fig. 1). Individual leaves or branches are placed in an enclosure and emissions are estimated from a mass balance calculation. The mass balance can be based on an increase in concentration with time (air is circulated within the chamber), or an increase in concentration between the inlet and outlet airstreams (air flows through and does not return to the chamber). An emission factor is estimated by dividing the emission rate by the amount of foliage and an emission activity factor. Dry weight foliar mass is typically used to quantify the amount of foliage. Alternatively, projected leaf area can be used. Most studies have used dry weight foliar mass because it is a relatively simple measurement. Foliage within the enclosure is cut, dried in an oven, and then weighed. It is not clear whether mass or area is the better of the two units for characterizing emission factors. As improved estimates of foliage distributions from satellite data become available, it will be important to determine whether these data are more representative of foliar mass or of leaf area in order to select which of these should be used for biogenic VOC emission factors.

There is a wide range of techniques and protocols that have been used for enclosure measurement studies. Some investigators only report that they observed an emission but provide no quantitative emission rate estimates. Other studies report emission rates but not the ancillary data (e.g., light and temperature, required to determine the activity factor). The uncertainty associated with most emission rate measurement studies cannot be assessed from the reported information. As a result, it is difficult to combine emission factors from various studies into a comprehensive emission factor database. If only one investigator has reported emissions of a particular compound from a plant species, or if several investigators report similar results, then assigning the best emission factor estimate is fairly straightforward. In other cases, there are somewhat conflicting estimates. The reason for this could be that (1) the emission factors are different, (2) some measurements were associated with activity factors (such as wounding) that were not accounted for, or (3) at least one of the estimates is inaccurate (analytical difficulties or other reasons). Table 2 contains isoprene emission factors for plants in North America and Europe based on the reviews of Guenther et al. (1994), Simpson et al. (1995), and Kesselmeier et al. (1996).

Both the top-down and bottom-up approaches are used to estimate landscape average emission factors that are associated with landcover types contained in regional and global landcover databases. The global 0.5° resolution landcover database compiled by Olson (1992), for example, has about 60 biomes including tropical rain forest, sand desert, and temperate mixed forest. While there are likely to be significant differences between the mean emission factors representative of these different landcover types, it is also likely that any one of these landcover types has a fairly wide range of possible emission factors. Loveland et al. (1991) developed a 1-km resolution landcover database for North America, based on satellite and ancillary data, that includes several hundred landcover types. This effort has been expanded to include all continents and is a potentially significant improvement for global biogenic emission modeling. Guenther et al. (1994) found that the larger number of landcover types found in this type of database could better represent the distribution of landscape average biogenic emission factors.

TABLE 2   Isoprene and Monoterpene Emission Factors for Common North American and European Vegetation[a]

Genus	Common Name	Isoprene[b]	Monoterpenes[b]
Quercus	North European oaks	70	0.2
Quercus	Mediterranean oaks	0	20
Quercus	North American oaks	70	0.2
Picea	Sitka spruce	10	3
Picea	Norway spruce	2	3
Abies	North American firs	0	3
Fagus	North American beech	0	0.6
Juniperus	North American junipers	0	0.6
Liquidambar	Sweetgum	70	3
Pinus	North American pines	0	3
Populus	Poplars, aspen	70	0
Ulmus	Elms	0	0

[a]Based on Guenther et al. (1994), Simpson et al. (1995), and Kesselmeier et al. (1996).
[b]Values in μg/g/h at leaf-level environmental conditions of 30°C and 1000 μmol/m^2/sec.

3. Activity Factors, γ

Photosynthetically active radiation (PAR) flux and leaf temperature have long been recognized as important factors in controlling isoprene emissions from tree foliage (Chapter 2). Early modeling efforts, including those of Zimmerman (1979) and Lamb et al. (1987), assumed an exponential increase in emissions with temperature and considered isoprene to be emitted only during daylight hours. Studies by Lamb et al. (1993) include methods for calculating hourly light intensity based on solar elevation angles and cloud cover, and a canopy environment model to calculate leaf temperature and light variations with canopy depth. Complex canopy environment models are available but field evaluations indicate that relatively simple models perform as well as the computationally expensive ones (Lamb et al., 1996).

Guenther et al. (1993) developed numerical algorithms that simulate the light and temperature dependence of isoprene emissions from a variety of vegetation types. The influence of light on emission activity is estimated as

$$C_L = (\alpha C_{L1} L)/[(1 + \alpha^2 L^2)^{0.5}], \qquad (2)$$

where α $(= 0.0027)$ and C_{L1} $(= 1.066)$ are empirical coefficients and L (μmol/m^2/s) is the PAR flux. The relationship between PAR flux and isoprene emis-

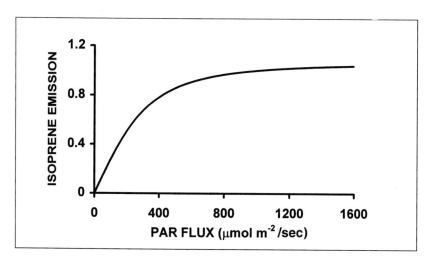

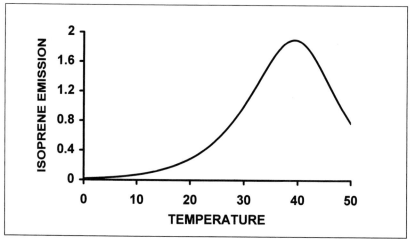

FIGURE 2 Influence of light (photosynthetically active radiation, PAR) and temperature on isoprene emissions predicted by Guenther *et al.* (1993).

sion is shown in Fig. 2. Variations in isoprene emissions due to temperature are estimated as

$$C_T = [\exp\{C_{T1}(T - T_s)/(R\,T_sT)\}]/[C_{T3} + \exp\{C_{T2}(T - T_M)/(R\,T_sT)\}], \quad (3)$$

where C_{T1} (= 95,000 J/mol), C_{T2} (= 230,000 J/mol), C_{T3} (= 0.961), and T_M (= 314 K) are empirical coefficients. T (K) is leaf temperature, T_S (K) is the leaf temperature at standard conditions, and R is the gas constant ($= 8.314$ J/K/mol). The behavior described by Eq. 3 is illustrated in Fig. 2.

Growth environment, leaf age, phenological events, leaf nitrogen content, water status, and stress can also influence isoprene emissions. These processes are particularly important for determining day-to-day and longer variations in emissions. Investigations of many of these processes have provided some insights into the regulation of isoprene emission rates but have not yet resulted in reliable numerical algorithms that can be used to describe seasonal variations from all landscapes. Several investigators (Monson et al., 1994; Kempf et al., 1996) have observed that isoprene emissions have a rapid rise in springtime and a rapid decrease in autumn. Annual average emissions will be overestimated unless an additional activity factor is applied. A general algorithm based on available data, such as the daily temperature record, is needed to estimate a seasonal activity factor.

4. Emission Estimates

Guenther et al. (1995) compared the results of seven global estimates of isoprene emission rates from vegetation that range from 175 to 500 Tg of carbon per year. The first global emission estimate, 350 Tg, was calculated by Zimmerman (1979) and is within about 30% of the estimate, about 500 Tg, reported by Guenther et al. (1995). Guenther et al. (1995) estimated that woodland landscapes, which are less than half of all land surfaces, contribute about 75% of all foliar isoprene emissions. Figure 3 illustrates the seasonal cycle of predicted isoprene emissions for each latitude. The global total emission rate is dominated by fluxes from tropical landscapes. The high emissions predicted for the tropics are due to the high foliar density and the year-round warm temperatures. Variations in monthly average emissions range from less than 20% in the tropics to more than three orders of magnitude in boreal and temperate regions.

B. MARINE PHYTOPLANKTON

Isoprene concentrations in surface layer seawater are often supersaturated relative to the atmosphere (Bonsang et al., 1992; Milne et al., 1995) which causes the ocean to be a source of isoprene in the atmosphere. McKay et al. (1996) found that isoprene in the air above laboratory diatom cultures coincided with chlorophyll peaks and concluded that isoprene was produced biotically either directly by the phytoplankton or indirectly from organics secreted by these organisms.

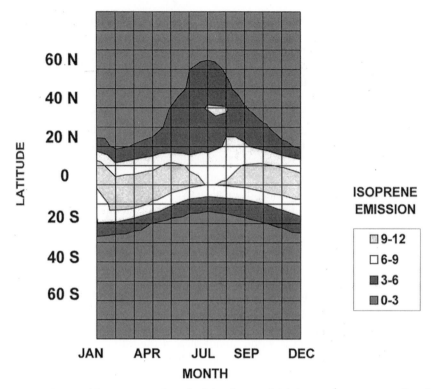

FIGURE 3 Monthly variations in latitudinal distribution of global isoprene emission rate (Tg of carbon per 10° latitude per month) from vegetation, based on estimates of Guenther *et al.* (1995).

Erickson and Eaton (1993) have developed methods for modeling trace gas emissions from oceans that have been used to predict global emissions of these VOCs from seawater (Guenther *et al.*, 1995). Fluxes are estimated using the model of Liss and Merlivat (1986) as

$$F = k_w (C_{so} - C_{eq}), \tag{4}$$

where k_w is a transfer velocity, C_{so} is the surface ocean VOC concentration, and C_{eq} is the equilibrium surface ocean VOC concentration expected from equilibrium with the ambient concentration. C_{eq} is typically at least an order of magnitude less than C_{so} and can be neglected. Guenther *et al.* (1995) assumed that marine VOC concentrations are controlled by biological activity that correlates with Coastal Zone Color Scanner (CZCS) satellite measurements and surface radiation fluxes. If we use the procedures described by

Guenther *et al.* (1995) to extrapolate the ocean data of Bonsang *et al.* (1992) and Milne *et al.* (1995), then the resulting annual global isoprene emission rate from oceans is about 1 Tg of carbon.

C. OTHER BIOGENIC SOURCES

Isoprene emissions have been reported from a variety of other biogenic sources, but there are few data available for developing quantitative emission estimates. The simple estimates described here are designed to give an initial indication of the importance of each source. This initial estimate provides guidance on whether a more accurate estimate is needed.

Isoprene is a dominant hydrocarbon in the breath exhaled by humans and some animals. Foster *et al.* (1996) observed emissions from 10 human subjects that range from 66 to 461 pmol/kg/min. An initial estimate of the global annual emission rate of isoprene in human breath can be calculated as the product of a source factor, the global human population of about 6 billion persons, and an emission factor of about 0.5 g per person per year. The resulting annual emission rate is approximately 0.003 Tg of carbon. There is evidence of isoprene emissions from other animals but the global total is probably less than that from humans.

Isoprene emissions have been observed from bacteria that inhabit soils (Kuzma *et al.*, 1995); some fungi (Berenguer *et al.*, 1991); and a variety of ground cover vegetation including mosses (Isidorov *et al.*, 1985), ferns, and some herbaceous plants (Evans *et al.*, 1982). Investigators using ground enclosures have observed isoprene emission rates ranging from 0 to 600 μg/m^2/h, with typical values of about 200 μg/m^2/h at 30°C in full sunlight (Zimmerman, 1979; Klinger *et al.*, 1994). If we assume an emission factor of 200 μg/m^2/h for all terrestrial surfaces—other than ice and sand desert—and the activity factors for isoprene emissions from tree foliage, the resulting annual global emission rate of isoprene from plant foliage other than trees is about 40 Tg. This source was included in the model of Guenther *et al.* (1995), which gave landscape average emissions. The results from Guenther *et al.* (1995) were separated into two components (canopy foliage and ground cover) in Table 1.

In addition to the light-dependent enzymatic production of isoprene, there is a nonenzymatic emission rate of isoprene from foliage that occurs even in the dark (see Chapter 2). This has been observed from many plant species but has not been investigated in detail because the emission rates have been considered negligible (0.001 to 0.1 μg/g/h). If we assume 0.01 μg/g/h as an

emission factor typical of all vegetation and assume an exponential temperature dependence, the resulting annual global emission is about 2 Tg.

D. ANTHROPOGENIC SOURCES

Although they are not biogenic sources, there are two additional sources of
isoprene emissions that should be considered when estimating global sources
of isoprene: vehicle emissions and industrial processes (e.g., synthetic rubber
production). Vehicular isoprene emissions are estimated from the number of
vehicles (the source density factor), the isoprene emissions per mile (the
emission factor), and the number of miles driven and the driving conditions
(activity factors). Isoprene emissions from industrial processes can be estimated from the number of industrial facilities (source density), and emission
and activity factors that are related to industrial output. Conner *et al.* (1995)
report that isoprene is typically about 0.04% of the total nonmethane volatile
organic compound (NMVOC) emissions from vehicles. By applying this percentage to an estimate of the global annual vehicle emission rate, 22 Tg (Singh
and Zimmerman, 1992), the estimated isoprene emission rate from vehicles is
about 0.01 Tg. Over 90% of the 15 Tg of total NMVOCs emitted by industrial
processes (Singh and Zimmerman, 1992) is associated with solvents. Isoprene
is a small fraction of the remainder and is likely to be less than 0.01 Tg.

E. RELATIVE IMPORTANCE OF ISOPRENE SOURCES

The global annual isoprene emission rate from all sources is about 500 Tg.
Table 1 shows that over 90% of this emission rate is attributed to canopy
foliage and that most of the remainder is associated with the foliage of ground
cover (ferns, mosses, etc.) and soil microbes. Additional sources (nonenzymatic emissions from foliage and marine phytoplankton) contribute less than
0.2% to the global total while other sources (vehicles, industrial processes,
humans, and other animals) contribute even less. Because of the short lifetime
of isoprene, emissions from sources that have a negligible annual global flux
may be the dominant source within a particular localized region of the atmosphere. Marine phytoplankton, in particular, may be an important source
since they could dominate over the open ocean. The relative importance of
each source is also expected to vary seasonally. Emissions from vegetation
vary with the season while emissions from automobiles and humans are

relatively constant and dominate in urban areas (at higher latitudes) in winter when biogenic emissions are at a minimum.

III. MONOTERPENES

Monoterpenes are a diverse group of natural products produced by many plant families (see Chapter 2). Monoterpenes are present in at least low concentrations above most terrestrial landscapes and are the dominant reactive VOCs in some regions. In addition to their importance for oxidant chemistry, some monoterpenes are of interest because of their high yield of particles following their oxidation in the atmosphere.

A. VEGETATION FOLIAGE

Monoterpene emissions from vegetation foliage can be estimated using Eq. 1. The source factor (foliar density) is described earlier. In some cases, monoterpene emissions are limited by light-dependent enzymatic production (Loreto et al., 1996) and the activity factor described earlier for isoprene emissions can be applied (Kesselmeier et al., 1996). In most cases, monoterpenes are emitted from stored pools and emissions are limited by volatilization and diffusion processes. The emission factor is related to the size of the stored pool, which is controlled by a number of factors, including genetics, growth conditions, and wounding. The activity factor must account for the observed exponential increase in emissions with temperature:

$$\gamma = \exp(\beta[T - T_s]). \qquad (5)$$

Guenther et al. (1993) reviewed empirical estimates of the coefficient β and recommended a value of 0.09 K^{-1}. This is significantly higher than the value that would be predicted from monoterpene vapor pressure curves. This demonstrates the importance of other factors that are presumably related to how the monoterpenes are transported from the stored pool to the atmosphere. Lerdau et al. (1995) report that there are ecological controls over longer term variations in monoterpene emission rates that should also be accounted for in emission models.

Estimates of global annual monoterpene emissions from tree foliage range from 115 to 480 Tg (Guenther et al., 1995). A best estimate of 115 Tg is used in Table 1. The higher estimates are based on enclosure measurement studies that may have overestimated monoterpene emissions due to disturbances caused by the enclosures.

B. TRUNKS, STEMS, AND GROUND SURFACES

Terpenes stored in the woody parts of many trees and shrubs can also be emitted into the atmosphere. Steinbrecher (1997) measured monoterpene emissions from various components of a Norway spruce forest and found that about 10% of the total emission was associated with trunks, stems, and the forest floor. The ecosystem total monoterpene emission rate (127 Tg) estimated by Guenther et al. (1995) has been separated into two components in Table 1: from foliage (115 Tg), and from trunks, stems, and the forest floor (13 Tg).

C. FLOWERS

The volatiles that comprise floral scents are perhaps the best known, and most appreciated, emissions from vegetation. Scientists interested in understanding their role as pollinator attractants have investigated the composition of floral scents but there has been little interest in quantifying emission rates. A wide range of VOCs is found in varying amounts in floral scents and includes alkanes, alcohols, esters, aromatics, nitrogen compounds, monoterpenes, and sesquiterpenes (Borg-Karlson et al., 1994). Arey et al. (1991) observed 17 μg/m^3 of the oxygenated monoterpene, linalool, in ambient air and used enclosure measurements to identify the source as orange blossoms and estimated emission rates (about 13 μg/g/h) that were considerably higher than the monoterpene emission rates from foliage. If we assume that floral emissions are a significant component of the total monoterpene emissions from about 10% of all plants and that these floral emissions occur on average during 12% of the year, then the annual global monoterpene emission rate associated with flowering would be about 2 Tg. The emissions of other VOCs from flowers is of a similar magnitude.

D. TIMBER HARVESTING AND LUMBER PRODUCTION

Timber harvesting and lumber production are likely to be the major direct anthropogenic impact on global monoterpene emissions. Large quantities of monoterpenes can be released when plants containing stored monoterpenes are disturbed. If we assume that approximately 70% of all monoterpene emitters are commercially harvested on an average 50-year rotation, and that 50% of the potential annual emission rate is released during the harvesting and production process, then the annual global emission rate from this source is about 1 Tg.

E. RELATIVE IMPORTANCE OF MONOTERPENE SOURCES

Emissions from foliage are estimated to be responsible for about 90% of the annual global monoterpene emission rate. Other sources are very uncertain but may make a significant contribution. The spatial and temporal distributions of the various sources are not expected to be uniform. As a result, some of the minor sources could be the dominant monoterpene source at a particular place and time.

IV. OTHER VOLATILE ORGANIC COMPOUNDS

There has been little effort directed at estimating biogenic emissions of VOC other than isoprene and monoterpenes. In addition, little is known about the biological processes responsible for their production and emission (Chapter 2). One reason for this is that many of these compounds are difficult to identify and quantify. Another reason is that photochemical models do not include reaction mechanisms for many of these compounds so there has been less demand for their emission rates. However, as indicated in Table 1, the sum of these emission rates is estimated to be about a third of the total global VOC emission rate.

A. TREE FOLIAGE

Enclosure measurements have demonstrated that many VOCs are emitted from tree foliage, but there are few data for developing emission models. Some of these compounds are important because of the predicted large magnitude of their flux. Methanol, for example, has been reported to be widespread in the troposphere and has high emission rates from most plant species (MacDonald and Fall, 1993b). Other compounds, including acetone, may have a relatively small global total emission rate (MacDonald and Fall, 1993a), but accurate emission estimates are needed because of their potential importance in global atmospheric chemistry.

Based on a review of enclosure measurements, Guenther *et al.* (1995) estimated emissions of biogenic VOCs other than isoprene and monoterpenes using an emission factor of 1.5 μg/g/h for reactive compounds (e.g., hexenol, formaldehyde, and acetic acid) and an additional emission factor of 1.5 μg/g/h for less reactive compounds (e.g., acetone and methanol). The annual global emission estimate of 260 Tg each for reactive (lifetime of less than one day) and less reactive (lifetime greater than one day) compounds demonstrates the potential importance of this source. These estimates are extremely uncertain and considerable effort is required to reduce these uncertainties.

B. MARINE ORGANISMS

Ocean surface waters are a significant source of ethane, propane, ethene, and propene in the atmosphere (Lamontagne *et al.*, 1974). The source of these compounds in the ocean is probably dissolved organic matter and their source strength is probably correlated with primary productivity. By using the model techniques described earlier for the marine isoprene source, Guenther *et al.* (1995) estimated a global annual emission rate of 4 Tg that was assumed to be 50% alkanes and 50% alkenes. This is at the low end of a wide range of previous estimates but similar to other estimates.

C. SOIL MICROBES

Geogenic VOCs (natural gas, which includes methane, ethane, propane, and other compounds) can migrate through soils and be emitted into the atmosphere. Lamb *et al.* (1987) estimate that local fluxes of nonmethane VOCs from this source range from 0.06 to 2.6 μg/m^2/h. An extrapolation of their U.S. estimate to ice-free global land surfaces results in an annual flux of about 0.01 Tg. Isidorov (1990) reviewed emission measurements in Russia and concludes that VOCs (primarily methane, ethane, and propane) have a much larger microbial source in soils. From these measurements, Isidorov (1990) estimates that the microbial soil source (for a region covering about 80% of the area of soil cover on land) is 46 to 101 Tg of ethane and 9 to 23 Tg of propane. Zimmerman (1979) measured emissions in the United States ranging between 100 and 260 μg/m^2/h from leaf litter and pasture surfaces. A global extrapolation of these measurements results in a total global VOC emission rate of about 60 Tg per year.

The annual loss rate of longer lived VOCs, including ethane and propane, can be estimated from their global removal rate due to OH. After accounting for any increase in the global average concentration of these compounds, we can estimate the total global source from the loss rate. Singh and Zimmerman (1992) estimated total annual global sources of 10 to 15 Tg ethane and 15 to 20 Tg propane. They concluded that biomass burning, oil combustion, natural gas production and distribution, and the oceans together contribute about 3 to 5.5 Tg of propane and 7 to 15 of ethane to the total global emission rate. The difference between the estimated sources and sinks, 9.5 to 17 Tg of propane and 0 to 8 Tg of ethane, is the emission range associated with biogenic sources (or some unknown anthropogenic source).

V. CONCLUSIONS

Although our understanding of the processes controlling biogenic VOC emissions is very uncertain, we can use the current state of knowledge to develop best estimates of the magnitude and global distributions of biogenic VOCs. The task is particularly challenging due to the large number of biogenic VOCs and wide range of sources. Numerical models have been developed to predict emissions of a few of the larger sources, including vegetation foliage. The limited understanding of other sources does not warrant a detailed numerical model but an order of magnitude flux estimate is worthwhile to demonstrate whether a particular source should be the subject of further study. Some VOC sources have a negligible annual global flux but may be important because they dominate in certain regions and in certain seasons.

Most investigations of biogenic VOC have focused on isoprene and, to a lesser extent, on monoterpenes. The large fraction of total VOC emissions associated with these compounds, especially isoprene, demonstrates that these studies have been needed. Uncertainties in isoprene emissions have been reduced considerably for specific regions (mostly the United States) and seasons (midsummer) but are more than a factor of three for most regions and seasons. Improved estimates require better databases of vegetation distributions and data relating isoprene emissions to vegetation species or type. Of particular importance for global models are studies of tropical vegetation that are thought to be responsible for a majority of the global total but have been investigated very little. The uncertainties associated with oxygenated VOCs are a factor of five or more for all regions and seasons. A better understanding of the processes controlling these emissions is needed to develop realistic numerical models. These compounds are likely to have a significant impact on atmospheric chemistry. Studies of the potential impact of these compounds on atmospheric trace gas distributions are needed to determine the level of effort that is required to produce emission estimates with an appropriate level of accuracy.

REFERENCES

Anastasi, C., Hopkinson, L., and Simpson, V. (1991). Natural hydrocarbon emissions in the United Kingdom. *Atmos. Environ.* **25A**, 1403–1408.

Arey, J., Corchnoy, S. B., and Atkinson, R. (1991). Emission of linalool from Valencia orange blossoms and its observation in ambient air. *Atmos. Environ.* **25A**, 1377–1381.

Berenguer, J., Calderon, V., Herce, M., and Sanchez, J. (1991). Spoilage of a bakery product by isoprene-producing molds. *Rev. Agroquim. Technol. Aliment.* **31**, 580–583.

Bonsang, B., Polle, C., and Lambert, G. (1992). Evidence for marine production of isoprene. *Geophys. Res. Lett.* **19**, 1129–1132.

Borg-Karlson, A., Valterova, I., and Nilsson, L. (1994). Volatile compounds from flowers of six species in the family apiaceae: bouquets for different pollinators? *Phytochemistry.* **35**, 111–119.

Chameides, W. L., Lindsay, R. W., Richardson, J., and Kiang, C. S. (1988). The role of biogenic hydrocarbons in urban photochemical smog: Atlanta as a case study. *Science.* **241**, 1–10.

Conner, T., Lonneman, W., and Seila, R. (1995). Transportation-related volatile hydrocarbon source profiles measured in Atlanta. *J. Air Waste Manage. Assoc.* **45**, 383–394.

EDC-NESDIS (1992). Monthly global vegetation index from Gallo bi-weekly experimental calibrated GVI (April 1985–December 1990). Digital raster data on a 10 minute geographic (lat/lon) 1080 × 2160 grid, *In* Global Ecosystems Database, Version 1.0, Disc A, National Geophysical Data Center (ed.), National Oceanic Atmospheric Administration Boulder, CO.

Erickson, D., and Eaton, B. (1993). Global biogeochemical cycling estimates with CZCS data and general circulation models. *Geophys. Res. Lett.* **20**, 683–686.

Evans, R. C., Tingey, D. T., Gumpertz, M. L., and Burns, W. F. (1982). Estimates of isoprene and monoterpene emission rates in plants. *Bot. Gaz.* **143**, 304–310.

Foster, W. M., Jiang, L., Stetkiewicz, P. T., and Risby, T. H. (1996). Breath isoprene: temporal changes in respiratory output after exposure to ozone. *J. Appl. Physiol.* **80**, 706–710.

Geron, C., Guenther, A., and Pierce, T. (1994). An improved model for estimating emissions of volatile organic compounds from forests in the eastern United States. *J. Geophys. Res.* **99**, 12773–12792.

Geron, C., Pierce, T., and Guenther, A. (1995). Reassessment of biogenic volatile organic compound emissions in the Atlanta area, *Atmos. Environ.* **29**, 1569–1578.

Geron, C., Nie, D., Arnts, R., Sharkey, T., Singsaas, E., Vanderveer, P., Guenther, A., Katul, G., Sickles, U., and Kleindienst, T. (1997). Biogenic isoprene emission: model evaluation in a southeastern U.S. bottomland deciduous forest. *J. Geophys. Res.* **102**, 18889–18901.

Guenther, A. (1997). Seasonal and spatial variations in natural volatile organic compound emissions. *Ecol. Appl.* **7**, 34–45.

Guenther, A., Baugh, W., Davis, K., Hampton, G., Harley, P., Klinger, L., Zimmerman, P., Allwine, E., Dilts, S., Lamb, B., Westberg, H., Baldocchi, D., Geron, C., and Pierce, T. (1996a). Isoprene fluxes measured by enclosure, relaxed eddy accumulation, surface-layer gradient, mixed-layer gradient, and mass balance techniques. *J. Geophys. Res.* **101**, 18555–18568.

Guenther, A., Hewitt, C. N., Erickson, D., Fall, R., Geron, C., Graedel, T., Harley, P., Klinger, L., Lerdau, M., McKay, W., Pierce, T., Scholes, B., Steinbrecher, R., Tallamraju, R., Taylor, J., and Zimmerman, P. (1995). A global model of natural volatile organic compound emissions. *J. Geophys. Res.* **100**, 8873–8892.

Guenther, A., Otter, L., Zimmerman, P., Greenberg, J., Scholes, R., and Scholes, M. (1996b). Biogenic hydrocarbon emissions from southern African savannas. *J. Geophys. Res.* **101**, 25859–25865.

Guenther, A., Zimmerman, P., Harley, P., Monson, R., and Fall, R. (1993). Isoprene and monoterpene emission rate variability: model evaluation and sensitivity analysis. *J. Geophys. Res.* **98**, 12609–12617.

Guenther, A., Zimmerman, P., and Wildermuth, M. (1994). Natural volatile organic compound emission rate estimates for U.S. woodland landscapes. *Atmos. Environ.* **28**, 1197–1210.

Hewitt, C. N., and Street, R. A., (1992). A qualitative assessment of the emission of non-methane hydrocarbon compounds from the biosphere to the atmosphere in the U.K.: present knowledge and uncertainties. *Atmos. Environ.* **26A**, 3069–3077.

Isidorov, V. (1990). "Organic Chemistry of the Earth's Atmosphere," Springer-Verlag: Berlin.

Isidorov, V. A., Zenkevich, I. G., and Ioffe, B. V. (1985). Volatile organic compounds in the atmosphere of forests. *Atmos. Environ.* **19**, 1–8.

Kempf, K., Allwine, E., Westberg, H., Claiborn, C., and Lamb, B. (1996). Hydrocarbon emissions from spruce species using environmental chamber and branch enclosure methods. *Atmos. Environ.* **30**, 1381–1389.

Kesselmeier, J., *et al.* (1996). Emission of monoterpenes and isoprene from a Mediterranean oak species *Quercus ilex* L. measured within the BEMA (biogenic emissions in the Mediterranean Area) project. *Atmos. Environ.* **30**, 1841–1850.

Klinger, L., Zimmerman, P., Greenberg, J., Heidt, L., and Guenther, A. (1994). Carbon trace gas fluxes along a successional gradient in the Hudson Bay lowland. *J. Geophys. Res.* **99**, 1469–1494.

Kuzma, J., Nemecek-Marshall, M., Pollock, W. H., and Fall, R. (1995). Bacteria produce the volatile hydrocarbon isoprene. *Curr. Microbiol.* **30**, 97–103.

Lamb, B., Gay, D., Westberg, H., and Pierce, T. (1993). A biogenic hydrocarbon emission inventory for the U.S. using a simple forest canopy model. *Atmos. Environ.* **27**, 1673–1690.

Lamb, B., Guenther, A., Gay, D., and Westberg, H. (1987). A national inventory of biogenic hydrocarbon emissions. *Atmos. Environ.* **21**, 1695–1705.

Lamb, B., Allwine, E., Dilts, S., Westberg, H., Pierce, T., Geron, C., Baldocchi, D., Guenther, A., Klinger, L., Harley, P., and Zimmerman, P. (1996). Evaluation of forest canopy models for estimating isoprene emissions. *J. Geophys. Res.* **101**, 22787–22798.

Lamontagne, R. A., Swinnerton, J. W., and Linnenbom, V. J. (1974). C_1–C_4 hydrocarbons in the North and South Pacific. *Tellus.* **26**, 71–77.

Lerdau, M., Matson, P., Fall, R., and Monson, R. (1995). Ecological controls over monoterpene emissions from Douglas-fir. *Ecology.* **76**, 2640–2647.

Liss, P., and Merlivat, L. (1986). Air-sea gas exchange rates: introduction and synthesis, *In* "The Role of Air-Sea Exchange in Geochemical Cycling" (P. Buat-Menard, ed.), D. Reidel: Norwood, MA.

Loreto, F., Ciccioli, P., Cecinato, A., Brancaleoni, E., Frattoni, M., Fabozzi, C., and Tricoli, D. (1996). Evidence of the photosynthetic origin of monoterpenes emitted by *Quercus ilex* L. leaves by C^{13} labelling. *Plant Physiol.* (in press).

Loveland, T. R., Merchant, J. W., Ohlen, D. O., and Brown, J. F. (1991). Development of a land-cover characteristics database for the conterminous U.S. *Photogrammetric Eng. Remote Sensing.* **57**, 1453–1463.

MacDonald, R. C., and Fall, R. (1993a). Acetone emission from conifer buds. *Phytochemistry.* **34**, 991–994.

MacDonald, R. C., and Fall, R. (1993b). Detection of substantial emissions of methanol from plants to the atmosphere. *Atmos. Environ.* **27A**, 1709–1713.

McKay, W., Turner, M., Jones, B., and Halliwell, C. (1996). Emissions of hydrocarbons from marine phytoplankton—some results from controlled laboratory experiments. *Atmos. Environ.* **30**, 2583–2593.

Milne, P. J., Riemer, D. D., Zika, R. G., and Brand, L. E. (1995). Measurement of vertical distribution of isoprene in surface seawater, its chemical fate, and its emission from several phytoplankton monocultures. *Marine Chem.* **48**, 237–244.

Molnar, A. (1990). Estimation of volatile organic compounds (VOC) emissions for Hungary. *Atmos. Environ.* **24A**, 2855–2860.

Monson, R., Harley, P., Litvak, M., Wildermuth, M., Guenther, A., Zimmerman, P., and Fall, R. (1994). Environmental and developmental controls over the seasonal pattern of isoprene emission from aspen leaves. *Oecologia.* **99**, 260–270.

Monson, R. K., Kaeger, C. H., Adams III, W. W., Driggers, E. M., Silver, G. M., and Fall, R. (1992). Relationships among isoprene emission rate, photosynthesis, and isoprene synthase activity as influenced by temperature. *Plant Physiol.* **98**, 1175–1180.

Olson, J. (1992). World ecosystems (WE 1.4): digital raster data on a 10 minute geographic 1080 × 2160 grid, *In* Global Ecosystems Database, Version 1.0, Disc A, National Geophysical Data Center (ed.), National Oceanic Atmospheric Administration, Boulder, CO.

Rasmussen, R. A. (1972). What do the hydrocarbons from trees contribute to air pollution? *J. Air Pol. Control Assoc.* **22**, 537–543.

Rasmussen, R., and Went, F. (1965). Volatile organic material of plant origin in the atmosphere. *Proc. Natl. Acad. Sci. U.S.A.* **53**, 215–220.

Roselle, S. J., Pierce, T. E., and Schere, K. L. (1991). The sensitivity of regional ozone modeling to biogenic hydrocarbons. *J. Geophys. Res.* **96**, 7371–7394.

Running, S. W., Nemani, R. R., Peterson, D. L., Band, L. E., Potts, D. F., Pierce, L. L., and Spanner, M. A. (1989). Mapping regional forest evapotranspiration and photosynthesis by coupling satellite data with ecosystem simulation. *Ecology.* **70**, 1090–1101.

Silver, G., and Fall, R. (1991). Enzymatic synthesis of isoprene from dimethylallyl diphosphate in aspen leaf extracts. *Plant Physiol.* **97**, 1588–1591.

Simpson, D., Guenther, A., Hewitt, C. N., and Steinbrecher, R. (1995). Biogenic emissions in Europe 1. Estimates and uncertainties. *J. Geophys. Res.* **100**, 22,875–22, 890.

Singh, H. B., and Zimmerman, P. (1992). Atmospheric distributions and sources of nonmethane hydrocarbons. *In* "Gaseous Pollutants: Characterization and Cycling" (J.O. Nriagu, ed.), pp. 235. John Wiley & Sons: New York.

Steinbrecher, R. (1997). VOC emission in Norway spruce ecosystems: compartimentation, source profile, source strength, and source strength parameterization. Proceedings of Workshop on biogenic hydrocarbons in the atmospheric boundary layer, American Meteorological Society, Charlottesville, VA.

Turner, D. P., Baglio, J. V., Wones, A. G., Pross, D., Vong, R., McVeety, B. D., and Phillips, D. L. (1991). Climate change and isoprene emissions from vegetation. *Chemosphere.* **23**, 37–56.

Watson, R., Filho, L., Sanhueza, E., and Janetos, A. (1992). Greenhouse gases: sources and sinks. "Climate Change 1992: The Supplementary Report to the IPCC Scientific Assessment" (J. Houghton, B. Callander, and S. Varney, eds.), Cambridge Universtiy Press: Cambridge.

Zimmerman, P. (1979). Testing of hydrocarbon emissions from vegetation, leaf litter and aquatic surfaces, and development of a methodology for compiling biogenic emission inventories. Report number EPA-450-4-70-004 to U.S. Environmental Protection Agency, Research Triangle Park, NC.

CHAPTER 4

The Sampling and Analysis of Volatile Organic Compounds in the Atmosphere

XU-LIANG CAO* AND C. NICHOLAS HEWITT[†]
*c/o Environmental Health Centre, Ottawa, Ontario, Canada
[†]Institute of Environmental and Natural Sciences, Lancaster University, Lancaster, United Kingdom

*Recipient of a Natural Sciences and Engineering Research Council Visiting Fellowship at Health Canada.

Accurate measurements of volatile organic compound (VOC) concentrations in the atmosphere are essential in assessing the risk of these compounds to the human environment, and particularly in controlling the concentrations of ozone and other secondary pollutants in the lower troposphere. In designing a VOC sampling program, suitable sampling and analytical methods should be selected according to the properties of the target compounds, to ensure representative sampling, precise and accurate quantification, and the best sensitivity. In this chapter, the application of commonly used sampling and analytical methods for airborne VOC monitoring and their range and limitations are summarized. The possible applications of several new potential sampling and analytical methods and developments in the use of new methods are also discussed.

I. INTRODUCTION

Volatile organic compounds (VOCs) may have direct detrimental effects, particularly on human health. In addition, VOCs, especially nonmethane hydrocarbons (NMHCs), together with nitrogen oxides (NO_x), are important precursors in the formation of photochemical oxidants including ozone. Although ozone in the stratosphere plays an important role in protecting the earth's surface from exposure to ultraviolet (UV) radiation, its presence in the lower troposphere has detrimental effects on human beings, animals, and plant species. Thus, measures are being taken to control ozone in ground-level ambient air.

The earliest ozone abatement strategies focused on reduction of anthropogenic hydrocarbon emissions, but urban and rural areas in the United States continue to be noncompliant with the ozone standard set by the National Ambient Air Quality Standards (NAAQS). This is in part due to a lack of understanding of the significant contributions made by biogenic emissions to the global VOC budget. As discussed in Chapters 1 and 2, VOCs in ambient air can be emitted from both anthropogenic and biogenic sources. Although there are still considerable uncertainties in the estimates of emissions of VOCs from the biosphere, it is believed that the majority of global VOC emissions are from biogenic sources (Fehsenfeld et al., 1992). The emissions of VOCs from plant foliage into the atmosphere account for about half of the estimated total VOC emissions in the United States (Lamb *et al.*, 1987) and two-thirds of global VOC emissions (Muller, 1992). Since the emissions of biogenic VOCs are generally considered to be uncontrollable, much greater reductions in anthropogenic VOC emissions than previously predicted are needed to reduce ozone levels. Therefore, accurate measurements of VOC concentrations in air and of their rates of emission from both anthropogenic and biogenic sources into the atmosphere are essential. In addition, accurate measurements of atmospheric concentrations of toxic organics are also necessary to assess their risk to the human environment.

Sampling is the first and very critical step in the whole process of obtaining the final results. Many different sampling methods are available, each of which has its own application range and limitations. Thus, suitable methods should be selected according to the properties (vapor pressure, boiling point, polarity, etc.) of the target compounds to obtain valid results. Since the concentrations of many VOCs in ambient air are very low, typically at levels in parts per billion volume (10^{-9}) or below, both sample preconcentration and use of sensitive detection methods are usually necessary. In this chapter, the sampling and analytical methods commonly used for the measurements of volatile organic compounds in air are summarized. The possible application of several potential detection methods in this area and recent development of new methods are also discussed.

II. SAMPLING METHODS

Several methods are available for collecting VOCs from the atmosphere, either indirect (e.g., reaction with a treated substrate, condensation in a cold trap) or direct using a continuous air-sampling device of which the detector is an integral part. Although the techniques of direct continuous real-time (or near real-time) analysis of VOCs in air have distinct advantages because they eliminate the need for the sample preconcentration step required with the other methods, they have not been widely applied. The two most commonly used methods for sampling VOCs in air are still whole air sampling and adsorbent sampling.

A. WHOLE AIR SAMPLING

Whole air sampling involves the direct collection and isolation of the test atmosphere in an impermeable and evacuated container. Different types of containers are available. Nonrigid Teflon or Tedlar bags have disadvantages (e.g., are difficult to clean and fragile, permit sample loss and contamination through permeation, and have photo-induced chemical effects) despite their relatively low cost. It has been claimed that permeation is not a problem for Tedlar bags containing air samples with organic compounds of molecular size as small as butadiene, and that they are suitable for halogenated hydrocarbons, such as chloroform and carbon tetrachloride; however, loss of polar organic compounds, such as ethanol and formaldehyde, and of higher boiling point nonpolar organic compounds has been seen in the Tedlar bags. Gholson *et al.* (1990) investigated using aluminum canisters to collect ambient-level air samples. The selected organic compounds showed no increase in stability when collected and stored in aluminum as compared to those collected and stored in stainless steel canisters. For polar oxygenated compounds, such as acetone and 1,4-dioxane, aluminum canisters are more reactive than those of passivated stainless steel. Stainless steel canisters are the best overall for whole air sampling. They have several advantages over the other approaches: they are not subject to sample permeation or photo-induced chemical effects, can be thoroughly cleaned, have good sample integrity, are rugged, and can be pressurized to increase sample volume. Despite their disadvantages (limited sample volume and relatively high cost), canister-based methods have been widely used for the sampling of VOCs in air (e.g., McClenny *et al.*, 1991; Evans *et al.*, 1992; Farmer *et al.*, 1994). The canister-based method has been summarized in the U.S. Environmental Protection Agency (EPA) *Compendium of Methods for the Determination of Toxic Organic Compounds in Ambient Air,* Method TO-14A (1997a), and also by ASTM (1996a).

Air samples can be collected either by subatmospheric pressure sampling (i.e., collection of an air sample in an evacuated canister at a final canister pressure below atmospheric pressure without assistance of a sampling pump), or by pressurized sampling (i.e., collection of an air sample in a canister with a final canister pressure above atmospheric pressure using a sampling pump). The subatmospheric pressure sampling method may be used to collect grab samples (duration of 10 to 30 sec) or time-integrated samples (duration of 12 to 24 h) taken through a flow restrictive inlet (e.g., mass flow controller, critical orifice). However, with a critical orifice flow restrictor, the sampling system may experience a decreased flow rate when atmospheric pressure is approached. Thus, a mass flow controller is required, with which the subatmospheric sampling system can maintain a constant flow rate from full vacuum to within about 7 kPa or less below ambient pressure. Pressurized sampling is used when longer term integrated samples or higher volume samples are required. The sample is collected in a canister using a metal bellows-type pump and flow control arrangement to achieve a typical final canister pressure of 100 to 200 kPa.

Canister-based sampling is only suitable for collection of truly gas-phase

VOCs (i.e., organic compounds having a vapor pressure greater than 10^{-1} torr at 25°C and 101 kPa). Compounds greater than C_{10} are more likely to adsorb onto the walls of the canister on storage, leading to an underestimate of their ambient concentrations. For example, Zielinska *et al.* (1996) compared the concentrations of individual hydrocarbons obtained from parallel Tenax and canister samples. The Tenax/canister concentration ratio was around one for most of the compounds in the range C_8 to C_{10}, but this ratio increases substantially for compounds eluting after 1,2,3-trimethylbenzene. Concentrations of compounds with a carbon number greater than C_{10} measured from Tenax cartridges were much higher than those measured from canisters.

The stability of samples during storage is an important factor to consider in VOC analysis. A number of studies (e.g., McClenny *et al.*, 1991; Oliver *et al.*, 1986) have found that a wide range of VOCs are stable in humidified passivated stainless steel canisters for at least 7 days, and some of these compounds are stable for up to 30 days. Although polar VOCs present more problems than nonpolar VOCs with the canister-based sampling method, Kelly *et al.* (1993) found that canister sampling for most of their target polar VOCs could be successful, provided that sample analysis was performed within 4 days after sample collection, while Pate *et al.* (1992) found that their target polar VOCs were stable in humidified stainless steel canisters for as long as 30 days. The difference between the two data sets may be due to the fact that the latter supplied a much larger amount of water to their canisters than was present in the canisters used by the former. The stability of VOCs, especially polar VOCs, in a dry canister is very poor since the inside of the passivated canister surface is coated with a nickel–chromium oxide layer on which compounds will be lost due to either physical adsorption or chemical interaction. Thus, the presence of water vapor is essential in maintaining sample integrity because water has a stronger affinity for the active sites than the nonpolar VOCs, and the formation of a layer of water over the metallic surface will prevent interaction with the organic compounds.

Although its presence in the stainless steel sampling canister is essential for sample integrity, water can cause problems in the subsequent analysis. The variability of concentration of ambient water vapor can cause detector baseline shifts, and co-collected water can cause blockage on injection into capillary columns in separation techniques that require subfreezing initial gas chromatographic (GC) oven temperatures. In extreme cases, the excess water when eluted can even extinguish the hydrogen flame of a flame ionization detector (FID). Thus, water should be removed from the sample as much as possible prior to being injected into the GC column, or a minimal volume of sample should be used for analysis. A Nafion permeable membrane dryer is commonly used to remove water selectively from the sample stream. This dryer consists of Nafion tubing (a copolymer of tetrachloroethylene and fluorosulfonyl monomer) that is coaxially mounted within larger tubing. The sample stream is passed through the interior of the Nafion tubing, allowing water to permeate through the walls into the dry purge stream flowing through the annular space between the Nafion and outer tubing. However, since polar compounds can also permeate through the membrane like water, use of this drying device results in the loss of these compounds, thus preventing satisfactory measurements of these species in air samples when a Nafion dryer is used. It has also been found that concentrations of some paraffins, olefins, and aromatics may decrease and some contaminants may be introduced into the system by the use of a Nafion dryer (Zielinska et al., 1996).

Since the principal limitation of whole air sampling is that it does not provide sample preconcentration, which will affect both sampling volume and detection limit, samples have to be concentrated prior to being analyzed, especially if the concentrations of VOCs in the sample air are very low. The whole air sample from the canister can be concentrated by passing through a single adsorbent packing or a multiadsorbent packing contained within a metal or glass tube maintained at or above the surrounding air temperature. Depending on the water retention properties of the packing, some or most of the water vapor could pass completely through the trap during this process. The concentrated sample is then thermally desorbed from the packing into a GC column. Quite often, a "refocusing" trap is placed between the primary trap and the GC column to obtain better resolutions of the eluted peaks. The other method commonly used for concentrating the whole air sample is by passing through a cold trap where the VOCs are condensed cryogenically. The condensed samples are subsequently thermally desorbed and back flushed from the trap with an inert gas into a GC column. It can be seen that for the sampling and analysis of VOCs at low concentrations in air, whole air sampling is not a stand-alone method if sensitive detection methods are not available; it has to be combined with some features of the adsorbent sampling method.

B. ADSORBENT SAMPLING

Collection of gas-phase organic compounds from the atmosphere onto solid adsorbents is the most widely used air-sampling methodology. This technique can be further classified as active (using a pump) and passive or diffusive (by diffusion) sampling on the basis of sampling mode. Active sampling onto solid adsorbent tubes followed by thermal desorption is an alternative to the canister-based sampling method, and has been summarized in the U.S. EPA

Compendium of Methods for the Determination of Toxic Organic Compounds in Ambient Air, Method TO-17 (1997b) and the standard test method of the ASTM (1996b). It has many advantages including: (1) the small size and light weight of the adsorbent packing and attendant equipment; (2) the placement of the adsorbent packing as the first element in the sampling train so as to reduce the possibility of contamination from upstream elements; (3) the availability of a large selection of adsorbents to match the target set of compounds including polar VOCs; and (4) the possibility of water management using a combination of hydrophobic adsorbents, dry gas purge of water from the adsorbent after sampling, and splitting of the sample during analysis. Thus, this method has been widely used for the sampling of VOCs in the atmosphere at low concentrations (e.g., Ciccioli *et al.,* 1993; Konig *et al.,* 1995).

Passive sampling has the advantages of having no requirement for power and pump, being very simple to use, being lightweight, and having greater wearer acceptability when determining personal human exposure; and thus offers a most attractive alternative to the active sampling technique. Passive samplers were originally developed for the measurement of time-weighted-average exposure to airborne pollutants at relatively high concentrations (parts per million volume level) in the occupational workplace, and thus have been widely used in this area. Following their greater development, they are now used occasionally for the measurement of VOCs at parts or subparts per billion volume levels in outdoor and indoor air (e.g., Cohen *et al.,* 1989; 1990; Otson *et al.,* 1994). However, several disadvantages remain; each passive sampler type has to be fully validated for each chemical to be monitored according to validation protocols, which is very time consuming and expensive to administer. The other reason is that the passive sampler has very low sampling rates (ca. 0.1 to 50 ml/min). During passive sampling, organic molecules in air will diffuse onto the sampling adsorbent at rates (SR) defined by Fick's first law of diffusion as:

$$SR = D \times A/L, \tag{1}$$

where D is the diffusion coefficient of the compound, A is the cross-sectional area of the sampler, and L is the length of the diffusion path.

Although samples collected onto tube-type passive samplers (e.g., the stainless steel Perkin–Elmer diffusion tube sampler) can be thermally desorbed in splitless mode, long sampling periods (>24 h) are required to collect enough mass to be detected by the detector. The sampling rates of the badge-type passive samplers (e.g., the 3M OVM3500 and the SKC badge) are much higher, but a long sampling period is still essential since solvent extraction has to be used instead of thermal desorption due to both the geometry of the sampler and the adsorbent (activated charcoal) used. Thus, passive samplers

TABLE 1   Characteristics of Commonly Used Adsorbents for Sampling VOCs from the Atmosphere[a]

Adsorbent	Maximum temperature (°C)	Specific area (m²/g)	Hydrophobicity	Approximate analyte volatility range
Carbotrap C	>400	12	Yes	C_8-C_{20}
Carbopack C	>400	12	Yes	C_8-C_{20}
Carbotrap	>400	100	Yes	C_4-C_{14}
Carbopack B	>400	100	Yes	C_4-C_{14}
Carbosieve S-III	400	800	No	bp $-60-80°C$
Carboxen 1000	400	800	No	bp $-60-80°C$
Chromosorb 102	250	350	Yes	bp 50–200°C
Chromosorb 106	250	750	Yes	bp 50–200°c
Porapak Q	250	550	Yes	C_5-C_{12}
Porapak N	180	300	Yes	C_5-C_8
Tenax-TA	350	35	Yes	C_7-C_{26}
Tenax-GR	350	35	Yes	C_7-C_{30}

[a]U.S. EPA (1997b).

are not suitable for the collection of air samples at short intervals (e.g., every 30 min) due to the low concentrations of VOCs in ambient air. In addition, passive samplers with activated charcoal as the adsorbent (e.g., the 3M OVM3500 and the SKC badge) may not be suitable for outdoor applications because deactivation of the charcoal adsorbent may occur at relative humidities above 80% (Lewis and Gordon, 1996).

There are generally three categories of solid adsorbents used for sampling VOCs in air: (1) organic polymeric adsorbents (e.g., Tenax, Chromosorb, and Porapak), (2) inorganic adsorbents (e.g., silica gel, alumina, Florisil, and molecular sieves), and (3) activated charcoal and various carbon-based adsorbents such as graphitized carbon black (e.g., Carbotrap) and carbon molecular sieves (e.g., Carbosieve). Some characteristics of the most commonly used adsorbents are summarized in Table 1. The inorganic adsorbents have relatively high polarity, and thus can collect polar VOCs efficiently. However, since they are hydrophilic and have a strong affinity for water, these adsorbents may become deactivated over time and cease to sample efficiently. Activated charcoal is a very strong adsorbent, and is especially suitable for sampling of low-molecular-weight organic compounds. However, its advantage (strong adsorbate bonding) is also its disadvantage (i.e., it is difficult to desorb the organic compounds from it after being collected, and its irreversible adsorption for the organic compounds leaves no leeway at all for the thermal desorption technique). Instead, highly polar solvents (e.g., carbon disulfide) have to be used to extract the adsorbed compounds, and the sensitivity of this method is consequently severely reduced since only a small portion of the extract (ca. 0.1%) can be analyzed. Therefore, although activated charcoal has been a very popular adsorbent in occupational hygiene VOC monitoring where VOC concentrations are high, around parts per million volume levels, and analytical sensitivity is not a problem, it has rarely been used for the measurement of VOCs in ambient air at low concentrations (parts or subparts per billion volume levels). However, graphitized carbon and carbon molecular sieve adsorbents are less hydrophilic, are easier to desorb than activated charcoal, and have larger capacities than organic polymer adsorbents.

The organic polymeric resin Tenax (poly-2,6-diphenyleneoxide) has the advantages of high thermal stability (to about 300°C) and low affinity for water, and thus large volumes of air can be sampled without collection of significant amounts of water. Tenax-TA and the earlier Tenax-GC, either alone or in combination with other adsorbents, have been widely used for the sampling of VOCs from the atmosphere. Tenax-GR, which is a Tenax matrix with 23% graphitized carbon, appears to be no better than Tenax-TA with regard to breakthrough volume. Tenax adsorbents also have several disadvantages, including low capacity for very volatile organic compounds (vapor pressure > ca. 15 kPa) and polar VOCs, and artifact formation due to chemical reactions that occur during sampling or thermal desorption (Walling *et al.*, 1986). Ozone reacts with Tenax adsorbents to form many artifact compounds, with benzaldehyde and acetophenone being the most abundant (Cao and Hewitt, 1994a). Tenax adsorbents should therefore not be used for the sampling of these two compounds from ambient air. For the sampling of other VOCs at low levels in air when ozone is present at relatively high concentrations, the use of an ozone scrubber is necessary. A potassium–iodide-based ozone trap has been used to remove ozone efficiently from the ambient air (Greenberg *et al.*, 1994), but reactive iodine compounds may be formed and lead to the formation of organic iodine species (Helmig and Greenberg, 1995). Pretreatment of the Tenax adsorbent with thiosulfate as an antioxidant and hydrogen carbonate as a buffer has been found to prevent decomposition or acid rearrangements of monoterpene compounds (Stromvall and Petersson, 1992).

Suitable adsorbents should be selected for the sampling of target VOCs in a particular study to ensure efficient sampling and ease of desorption. The following factors are important considerations during this selection: characteristics of the adsorbents, such as the capacity or specific surface area, hydrophobicity, breakthrough volume or safe sampling volume (SSV), polarity, maximum temperature; and properties of the target VOCs, such as polarity,

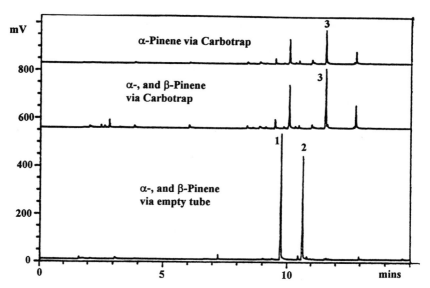

FIGURE 1 GC-FID chromatograms of α- and β-pinene after thermal desorption at 280°C for 8 min. Peak number 1: α-pinene, 2: β-pinene, and 3: 1,2,3-trimethylbenzene (Cao and Hewitt, 1993a).

vapor pressure, and boiling point. In addition, it is also very important to consider the stability of the target VOCs on the selected adsorbents to ensure both their high recovery and thus high sensitivity of the method and representativeness of the sampling and subsequent analytical results. The recoveries of some VOCs, especially more reactive VOCs, from some adsorbents are very low because of partial or complete decomposition, and the decomposition products could be the same as some of the target VOCs, leading to both over- and underestimation of VOC concentrations in the atmosphere. For example, recoveries of monoterpene compounds from the carbonaceous adsorbents are very low. Cao and Hewitt (1993) showed that α-pinene and β-pinene were decomposed completely on Carbotrap, and several new compounds were consequently formed (Fig. 1). Recovery of β-pinene from Tenax-GR was also relatively low due to the presence of 23% graphitized carbon in the Tenax matrix. The recovery of β-pinene from Chromosorb 106 was 79%, while recovery of α-pinene from Chromosorb 106 was very high (124%), which suggests the rearrangement of the monoterpene isomers at elevated temperature on this adsorbent.

A single adsorbent can rarely efficiently collect a wide volatility range of VOCs. Pollack *et al.* (1993) showed that in many cases a compound was not collected quantitatively by a single adsorbent but could be efficiently collected using a combination of several adsorbents with different characteristics (capacity, polarity, hydrophobicity, etc.). Multi-adsorbent sampling is therefore the preferred method for sampling VOCs over a wide range of volatility and polarity, especially in unknown atmospheres such as ambient air.

In multi-adsorbent sampling tubes, adsorbents are packed in order of increasing adsorbent strength, with the higher molecular weight VOCs retained on the front, least retentive adsorbents; and the more volatile compounds are retained farther into the packing on a stronger adsorbent. The higher molecular weight compounds never encounter the stronger adsorbents, thus improving the efficiency of the thermal desorption process. In the U.S. EPA Method TO-17 (1997b), three styles of multi-adsorbent tubes are suggested. Tube style 1 consists of 30 mm of Tenax-GR plus 25 mm of Carbopack B separated by 3 mm of unsilanized, preconditioned glass or quartz wool, which is suitable for compounds ranging in volatility from n-C_6 to n-C_{20} for air volumes of 2 liters at any humidity. Tube style 2 consists of 35 mm of Carbopack B plus 10 mm of Carbosieve S-III or Carboxen 1000 separated by glass or quartz wool as in tube style 1, and it is suitable for compounds ranging in volatility from n-C_3 to n-C_{12} for air volumes of 2 liters at relative humidities below 65% and temperature below 30°C. Tube style 3 consists of 13 mm of Carbopack C, 25 mm of Carbopack B plus 13 mm of Carbosieve S-III or Carboxen 1000 all separated by 3-mm plugs of glass or quartz wool, and it is suitable for compounds ranging in volatility from n-C_3 to n-C_{16} for air volumes of 2 liters at relative humidities below 65% and temperature below 30°C. However, since a hydrophilic adsorbent may be used in multi-adsorbent tubes, more water vapor will be retained than by the single hydrophobic (e.g., Tenax) adsorbent tube. Thus, suitable water management techniques should be used to reduce the effect of water vapor on the analysis system.

The success of using adsorbent tubes for the sampling of airborne VOCs at parts and subparts per billion volume levels depends on the mass of the artifact of the adsorbent being significantly lower ($< 10\%$) than the masses of analytes collected during air monitoring. Therefore, adsorbents with the lowest artifact levels should be selected, and adsorbent tubes should be conditioned as much as possible using a flow of inert gas at a maximum temperature. The conditioned adsorbent tubes should be capped tightly with Swagelok or other airtight fittings and stored in a solvent-free refrigerator. Adsorbent tubes should be reconditioned prior to use if they have been stored for a relatively long time (e.g., weeks), since artifacts could build up on adsorbents during storage over time, especially for Chromosorb 106 (Cao and Hewitt, 1994b).

As mentioned Section II.A (whole air sampling), water present in the air collected by a sample canister can cause problems for subsequent analysis (e.g., detector baseline drifts, extinguishing the hydrogen flame of FID). Water is also a severe problem for the analysis of samples collected on adsorbent tubes. It can stop the flow of carrier gas by forming ice in the secondary cold trap, especially the single path capillary trap used in some thermal desorption units (e.g., Chrompack TCT unit), during thermal desorption, and this will certainly affect the quantitative transfer of the desorbed analytes to the cold trap. Although this problem can be eased by using a multipath cold trap packed with adsorbent, as used in some thermal desorption units (e.g., the Perkin–Elmer ATD-400 thermal desorber), water is still a problem when it reaches the analytical detector. To minimize the amount of water collected on the adsorbent tubes, hydrophobic adsorbents should be used whenever possible. It is also essential to ensure that the temperature of the adsorbent tube is at least the same and certainly not lower than ambient temperature at the start of sampling or moisture will be retained via condensation, however hydrophobic the adsorbent is. Although water collected on the sampling tubes can be partly removed during thermal desorption in split mode, the sensitivity of the method will be decreased significantly; and thus this option is not suitable for the sampling and analysis of airborne VOCs at low concentrations. Water can also be partly eliminated by dry purging the sampling tube. This involves passing a volume of pure, dry, inert gas through the tube from the sampling end. Tubes can be heated while purging at slightly elevated temperatures at which the target analytes are still quantitatively retained while water is purged to vent from the tube (U.S. EPA, 1997b).

C. Sampling of Oxygenated Compounds

Compared with the hydrocarbon compounds, carbonyl compounds, aldehydes (RCHO), and ketones (R_1COR_2) are more reactive, and thus more difficult to determine in ambient air (although several methods are available for formaldehyde, other aldehydes, and ketones). The spectrometric methods, which are now considered obsolete for determination of atmospheric carbonyls other than formaldehyde, are usually based on functional group detection and lack adequate sensitivity for ambient air measurements (Vairavamurthy et al., 1992). The requirements of large, complex, and expensive instrumentation make these methods unsuitable for routine applications. Colorimetric techniques (e.g., the chromotropic acid method), and chemiluminescent detection also have limited applicability due to the lack of adequate sensitivity and the negative effect caused by an unknown interferent. Chromatographic methods, which allow the analysis of many carbonyls at the same time, have been widely used for the measurements of carbonyl compounds in air. Several methods have been developed using GC or liquid chromatographic (LC) analysis after derivatization, while carbonyls can also be detected directly by GC method.

1. Gas Chromatography

Oxygenated compounds may be collected by solid adsorbent sampling, then thermally desorbed into the gas chromatograph, and with subsequent detection by flame ionization or by mass spectrometry (MS). This method has been used for the measurement of biogenic carbonyl and other oxygenated compounds in air (e.g., Winer *et al.*, 1992; Ciccioli *et al.*, 1993). The commonly used adsorbents for carbonyl sampling are Tenax-GC, Tenax-TA, Carbotrap, Carbosieve S-III, and 13X molecular sieve. However, the sampling capacity of 13X molecular sieve may be affected by humidity. Tenax adsorbents may not be suitable for the sampling of some carbonyls in air because they react with ozone, generating several carbonyl compounds (e.g., hexanal, heptanal, octanal, and nonanal). Thus, there may be some question about the accuracy of the carbonyl data obtained using Tenax adsorption methods, because at least some of the carbonyls observed could be due to sampling artifacts. Nevertheless, despite the occasional application of this method, there has been no detailed investigation of the use of solid adsorbents for sampling and analysis of the wide range of carbonyls in air.

The other method for determination of atmospheric carbonyls with GC is based on chemical derivatizations (e.g., *o*-alkyloxime derivatives and oxazolidine derivatives). However, this method is less popular mainly because of the low volatility of the derivatives, especially for the high-molecular-weight carbonyls, and thus high oven temperatures required for elution, and the operational success of the DNPH derivatization with LC.

2. Chemical Derivatization and Liquid Chromatography

Although many other chemical derivatization methods are available [e.g., the use of 3-methyl-2-benzothiazolinon hydrazone (MBTH)], the acid-catalyzed condensation reaction of carbonyl compounds with 2,4-dinitrophenylhydrazine (DNPH) is widely used for characterizing carbonyl compounds. The reaction proceeds by nucleophilic addition to the carbonyl followed by 1,2-elimination of water to form the 2,4-dinitrophenylhydrazone. Although the derivatives can be separated by GC and detected by some suitable detector, LC separation of hydrazones combined with UV detection has become the most popular method for determination of carbonyls in air.

The air samples can be collected either by the DNPH-based impinger sampling technique or with DNPH-coated solid adsorbents. The impinger sampling method is cumbersome and not well suited to large field studies or to sampling at remote locations when samples have to be stored and transported to a central laboratory for analysis. The solid adsorbent method, on the other hand, has much higher sensitivity than the impinger method because the derivatives are usually preconcentrated to a high degree in the sample. Thus, DNPH-coated solid adsorbents have been increasingly used for the sampling of carbonyls in atmosphere (e.g., Shepson et al., 1991). Many solid adsorbents are available, with silica gel and C_{18} (octadecylsilane-bonded silica) being the most commonly used. Silica gel is a reactive polar material with surface OH groups, and is known to promote surface reactions with ozone at the parts per billion volume level. The C_{18} cartridge can provide nonpolar, hydrophobic, and relatively inert surface characteristics since its surface may be passivated with nonpolar paraffinic groups (e.g., n-octadecyl).

The major problem with the DNPH LC method is coelution or poor resolution of some compounds including saturated and corresponding unsaturated aldehydes. Ozone can also affect the DNPH-based method by formation of carbonyls as artifacts from reaction with sampling substrates and degradation of DNPH hydrazones. In addition, the DNPH-based method provides time-weighted-average concentrations because of the lengthy collection time required to achieve detection limits of subparts per billion volume. Thus, this method is not ideal for applications requiring good time resolutions.

D. Solid-Phase Microextraction

Solid-phase microextraction (SPME) is a new sampling technique. It integrates sampling, extraction, concentration, and sample introduction into a single step. It is rapid, inexpensive, solvent free, portable, relatively independent of the instrument design, and completely automated. The early work focused on its application for the sampling of organic compounds from liquid samples, but the SPME technique has been described for the sampling of VOCs in air by Chai et al. (1993) and Chai and Pawliszyn (1995).

The SPME device consists of a 1-cm length of fused silica fiber coated with an organic stationary phase that is contained in the needle of a modified syringe. Either the nonpolar stationary phase [poly(dimethylsiloxane)] or the more polar stationary phase [poly(acrylate)] can be coated on the fiber, which allows the SPME technique to be used for the measurements of both nonpolar and polar VOCs in air. To sample with the SPME device, the fiber is lowered into the atmosphere to be sampled by depressing the syringe plunger. The VOCs in the atmosphere then partition into the coating of the fiber until equilibrium is reached. The sampling time required for atmospheric sampling is short, as low as 2 min, due to the higher diffusion coefficients of compounds in air than in other media (e.g., water). The plunger is then with-drawn and the fiber is retracted into the needle. The syringe needle is used to pierce the septum of the injector of a gas chromatograph, and the samples collected on the fiber are thermally desorbed at temperatures between 150 and 250°C. A coated fiber can typically be used for up to 100 injections (Boyd-Boland *et al.*, 1994).

The major limitation of this technique is the storage stability of compounds collected on the coatings. Chai and Pawliszyn (1995) found that the analytes on the fiber were relatively stable when a sample fiber was withdrawn into the syringe needle with no plug on the needle tip for a time lapse of 2 min between sampling and desorbing the analytes. Longer storage time caused a significant concentration drop for all target compounds, especially for highly volatile compounds. Even with a Thermogreen septum cap on the needle tip, an average of 24% of the analytes was lost after 1 h of storage at room temperature. Maintaining a sampled syringe at low temperature yielded a notable increase in the stability of the analytes on the fiber due to the slower diffusion of analytes at lower temperatures. Most analytes studied can be stored in a refrigerator for 1 h without significant loss. The storage time for the less volatile analytes can be extended for up to 2 days when the sampled fiber is kept on dry ice at about $-70°C$. Therefore, this method is unlikely to be used for ambient air sampling in the field unless on-site analysis is avail-able. However, this limitation could be significantly eased if this method is combined with other sampling methods (e.g., the canister sampling method). That is, air samples are collected using the stainless steel canister, and are analyzed in the lab using the SPME technique.

The precision of the SPME technique is very good, typically about 5% relative standard deviation (RSD) for manual operation and as low as 1% using an autosampler, because it is a single-step method and thus the random sources of error associated with transfer of analytes are minimized. However, accuracy of this method is still uncertain, since this method has not been compared with any of the established methods. Thus, further validation work should be done before it can be considered as a useful sampling method for air monitoring.

E. MEASUREMENT METHODS FOR VOLATILE ORGANIC COMPOUND EMISSION FLUXES FROM SURFACES

The emission fluxes of VOCs from VOC-emitting surface areas (e.g., vegeta-tion) can be measured by several different methods, such as the bag enclosure method, the gradient profile method, and the conditional sampling or relaxed

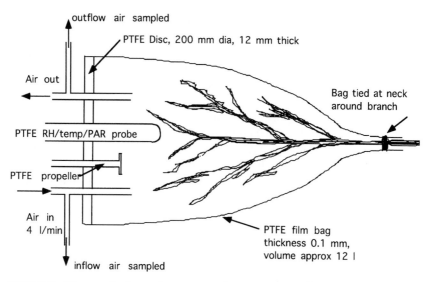

FIGURE 2   Diagram of a bag enclosure system for measurement of VOC emission rates from vegetation.

eddy accumulation method. The enclosure technique is indirect and measures the VOC fluxes from a relatively small sample of plant material, whereas the other techniques determine the average compound flux over a large surface area (typically 10^5 m² or more).

1. Bag Enclosure

In a dynamic flow-through branch enclosure system as shown in Fig. 2, a Teflon bag is placed around a branch of vegetation, and ambient air is pumped into the chamber. The emission rate (E, ng/gdw/h) and the corresponding emission flux (F, ng/m²/h) can then be calculated according to the following equations:

$$E = [(m_{out}/f_{out} - m_{in}/f_{in})F_{in}]/(T \times M) \qquad (2)$$

and

$$F = E \times B, \qquad (3)$$

where m_{out} and m_{in} are the masses (ng) of VOC in the outflow and inflow samples, f_{out} and f_{in} are the sampling flow rate (cm³/min) for the outflow and inflow, F_{in} is the total flow rate (cm³/min) into the chamber, T is the sampling time (h), M is the dry weight (g) of the leaves or needles on the branch, and B is the biomass factor (g/m²) appropriate to the particular forest site.

This method is very simple and easy to perform, can sample different vegetation species individually, and does not require highly sensitive or fast response chemical detectors or the meteorological data required by the other techniques. It can be employed in the field or in the laboratory where the effects of different environmental conditions can be investigated systematically. Thus, this method has been widely used for the measurement of emission fluxes of VOCs (mainly isoprene and monoterpenes) from vegetation (e.g., Cao *et al.*, 1997). However, since physical confinement of the plant under investigation is required, the enclosure technique may perturb the normal biological function of the plant and hence yield unrepresentative emission rates. In addition, a detailed biomass survey is required to allow extrapolation from a single branch to a forest or region to use these measurements for calculation of an emission inventory.

2. Gradient Profile

This method is based on micrometeorological surface layer theory, and can be used to obtain fluxes from plants distributed over much larger areas. The VOC concentration gradients (dC/dz) above an essentially infinite, uniform plane source (e.g., an ideal forest canopy) can be obtained by measuring VOC concentrations at several different heights. Temperature and wind speed or water vapor concentration gradients must also be measured. The meteorological data are used to determine the eddy diffusivity (K_z) so that the VOC emission flux can be calculated from the concentration gradient according to the following equation:

$$F = K_z(dC/dz). \tag{4}$$

However, since the method is difficult to set up and has extremely stringent sensor and site requirements, this method has been used only occasionally, primarily as an independent check on enclosure measurements. For example, in a field study by Cao *et al.* (1997), emission fluxes of isoprene and monoterpenes from gorse plants were measured by both bag enclosure and gradient methods, and the results from these two methods agreed very well (see later section for more information). In addition, this method may also be affected by chemical reactions between VOCs and oxidizing species (e.g., O_3, and OH and NO_3 radicals) during their upward transport, which may result in unrepresentative emission fluxes.

3. Conditional Sampling or Relaxed Eddy Accumulation

The most direct approach to flux measurement is the eddy correlation (EC) technique. This technique is based on the mean product of the fluctuations of vertical wind velocity (w) and concentration of the gas of interest (c). However, since this technique requires continuous fast response sensors with sufficient resolution to accurately measure the covariances of vertical wind velocity and concentrations of the gas of interest, it has not been used for organic compounds to date.

The eddy accumulation technique (EA) overcomes the need for fast response gas sensors without adding other uncertainties, since it is based on the same physical principles as EC. In this method, air is drawn from the immediate vicinity of an anemometer measuring vertical wind speed (w) and diverted into one of two "accumulators" on the basis of the sign of w, at a pumping rate proportional to the magnitude of w. Gas samples can be collected from the two accumulators and analyzed with a slow response detector.

Businger and Oncley (1990) suggested that the demands of eddy accumulation might be "relaxed" by sampling air at a constant rate for updrafts and downdrafts, rather than proportionally [relaxed eddy accumulation (REA) or conditional sampling]. They proposed that the flux (F) of the compound of interest should be given by

$$F = \beta \sigma_w (c^+ - c^-), \tag{5}$$

where β is an empirical constant (about 0.6), σ_w is the standard deviation of the vertical wind speed (m/sec), and c^+ and c^- are the mean concentrations ($\mu g/m^3$) of the gas in the upward- and downward-moving eddies.

Due to their technical difficulties, the EA and REA techniques have mainly been used only to measure fluxes of nonreactive species such as methane, CO_2, and H_2O (e.g., Baker et al., 1992; Oncley et al., 1993). Efforts have been made to use these techniques to determine the emission fluxes of biogenic VOCs from vegetation (Beverland et al., 1996).

III. ANALYTICAL METHODS

Because of the complexity of the mixture of organic compounds present in air, an analytical method that can resolve one compound from another is required. GC, particularly combined with the use of a high-resolution capillary column, offers excellent possibilities of speciation.

A. COLUMN CHROMATOGRAPHY

Chromatography is a collective term referring to a group of separation processes whereby a mixture of solutes are separated from one another by a differential distribution of the solutes between two phases, that is, the mobile phase and the stationary phase. Chromatographic methods are generally classified according to the physical state of the mobile phase. Thus, LC refers to the liquid state of the mobile phase, while GC refers to the gaseous state of the mobile phase. LC is further divided into flat and column methods, depending on whether the stationary phase is a thin layer mechanically supported on a sheet or is packed into a column. Flat chromatography has rarely been used for environmental analysis due to its limitations in separation and quantification, while column chromatographic techniques have played an extremely important role in our understanding of the environment.

Although column methods are classically referred to as LC, it should be noted that GC is strictly a column method because a column must be used for containment of the stationary phase. LC is suitable for the analysis of high-molecular-weight organic compounds and compounds that are reactive and unstable under high temperatures (e.g., aldehydes). GC, especially with the high-resolution capillary column, is the best method available for the separation and analysis of VOCs.

A gas chromatograph consists of six components: (1) a supply of pressurized carrier gas with ancillary pressure and flow regulators; (2) a sample–injection port; (3) a column; (4) a detector; (5) an electrometer and signal recorder; and (6) thermostated compartments encasing the column, detector, and injection port. The pressure inside the injection port is usually well above atmospheric pressure, and the stream of carrier gas sweeps away the sample and aids in vaporization. The injection port temperature is usually set at 25 to 50°C higher than the boiling point of the highest boiling components in the sample to ensure immediate vaporization. After the sample (in liquid or gaseous phase) is injected by a syringe into the injection port, it is vaporized and swept by the carrier gas into the column. In the less frequently used gas–solid chromatography (GSC), where the inner wall of the capillary column is coated with an adsorptive material, the components in the sample vapor are separated based on their differences in attraction to the solid stationary phase. Gas–liquid chromatography (GLC), where the liquid stationary phase is coated on the inner wall of the capillary column, is more widely used since it offers more choices for the selection of stationary phases with different degrees of polarity according to the application. In GLC, the components in the sample vapors are separated according to the differences in their distribution between the gas and liquid phases, that is, different partition coefficients (K_D) for different components. The partition coefficient is the ratio of component concentration in the stationary phase (C_s) and its concentration in the mobile phase (C_m):

$$K_D = C_s/C_m = \beta \, k, \tag{6}$$

where β, the phase ratio, is the ratio of the mobile-phase (gas) and stationary-phase volumes; and k, the capacity ratio, is the ratio of sample weight in the stationary phase and mobile phase. The phase ratio and capacity ratio are characteristic for a particular column. However, their product, K_D, is independent of the particular column. Therefore, the higher the phase ratio, the smaller is the capacity ratio. In general, the smaller the capacity ratio, the more difficult it is to achieve a particular separation.

After being separated on the capillary column, the sample components are then measured by the detectors. Although many GC detection systems are available, only FID and MS have been widely used for the quantitative and qualitative analysis of VOCs in the atmosphere.

B. Flame Ionization Detection

The FID is highly nonselective, and can respond to almost all VOCs. In an FID, eluted components in the carrier gas from the GC column are mixed with hydrogen and burned in air to produce a very hot flame to ionize the organic compounds. A pair of electrodes, charged by a polarizing voltage, collect the ions and generate a current proportional to the number of ions collected. The resultant current is amplified by an electrometer, producing a response on the recorder. The FID has a very wide range of response linearity, and its response factors for true hydrocarbon compounds can be predicted from the number of carbon atoms in the molecule; thus, concentrations of all other hydrocarbons can be determined from calibration with only a single hydrocarbon. It is also less expensive than other chromatographic detectors (e.g., the MS). Therefore, the FID has been widely used for the determination of VOCs in air, and has undergone little change during the last two decades.

Due to the significant contributions of biogenic VOCs to global VOC emissions, their airborne concentrations in different forest and agricultural sites and their emission rates from different plant species have been measured, mainly by gas chromatography–flame ionization detection (GC-FID) analysis of air samples collected either by stainless steel canisters or, more frequently, on adsorbent tubes. In a study by Cao et al. (1997), diurnal air concentrations of biogenic hydrocarbons were measured every 2 h at two different heights above an area of gorse located at Kelling Heath, United Kingdom, for two consecutive days to measure their emission rates by the gradient method. Gorse (Ulex europaeus) is known to be a prolific emitter of isoprene and it can also emit many monoterpenes at lower rates. The air samples were collected by drawing air through adsorbent tubes packed with Tenax-TA and Carbotrap, and analyzed by GC-FID after thermal desorption. Since the isoprene emission rate from gorse is significantly higher than that of the monoterpenes, air samples were collected and analyzed separately for isoprene and monoterpenes. Some of the samples were also analyzed by gas chromatogra-

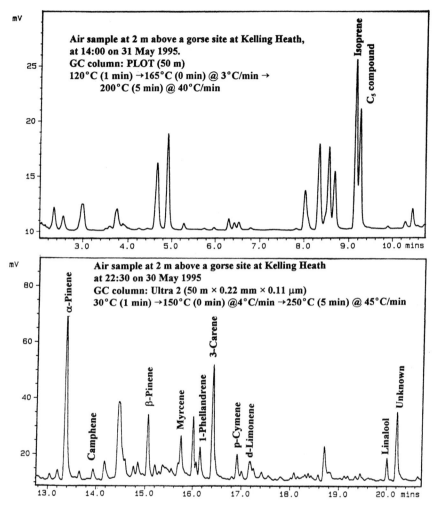

FIGURE 3   Typical GC-FID chromatograms of isoprene and monoterpene in air samples collected above gorse at Kelling Heath, United Kingdom.

phy–mass spectrometry (GC-MS) for identification of the dominant compounds emitted. Figure 3 shows typical GC-FID chromatograms of the isoprene and monoterpene samples, and Fig. 4 shows the mass spectra for isoprene and one of the monoterpenes (α-pinene). The diurnal variations of the air concentrations of isoprene at 2 and 6 meters and its emission rates are shown in Fig. 5. A very clear diurnal pattern of isoprene concentrations was observed, with maximum air concentrations (ca. 170 pptv at 2 m and

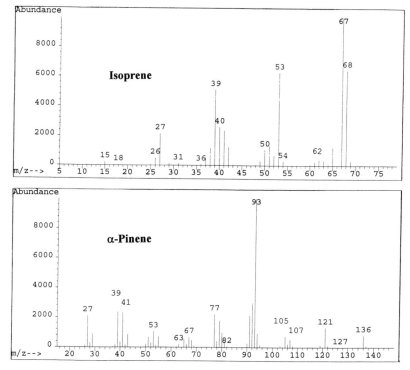

FIGURE 4   Mass spectra of isoprene and α-pinene detected in air.

150 pptv at 6 m) around midday and minimum concentrations (ca. 10 pptv at 2 and 6 m) during the night. The air concentrations at 2 m were generally higher than those at 6 m, except the point at 02:00 A.M. during the second day, which may be due to the effect of a surface temperature inversion. A clear diurnal pattern of isoprene emission fluxes was also observed, with maximum emission fluxes (ca. 100 $\mu g/m^2/h$) just before midday and late afternoon and minimum (zero) during the night. The decrease of emission fluxes after midday and before late afternoon for both sampling days was due to the changes of weather conditions (cloudy and raining). Table 2 shows the comparison of the isoprene emission fluxes measured by the gradient method with those measured simultaneously by the bag enclosure technique. It can be seen that for all time periods, the isoprene emission fluxes from the new grown and flowering branches were very close to the gradient results within a factor of two, while isoprene emission fluxes from the "mixed" branch showed lower results, especially during the first day. This is due to the fact that this branch was shaded by other parts of the gorse plant. The average

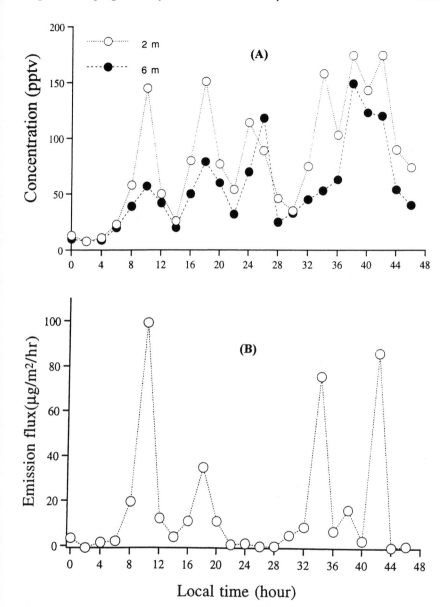

FIGURE 5   Diurnal variations of isoprene concentrations in air at two different heights above gorse at Kelling Heath, United Kingdom, and its emission fluxes measured by the gradient profile method (Cao *et al.*, 1997).

TABLE 2 Comparison of Average Emission Fluxes[a] of Isoprene Obtained from Enclosure and Gradient Methods[b]

Time	Gradient	Enclosure			
		New grown branch	Mixed branch	Flowering branch	Average from three branches
06:00–12:30, day 1	43.7	46.4	5.8	20.2	24.1
12:30–19:00, day 1	16.6	13.0	6.8	13.1	11.0
06:00–12:30, day 2	23.9	26.0	13.2	23.8	21.0
12:30–19:00, day 2	35.0	54.3	31.3	78.5	54.7
Daytime, day 1	26.1	36.2	6.7	18.1	20.3
Daytime, day 2	28.7	68.9	37.2	83.9	63.3
Daytime, 2 days	27.4	52.6	22.0	53.2	42.6

[a]Values in $\mu g/m^2/h$.
[b]Cao et al. (1997).

fluxes for the first and second days and the two-day average all showed good agreement.

Since manual sampling and operating instruments for subsequent analysis are extremely time and labor consuming, especially for frequent routine monitoring at remote field sites, efforts have been made to design automated and unattended sampling and analysis systems. The system designed by Mowrer and Lindskog (1991) was based on the thermal desorption of samples collected on adsorbent tubes with analysis by GC-FID. Air samples 2 liter in volume were taken every 4 h by pumping air through a silanized glass tube packed with Carbotrap and Carbosieve S-III. A Perma Pure dryer was used to remove water from the sample, though it was found to introduce contaminants to the system. The residual hydrocarbons and water from previous analysis were minimized by heating the adsorbent at 320°C, and the Perma Pure tubing at 90°C. This system can be left unattended for up to 2 weeks at a time, depending on the consumption of liquid nitrogen and the compressed gases, and has been used for the sampling and analysis of C_2 to C_5 hydrocarbons in air.

Automated sampling and analysis systems based on canister sampling have also been designed (Farmer et al., 1994; Greenberg et al., 1994; Castello et al., 1995). For example, in the work of Farmer et al. (1994), an automated integrated system was used successfully for the continuous hourly analysis of C_2 to C_{10} hydrocarbons in air down to the subparts per billion volume range. By using an independent mass flow controller, a time-integrated sample was accumulated into an electropolished SUMMA® canister over a 55-min period.

Next a subsample was cryogenically concentrated and analyzed by GC-FID. The integrating canister was then automatically evacuated, and the integration-fill cycle was recommenced. The whole cycle was repeated hourly. The sample-to-sample holdup within the integrating canister was shown to be insignificant.

The major problem with the previously mentioned automated sampling and analysis systems is the use of liquid nitrogen or other cryogens, which is expensive and can be difficult to supply reliably for long-term unattended operations at remote field sites. One solution is to use the small Peltier-cooled adsorbent trap that was originally designed for refocusing analytes in the Perkin–Elmer ATD-400 automatic thermal desorption system. Instead of using cryogens, analytes can be concentrated at $-30°C$ on this trap, cooled electronically.

The detection limit of GC-FID systems for VOCs depends on the operation modes (split or splitless) of the thermal desorption system used, and generally ranges from 5 to 50 pg. One of the limitations of the FID is that coeluted compounds cannot be measured separately. Misidentification of the eluted compounds is also a problem. In GC-FID, peak identifications are carried out by analyzing known composite VOC standard solutions and then comparing the retention time of a component in the sample with the retention time of the standard. Further peak identification by analyzing samples and standards using two different GC columns is useful, but still limited. Therefore, MS must normally be used to confirm the tentative identification of compounds obtained by GC-FID.

C. MASS SPECTROMETRY

MS has been used extensively as a GC detection method for the identification of organic compounds. A mass spectrometer includes an ionization source, a mass filter or analyzer, ion collection, amplification, and detection devices, and a system to create and maintain a vacuum. Electron impact (EI) is the most widely used method of ionization. As gaseous sample molecules enter the ion source, which is maintained in a high-vacuum environment to enhance collision efficiency and ion formation, they are bombarded by the electron beam originating from a heated rhenium or tungsten filament. A small positive potential on the repeller plate focuses and repels the positive ions generated through the exit slit toward the mass analyzer. Although negative ions are also formed in the source, and they can be analyzed by reversing the voltage potentials of the repeller and accelerating plates, the resulting negative ion mass spectrum has fewer ion fragments, and they are usually at relatively low masses. Consequently, negative-ion EI provides less structural information than its positive-ion counterpart, and as a result most applications center on positive-ion MS. The ions are separated according to their m/z ratios by the mass filter either electronically, as is the case in a quadrupole mass spectrometer, or magnetically, as is the case in magnetic sector mass spectrometers. These separated ions ultimately reach the detector.

MS has made a significant contribution to the understanding of the emissions of VOCs from vegetation. The early investigative work on VOC emissions from vegetation was focused mainly on hydrocarbons (isoprene and monoterpenes), because of their role as precursors to ozone formation, and also because the canister sampling methods used in the early investigations were not suitable for the oxygenated organic compounds because of wall adsorption–desorption problems and poor recovery. The results from the work of Arey et al. (1991) where samples were collected by adsorbent sampling followed by thermal desorption and GC-MS analysis have shown that oxygenated organic compounds, in addition to isoprene and monoterpenes, are also emitted by vegetation. Konig et al. (1995) identified more than 30 oxygenated organic compounds from the vegetation species studied in addition to isoprene and many monoterpene compounds, many for the first time (e.g., 2-methyl-1-propanol, 1-butanol, 2-butanol, 1-pentanol, 3-pentanol, 1-hexanol, 6-methyl-5-hepten-2-one, butanal, and ethylhexylacetate).

Efforts have been made to develop automated sampling and GC-MS analysis systems for the continuous monitoring of VOCs in the atmosphere. In the automated system designed by Yokouchi et al. (1993) for the trace determination of natural VOCs including isoprene and its products, a volume of 200 ml of air was collected on a small Tenax TA trap cooled electrically to 15°C. The samples were then thermally desorbed and analyzed by GC-MS. These processes were repeated automatically every 45 min, using a time controller. With a sampling volume of 200 ml, the detection limit was 1.9 pptv for isoprene at a signal-to-noise ratio of 3. This method was applied successfully to the monitoring of airborne VOCs in a field study.

An automated sampling and GC-MS analysis system has been designed by Simmonds et al. (1995) to operate automatically at a remote atmospheric research station for routine measurements of the chlorofluorocarbon (CFC) replacement compounds in air. A 2-liter air sample is collected onto a microtrap packed with adsorbents that are cooled to subambient temperatures ($-50°C$) using Peltier thermoelectric devices, thereby eliminating the need for liquid cryogens. The samples are then thermally desorbed, followed by GC-MS analysis. Detection limits of 0.5 pptv are achieved with sample air volume of 2 liter.

Efforts have been also made to develop mass spectrometer-based systems to monitor airborne VOCs in real time (Hansel et al., 1995; Cisper et al.,

1995; Gordon *et al.*, 1996). In these systems, samples can be introduced directly into the mass spectrometer without prior sample preparation, preconcentration, or chromatographic separation, so that all the analytes of interest in the sample are measured simultaneously in real time.

MS can be operated in two different modes: full-scan mode and selected ion monitoring (SIM) mode. In the full-scan mode, the entire mass range of interest is scanned; thus, full-scan analysis offers the highest specificity and is mainly used for the qualitative identification of organic compounds. However, the MS sensitivity in scan mode is limited since the detector spends a significant amount of time in regions where fragments give low-intensity adducts. By focusing on just a few preselected intense ion fragments, sensitivity is enhanced because the detector can spend all its time on those few ions instead of dividing its time on the several hundred ions scanned in the typical full-scan mode. Consequently, SIM techniques tend to enhance sensitivity by 10 to 100 times. However, this enhanced sensitivity is obtained at the expense of decreased specificity, since the vast amount of data from the mass spectrum is lost. Therefore, SIM techniques are only suitable for the analysis of preselected target compounds. In addition, data from the SIM analysis must be censored according to the criteria of retention time and ratios of qualifier ions over target ions to make positive identification. In the U.S. EPA TO-14A method (U.S. EPA, 1997a), the criterion of retention time is within $\pm$ 0.10 min of the library retention time of the compound, and the acceptance level for ratios of qualifier ions over target ions is determined to be $\pm 15\%$ of the ion ratios of the standard compound. However, since the retention time, and especially the ion ratios, vary depending on the amount of analyte in the ion source, criteria of retention time and ion ratios obtained from different levels of standard solutions should be applied to the target compounds at similar levels.

In addition to the advantage of compound identification, MS can also measure quantitatively compounds coeluted from the GC. A necessary condition is that the target ions of the coeluted compounds must be different, and the accuracy depends on the contribution of one of the coeluted compounds to the target ions of the other coeluted compound.

D. PHOTOIONIZATION DETECTION

In the photoionization detector (PID), gaseous sample molecules with ionization potentials less than the energy of a photon are ionized by absorbing photons from a far-UV lamp. The ions formed by UV ionization are then pushed through the accelerating electrode to the collection electrode where the current (proportional to concentration) is measured.

PID can almost be considered as a nondestructive detection system, since the ionization efficiency is about 0.1%. It is highly sensitive to most organic compounds, and the detection limits are typically 10 to 50 times lower than those of an FID for the same compounds, due to a larger response and a lower signal noise. Since a response is obtained only from compounds that have an ionization potential below the energy of the UV photons generated by the lamp, PID is a highly selective detection system, especially for alkenes, and aromatic and other reactive hydrocarbons that have lower ionization potentials. This is very advantageous for atmospheric monitoring programs focusing on photochemical ozone production in which the priority is the speciation and quantification of the more reactive hydrocarbons. PID may also give a less complex chromatogram than FID, and thus simplify peak identification.

PID is a very popular detector in the portable gas chromatographs (PGCs) that are now capable of near real-time ambient air monitoring. In the work of Berkley *et al.* (1991), the performance of a laboratory-tested portable GC equipped with a highly sensitive PID was compared with the U.S. EPA Method TO-14A canister-based method, and the results showed acceptable agreement between the two data sets. Despite the advantages mentioned earlier, PID has been used only occasionally for the analysis of VOCs in ambient air. This may be due to the fact that improvements to the design of PID systems with capillary GC are still in progress (Driscoll, 1992), and its response is compound specific, making its general application tedious.

E. REDUCTION GAS DETECTION

Reduction gas detection (RGD) was originally developed for detecting reducing inorganic gases, particularly CO and H_2. However, since the principle of detection relies only on the reduction of HgO to Hg vapor:

$$HgO \text{ (s)} + X \rightarrow Hg \text{ (vapor)} + \text{products}, \tag{7}$$

any reducing species (X) will, in principle, be detected, including organic molecules containing unsaturated bonds. O'Hara and Singh (1988) first described the use of the RGD for the determination of carbonyls, with good sensitivity. Since then it has also been used occasionally to determine isoprene concentrations at subparts per billion volume levels in air (Greenberg *et al.*, 1993; Monson *et al.*, 1994).

The responses of the RGD to C_2 through C_6 alkenes, C_2 through C_6 alkanes, isoprene, and benzene have been investigated under different conditions using packed column GC (Cao *et al.*, 1995). The results show that RGD is considerably more sensitive to alkenes than is FID, and it has much greater sensitivity to alkenes than alkanes. The sensitivity of RGD increases with increasing HgO bed temperature, but its selectivity toward alkenes decreases at the same time. An additional feature of this detector is that it does not require flammable support gases.

Although the RGD was initially engineered for use with packed GC columns, an interfaced capillary GC-RGD system has also been developed, and used for environmental analysis (Cao and Hewitt, 1993b; Cao et al., 1994). The detection limit of this system for hydrocarbons is still not adequate for ambient air sampling, being about 10 pg, due to peak tailing. However, it should be possible to improve the detection limit by heating the transfer line between the GC column and the detector, although the peak tailing problem may be also due to a dynamic equilibrium process (adsorption–desorption) between the mercury vapor and the wall of the detection cell.

F. Electron Capture Detection

After FID and MS, electron capture detection (ECD) is the most commonly used detector for GC. In an electron capture detector, a radioactive isotope (e.g., nickel-63) releases β particles that collide with the carrier gas molecules, producing many low-energy electrons. The electrons are collected on electrodes and produce a small, measurable, standing current. As sample components that contain chemical groups with high electron affinity (i.e., electrophilic species), particularly halogen atoms, are eluted from the GC column, they capture the low-energy electrons generated by the isotope to form negatively charged ions. The loss of current is then measured by the detector.

The sensitivity of this detector is extremely high, and the detection limits can be 10^4 times lower than with FID. It is also extremely selective, and it has been widely used for the analysis of airborne organic compounds having strong electron affinity, such as the CFC compounds. Its responses to hydrocarbons are very low. However, if highly electronegative atoms, such as halogens, can be added to the hydrocarbon molecules, then use of ECD, with much higher sensitivity than FID, should be possible.

Efforts have been made to use the GC-ECD for the analysis of volatile alkenes via on-column bromine addition reactions (Cao and Hewitt, 1995; Trigg et al., 1995). In the former, pyridinium bromide perbromide (PBPB) was used as the Br_2 source, and the excess Br_2 remaining after bromine addition to alkenes was removed by a methanol-treated cholesterol–glass bead mixture. This mixture also provides suitable polar conditions for bromine addition reactions with the alkenes in the gas phase. The conversion efficiencies of the individual alkenes to their brominated products are very low for ethene, but increase with carbon number, reaching 74% for 1-butene. The sensitivity of the ECD to brominated C_3 to C_5 alkenes is about 200 to 300 times higher than conventional FID, but poor peak shapes limit its applicability at present. However, the detection limits could be improved by using a two-oven system that would allow the temperatures for the brominating phase and the GC column to be maintained independently.

In the work of Trigg et al. (1995), copper(II) bromide coated onto a solid support was used as the Br_2 source. Because of the low bleed of Br_2 from $CuBr_2$, higher temperatures (up to 140°C) could be used, and a bromine bleed scrubbing phase was not required. In addition, the Cu^{2+} ion in $CuBr_2$ may have a catalytic effect on the bromination of alkenes, but at higher temperatures (above 80°C), substitution reactions may occur. The conversion rates for C_2 to C_4 alkenes were normally greater than 80%, and the detection limits of the GC–bromination–ECD system for alkenes (C_2 to C_5) were less than 5 pg.

This pioneering work shows the potential of this detection method in the real-time monitoring of reactive hydrocarbons in the atmosphere.

G. Ozone Chemiluminescence Detection

The reactions between alkenes and ozone produce electronically excited formaldehyde that subsequently chemiluminesces:

$$alkenes + O_3 \rightarrow HCHO^* + products \qquad (8)$$

and

$$HCHO^* \rightarrow HCHO + h\nu. \qquad (9)$$

Emission from HCHO* occurs in the region of 450 to 550 nm, and the intensity of this light can be measured and related to the concentrations of alkenes present.

The chemiluminescence of alkene–ozone reactions was first explored as a possible method of detecting ozone by Nederbragt et al. (1965), and as a selective GC detector for hydrocarbon gas analysis by Bruening and Concha (1975). It is selective, due to the relatively small number of compounds that chemiluminesce on reaction with a given reactant, and very sensitive since the chemiluminescence appears out of a near-zero light background. In principle, a single photon generated from a chemiluminescent reaction can be detected. Its selectivity depends very much on the detector temperature: at lower temperature (100°C), only alkenes can be detected; at higher temperature (250°C), alkanes can also be detected. The detection limit is frequently at the nanogram level and is temperature dependent. This detection method has the advantage of being based on a very fast, noncatalytic and flameless reaction, but the foremost advantage is that it is possible to monitor certain atmospheric species in real time.

Hills and Zimmerman (1990) constructed a continuous isoprene monitor, based on its reaction with ozone. It has a linear response over three orders of magnitude. With a response time of about 10 sec, the lower detection limit was less than 1 ppbv. This continuous analyzer has been used for the measurement of isoprene emission fluxes from different plant species (Guenther et al., 1991). In general, one would expect discrimination between isoprene and other alkenes to be poor since all alkenes react to some extent with ozone to produce HCHO*, but the rapid reaction of isoprene with ozone and the use of selected wave bands does allow discrimination of isoprene. Interference from propene is the major problem because responses to these two compounds are roughly the same, and if comparable amounts of each were present, the chemiluminescent signals could not be distinguished. Fortunately, this is rarely the case since the isoprene/propene ratio in air is usually > 10 in regions where biogenic isoprene fluxes are significant. Interferences from monoterpene compounds are also slight. The instrument has the rapid response necessary to measure isoprene fluxes using the micrometerological eddy correlation technique, although further development is certainly needed.

H. Combustion/Isotope Ratio Mass Spectrometry

Since the isotopic abundances of carbon may be different according to origin (natural and anthropogenic), it may be possible to establish the origin of a chemical compound in air and to evaluate the relative importance of different sources by measurement of the $^{13}C/^{12}C$ ratio of that compound. Isotopic data may also facilitate an understanding of the mechanisms of the production and consumption of compounds. GC–combustion–isotope ratio mass spectrometry (GC-C-IRMS) is a very useful method for measurement of the $^{13}C/^{12}C$ ratio. The GC-C-IRMS system consists of a gas chromatograph, an interface with a combustion furnace and an isotope ratio mass spectrometer. The chemical compounds of interest in a gaseous sample are first separated on a GC column. The compounds eluted from the GC column are then converted to carbon dioxide by burning in an electrically heated quartz tube packed with CuO particles. The products of the combustion, CO_2 and H_2O, are carried into a cryogenic trap to remove water, while CO_2 passes and flows through the capillary column into the ion source of the mass spectrometer.

GC-C-IRMS has been used for the determination of the $^{13}C/^{12}C$ ratios for CO_2 and CH_4 in the atmosphere (e.g., Zeng et al., 1994; Sugimoto, 1996), and for studying the biosynthetic pathway of isoprene by measuring the fractionation between stable carbon isotopes during biosynthesis (Sharkey et al., 1991). It has also been used in some other areas, for example, detection of vegetable oil adulteration (Woodbury et al., 1995), and detection of testosterone abuse (Aguilera et al., 1996a; 1996b). However, this method has not yet been successfully used for the isotope analysis of VOCs in the atmosphere, due to at least two reasons: (1) unlike CO_2 and CH_4 at parts per million volume levels in the atmosphere, concentrations of VOCs in the atmosphere are very low (at parts or subparts per billion volume levels), and (2) the high detection limits of present GC-C-IRMS systems (e.g., ca. 50 ng for isoprene). As can be seen from its principle mentioned earlier, the detection limits of this method for hydrocarbons depend greatly on the combustion efficiency of each organic compound and also on the efficiency of the subsequent process of transferring CO_2 produced to the MS for analysis. Since interest in isotopic studies of the biogenic emission of hydrocarbons is growing, improvements to the sensitivity of the GC-C-IRMS method are both necessary and likely.

I. DIFFERENTIAL OPTICAL ABSORPTION SPECTROSCOPY

Although not a chromatographic method, differential optical absorption spectroscopy (DOAS) has been used for more than 15 years to measure a wide range of gaseous air pollutants. DOAS is based on the absorption of light at specific wavelengths for different compounds according to Beer's law:

$$A = \log (P_o/P) = \epsilon bC, \tag{10}$$

where A is the absorbance, C is the concentration of the compound (mol/l), b is the length of the light path (cm), ϵ is the molar absorptivity (l/mol/cm), P_o is the radiant power when the absorbing medium is not present in the light path, and P is the radiant power when the absorbing medium is present in the light path. A DOAS instrument is composed of a broadband light source (emission between 200 and 1000 nm) and a receiver–spectrometer assembly. The distance between the light source and the receiver can range from 100 to 2000 m, depending on the pollutant to be monitored and the species concentrations. This method has been used more successfully for the real-time monitoring of inorganic gases (SO_2, NO_2, O_3, and HNO_2) and formaldehyde. For example, Stevens *et al.* (1993) measured the air concentrations of SO_2, NO_2, O_3, HNO_2, and HCHO simultaneously by both a DOAS instrument and the Federal Reference Methods (conventional methods) designated by the U.S. EPA. The results show excellent agreement between the DOAS and the conventional methods. DOAS has also been used for the real-time monitoring of aromatic hydrocarbons (e.g., Lofgren, 1992; Axelsson *et al.*, 1995; Barrefors, 1996), but the reliability of this method for such application is still questionable because the correlation between DOAS and GC data is very poor. This is mainly due to the severe interference effects of the presence of many hydrocarbon compounds with similar spectra.

J. INTERCOMPARISON EXPERIMENTS

Although most ambient VOC measurement methods are based on GC techniques, the details of the experimental procedures vary from one laboratory to another. For example, there are several methods available for sample preconcentration prior to injection. Some laboratories may prefer to trap the air samples cryogenically in a glass bead-filled, stainless steel or nickel loops immersed in liquid oxygen, argon, or nitrogen, while other laboratories may prefer to trap the air samples on adsorbent materials. Various standard VOCs or mixtures of VOCs are employed by investigators to calibrate their instruments for quantification. Some investigators use secondary standards from a commercial source, while others prepare their own standards. When using GC-FID, many investigators have taken advantage of the linear response of the detector to the molecular weight of the hydrocarbon and simply report concentrations of hydrocarbons in the air samples by referencing to a particular single standard compound. Other laboratories prefer to generate a standard for each compound to be analyzed, and calibrate their instrument accordingly. All these factors and other variations in the experimental procedures used by different laboratories raise doubts on the reliability and reproducibility of the various analytical techniques used in deriving VOC concentrations in the atmosphere.

To identify existing problems in analytical procedures, to correct these problems, and to ensure quality control of VOC analyses made by atmospheric scientists throughout the world, it is essential to participate in international intercomparison experiments. The Nonmethane Hydrocarbon Intercomparison Experiment (NOMHICE) has been carried out to evaluate current methods being used to determine the ambient levels of various atmospheric nonmethane hydrocarbons (Apel *et al.*, 1994). There are 40 laboratories participating in the NOMHICE program, and this program has a total of five tasks. Task 1 was designed to check on both the reliability of the standards used by each of the participating laboratories and the basic analytical procedures employed by each group. A two-component mixture of n-butane and benzene (parts per billion volume range) was used as a standard in task 1. Of the 36 laboratories that submitted their results for this task, 12 reported results that differed from the National Institutes of Standards and Technology (NIST)-reported concentrations by more than 20% for n-butane or benzene. These laboratories were asked to perform a follow-up analysis of the mixture provided in another canister. This time, for 8 out of the 12 laboratories, improvements were observed for the second analysis, and significant improvements were observed for 4 of the laboratories. However, the results from a few laboratories moved out of the "adequate range," indicating a problem of reproducibility of their analytical procedures.

Task 2 was designed to be a more stringent test for both identification and quantification of a range of compounds. A more complex 16-component mixture was used in this task. Of 28 laboratories that submitted their results, 12 were able to correctly identify all 16 components. Of these 28 laboratories, 12 reported their results within 10% of the NIST values for the sum of the compounds that they reported, and 3 of the laboratories were within 5%. Some constructive comments were provided by the organizers to those laboratories with poor performance, and improvements were indeed observed as a result of this effort.

Task 3 of this program involved the distribution of a still more complex prepared mixture. Any problems occurring in this task will trigger the supply of a second mixture, if necessary, followed by personal contact from the NOMHICE staff with subsequent investigation and evaluation of the methods and procedures employed. Task 4 will involve the analysis of ambient air samples collected by the NOMHICE scientists and analyzed by selected scientists from other laboratories who have shown their reliability in the previous tasks. A subset of the participants who have demonstrated their ability to analyze ambient levels of hydrocarbon in task 4 will be invited to participate in the final task, a field intercomparison study made at a common field location.

IV. SUMMARY

Accurate measurements of VOC concentrations in air and of their emission rates from both anthropogenic and biogenic sources into the atmosphere are essential to control the concentrations of ozone in the lower atmosphere and to assess their risk to the human environment. In designing a VOC sampling program, suitable sampling and analytical methods should be selected according to the properties of the target compounds, to ensure representative sampling, precise and accurate quantification, and the best sensitivity of the method.

Several methods are available for collecting VOCs from the atmosphere, the most commonly used methods being whole air sampling with stainless steel canisters and adsorbent sampling. The stainless steel canister method is only suitable for collection of truly gas-phase VOCs. Compounds greater than C_{10} are more likely to adsorb onto the walls of the canister on storage, leading to underestimation of their ambient concentrations. Collection of polar VOCs by this method presents more problems than for nonpolar VOCs. The stability of samples during storage depends greatly on the presence of water vapor in the canister.

Adsorbent sampling, especially with multi-adsorbents, has much broader applications due to the availability of a wide variety of adsorbents with different characteristics. Some organic compounds are partly or completely decomposed on some adsorbents, and artifact compounds can be formed on some adsorbents due to reactions of these adsorbents with oxidizing species in the air. Thus, suitable adsorbents should be selected for the sampling of target VOCs in a particular study to ensure efficient sampling, ease of desorption, and high recovery and thus high sensitivity of the method by taking into account the following factors: characteristics of the adsorbents (such as the capacity or specific surface area, hydrophobicity, breakthrough volume, polarity, maximal temperature, and others); and the properties of the target VOCs (such as polarity, vapor pressure, and boiling point).

GC with high-resolution capillary column and FID or MS detection is still the dominant method for the analysis of VOCs. Although many other detection methods are available, they have been used only occasionally for the analysis of VOCs due to their various limitations, including narrow range of application, poor resolution, low sensitivity, and others. Since the details of the analytical procedures used for VOC analysis differ from one laboratory to another, programs of international intercomparison experiment on VOC analysis should be organized regularly, and atmospheric scientists should participate in these programs to ensure the reliability and reproducibility of their data.

REFERENCES

ASTM (1996a). Standard test method for determination of volatile organic chemicals in atmospheres (canister sampling methodology). *In* "Annual Book of ASTM Standards," Vol. 11.03, pp. 528–545. American Society for Testing and Materials (ASTM): Washington, DC.

ASTM (1996b). Standard practice for analysis of organic compound vapours collected by the activated charcoal tube adsorption method. *In* "Annual Book of ASTM Standards," Vol. 11.03, pp. 208–212. American Society for Testing and Materials (ASTM): Washington, DC.

Aguilera, R., Grenot, C., Casabiance, H., and Hatton, C. K. (1996a). Detection of testosterone misuse: comparison of two chromatographic-combustion/isotope ratio mass spectrometric analysis. *J. Chromatogr. (B)* **687**, 43–53.

Aguilera, R., Becchi, M., Casabianca, H., Hatton, C. K., Catlin, D. H., Starcevic, B., and Pope, H. G., Jr. (1996b). Improved method of detection of testosterone abuse by gas chromatography/combustion/isotope ratio mass spectrometry analysis of urinary steroids. *J. Mass Spectrom.* **31**, 169–76.

Apel, E. C., Calvert, J. G., and Fesenfeld, F. C. (1994). The nonmethane hydrocarbon intercomparison experiment (NOMHICE): tasks 1 and 2. *J. Geophys. Res.* **99**, 16651–16664.

Arey, J., Winer, A. M., Atkinson, R., Aschmann, S. M., Long, W. D., and Morrison C. L. (1991). The emission of (Z)-3-hexen-1-ol, (Z)-3-hexenylacetate and other oxygenated hydrocarbons from agricultural plant species. *Atmos. Environ.* **25A**, 1063–1075.

Axelsson, H., Edner, H., Eilard, A., Emanuelsson, A., Galle, B., Kloo, H., and Ragnarson, P. (1995). Measurement of aromatic hydrocarbons with the DOAS technique. *Appl. Spectrom.* **49**, 1254–1260.

Baker, J. M., Norman, J. M., and Bland, W. L. (1992). Field-scale application of flux measurement by conditional sampling. *Agric. Forest Meteorol.* **62**, 31–52.

Barrefors, G. (1996). Monitoring of benzene, toluene and *p*-xylene in urban air with differential optical absorption spectroscopy technique. *Sci. Total Environ.* **189/190**, 287–292.

Berkley, R. E., Varns, J. L., and Pleil, J. (1991). Comparison of portable gas chromatographs and passivated canisters for field sampling airborne toxic organic vapours in the United States and the USSR. *Environ. Sci. Technol.* **25**, 1439–1444.

Beverland, I. J., Milne, R., Boissard, C., O'Neill, D. H., Moncrieff, J. B., and Hewitt, C. N. (1996). Measurement of carbon dioxide and hydrocarbon fluxes from a sitka spruce forest using micrometeorological techniques. *J. Geophys. Res.* **101**, 22807–22815.

Boyd-Boland, A. A., Chai, M., Luo, Y. Z., Zhang, Z., Yang, M. J., Pawliszyn, J. B., and Gorecki, T. (1994). New solvent-free sample preparation techniques based on fibre and polymer technologies. *Environ. Sci. Technol.* **28**, 569A–574A.

Bruening, W., and Concha, F. J. M. (1975). Selective detector for gas chromatography based on the chemiluminescence of ozone reactions. *J. Chromatogr.* **112**, 253–265.

Businger, J. A., and Oncley, S. P. (1990). Flux measurement with conditional sampling. *J. Atmos. Oceanic Technol.* **7**, 349–352.

Cao, X.-L., and Hewitt, C. N. (1993a). Thermal desorption efficiencies for different adsorbate/adsorbent systems typically used in air monitoring programmes. *Chemosphere.* **27**, 695–705.

Cao, X.-L., and Hewitt, C. N. (1993b). Passive sampling and capillary gas chromatographic determination of reactive hydrocarbons in ambient air with the reduction gas detector. *J. Chromatogr.* **648**, 191–197.

Cao, X.-L., Hewitt, C. N., and Waterhouse, K. S. (1994). Capillary gas chromatographic analysis of reactive hydrocarbons with the reduction gas detector. *J. Chromatogr.* **697**, 115–112.

Cao, X.-L., and Hewitt, C. N. (1994a). Study of the oxidation by ozone of the adsorbents and the organic compounds adsorbed during passive sampling. *Environ. Sci. Technol.* **28**, 757–762.

Cao, X.-L., and Hewitt, C. N. (1994b). Artifact build-up on adsorbents during storage and its effect on passive sampling and GC-FID analysis of low concentrations of VOCs in air. *J. Chromatogr.* **688**, 368–374.

Cao, X.-L., Hewitt, C. N., and Waterhouse, K. S. (1995). A study of the responses of gas chromatography-reduction gas detector system to gaseous hydrocarbons under different conditions. *Anal. Chim. Acta* **300**, 193–200.

Cao, X.-L., and Hewitt, C. N. (1995). Gas chromatographic analysis of volatile alkenes by on-column bromination and electron capture detection. *J. Chromatogr.* **690**, 187–195.

Cao, X.-L., Boissard, C., Juan, A. J., Hewitt, C. N., and Gallagher, M. (1997). Biogenic emissions of volatile organic compounds from gorse (*Ulex europaeus*). I. Diurnal emission fluxes obtained from bag enclosure and gradient methods over a gorse site at Kelling Heath, England. *J. Geophys. Res.* **102**, 18903–18915.

Castello, G., Benzo, M., and Gerbino, T. C. (1995). Automated gas chromatographic analysis of volatile organic compounds in air. *J. Chromatogr.* **710**, 61–70.

Chai, M., Arthur, C. L., Pawliszyn, J. (1993). Determination of volatile chlorinated hydrocarbons in air and water with solid-phase microextraction. *Analyst.* **118**, 1501–1505.

Chai, M., and Pawliszyn, J. (1995). Analysis of environmental air samples by solid-phase microextraction and gas chromatography/ion trap mass spectrometry. *Environ. Sci. Technol.* **29**, 693–701.

Ciccioli, P., Brancaleoni, E., Cecinato, A., and Sparapani, R. (1993). Identification and determination of biogenic and anthropogenic volatile organic compounds in forest areas of Northern and Southern Europe and a remote site of the Himalaya region by hi-resolution gas chromatography-mass spectrometry. *J. Chromatogr.* **643**, 55–69.

Cisper, M. E., Gill, C. G., Townsend, L. E., and Hemberger, P. H. (1995). On-line detection of volatile organic compounds in air at parts-per-trillion levels by membrane introduction mass spectrometry. *Anal. Chem.* **67**, 1413–1417.

Cohen, M. A., Ryan, P. B., Ozkaynak, H., and Epstein, P. S. (1989). Indoor/outdoor measurements of volatile organic compounds in the Kanawha Valley of West Virginia. *JAPCA* **39**, 1086–1093.

Cohen, M. A., Ryan, P. B., Yanagisawa, Y., and Hammond, S. K. (1990). The validation of a passive sampler for indoor and outdoor concentrations of volatile organic compounds. *J. Air Waste Manage. Assoc.* **40**, 993–997.

Driscoll, J. N. (1992). Far-UV ionization (photoionization) and absorbance detectors. In "Detectors for Capillary Chromatography" (H. H. Hill, and D. G. McMinn, eds.), pp. 51–82. John Wiley & Sons, New York.

Evans, G., Lumpkin, T. A., Smith, D. L., and Somerville, M. C. (1992). Measurements of VOCs from the TAMS Network. *J. Air Waste Manage. Assoc.* **42**, 1319–1323.

Farmer, C. T., Milne, P. J., Riemer, D. D., and Zika, R. G. (1994). Continuous hourly analysis of C_2–C_{10} non-methane hydrocarbon compounds in urban air by GC-FID. *Environ. Sci. Technol.* **28**, 238–245.

Fehsenfeld, F., Calvert, J., Fall, R., Goldan, P., Guenther, A. B., Hewitt, C. N., Lamb, B., Liu, S., Trainer, M., Westberg, H., and Zimmerman, P. R. (1992). Emissions of volatile organic compounds from vegetation and the implications for atmospheric chemistry. *Global Biogeochem. Cycles* **6**, 389–430.

Gholson, A. R., Jayanty, R. K. M., and Storm, J. F. (1990). Evaluation of aluminum canisters for the collection and storage of air toxics. *Anal. Chem.* **62**, 1899–1902.

Gordon, S. M., Callahan, P. J., Kenny, D. V., and Pleil, J. D. (1996). Direct sampling and analysis of volatile organic compounds in air by membrane introduction and glow discharge ion trap mass spectrometry with filtered noise fields. *Rapid Commun. Mass Spectrom.* **10**, 1038–1046.

Greenberg, J. P., Zimmerman, P. R., Taylor, B. E., Silver, G. M., and Fall, R. (1993). Sub-parts per billion detection of isoprene using a reduction gas detector with a portable gas chromatograph. *Atmos. Environ.* **27A**, 2689–2692.

Greenberg, J. P., Lee, B., Helmig, D., and Zimmerman, P. R. (1994). Fully automated gas chromatograph-flame ionization detector system for the *in situ* determination of atmospheric non-methane hydrocarbons at low parts per trillion concentration. *J. Chromatogr.* **676**, 389–398.

Guenther, A. B., Monson, R. K., and Fall, R. (1991). Isoprene and monoterpene emission rate variability: observations with Eucalyptus and emission rate algorithm development. *J. Geophysical Research.* **96**, 10799–10808.

Hansel, A., Jordan, A., Holzinger, R., Prazeller, P., Vogel, W., and Lindinger, W. (1995). Proton transfer reaction mass spectrometry: on-line trace gas analysis at the ppb level. *Int. J. Mass Spectro. Ion Proc.* **149/150**, 609–619.

Helmig, D. and Greenberg, J. (1995). Artifact formation from the use of potassium-iodide-based ozone traps during atmospheric sampling of trace organic gases. *J. High Resolution Chromatogr.* **18**, 15–18.

Hills, A. J., and Zimmerman, P. R. (1990). Isoprene measurement by ozone-induced chemiluminescence. *Anal. Chem.* **62**, 1055–1060.

Kelly, T. J., Callhan, P. J., Pleil, H., and Evans, G. F. (1993). Method development and field measurements for polar volatile organic compounds in ambient air. *Environ. Sci. Technol.* **27**, 1146–1153.

Konig, G., Brunda, M., Puxbaum, H., Hewitt, C. N., and Duckham, S. C. (1995). Relative contribution of oxygenated hydrocarbons to the total biogenic VOC emissions of selected mid-European agricultural and natural plant species. *Atmos. Environ.* **29**, 861–874.

Lamb, B., Guenther, A., Gay, D., and Westberg, H. (1987). A national inventory of biogenic hydrocarbon emissions. *Atmos. Environ.* **21**, 1695–1705.

Lewis, R. G., and Gordon, S. M. (1996). Sampling for organic chemicals in air. *In* "Principles of Environmental Sampling (L. H. Keith, ed.), pp. 401–407. American Chemical Society: Washington, DC.

Lofgren, L. (1992). Determination of benzene and toluene in urban air with differential optical absorption spectroscopy. *Intern. J. Environ. Anal. Chem.* **47**, 69–74.

McClenny, W. A., Pleil, J. D., Evans, G. F., Oliver, K. D., Holdren, M. W., and Winberry, W. T. (1991). Canister-based method for monitoring toxic VOCs in ambient air. *J. Air Waste Manage. Assoc.* **41**, 1308–1318.

Monson, R. K., Harley, P. C., Litvak, M. E., Wildermuth, M., Guenther, A. B., Zimmerman, P. R., and Fall, R. (1994). Environmental and developmental controls over the seasonal pattern of isoprene emission from aspen leaves. *Oecologia.* **99**, 260–270.

Mowrer, J., and Lindskog, A. (1991). Automated unattended sampling and analysis of background levels of C_2–C_5 hydrocarbons. *Atmos. Environ.* **25A**, 1971–1979.

Muller, J.-F. (1992). Geographical distribution and seasonal variation of surface emissions and deposition velocities of atmospheric trace gases. *J. Geophys. Res.* **97**, 3787–3804.

Nederbragt, G. W., van der Horst, A., and van Duijn, J. (1965). Rapid ozone determination near an electron accelerator. *Nature* **206**, 87–88.

O'Hara, D., and Singh, H. B. (1988). Sensitive gas chromatographic detection of acetaldehyde and acetone using a reduction gas detector. *Atmos. Environ.* **22**, 2613–2615.

Oliver, K. D., Pleil, J. D., and McClenny, W. A. (1986). Sample integrity of trace level volatile organic compounds in ambient air stored in Summa polished canisters. *Atmos. Environ.* **20**, 1403–1411.

Oncley, S. P., Delany, A. C., Horst, T. W., and Tans, P. P. (1993). Verification of flux measurement using relaxed eddy accumulation. *Atmos. Environ.* **27A**, 2417–2426.

Otson, R., Fellin, P., and Tran, Q. (1994). VOCs in representative Canadian residences. *Atmos. Environ.* **28**, 3563–3569.

Pate, B., Jayanty, R. K. M., Peterson, M. R., and Evans, G. F. (1992). Temporal stability of polar organic compounds in stainless steel canisters. *J. Air Waste Manage. Assoc.* **42**, 460–462.

Pollack, A. J., Gordon, S. M., and Moschandreas, D. J. (1993). Evaluation of portable multisorbent air samplers for use with an automated multitube analyser, U.S. Environmental Protection Agency: Research Triangle Park, NC; EPA/600/R-93/053.

Roberts, J. M., Fehsenfeld, F. C., Albritton, D. L., and Sievers, R. E. (1983). Measurements of monoterpene hydrocarbons at Niwot Ridge, Colorado. *J. Geophys. Res.* **88**, 10667–10678.

Shepson, P. B., Hastie, D. R., Schiff, H. I., Polizzi, M., Bottenheim, J. W., Anlauf, K., Mackay, G. I., and Karecki, D. R. (1991). Atmospheric concentrations and temporal variations of C_1–C_3 carbonyl compounds at two rural sites in central Ontario. *Atmos. Environ.* **25A**, 2001–2015.

Sharkey, T. D., Loreto, F., Delwiche, C. F., and Treichel, I. W. (1991). Fractionation of carbon isotopes during biogenesis of atmospheric isoprene. *Plant Physiol.* **97**, 463–466.

Simmonds, P. G., O'Doherty, S., Nickless, G., Sturrock, G. A., Swaby, R., Knight, P., Ricketts, J., Woffendin, G., and Smith, R. (1995). Automated gas chromatography/mass spectrometer for routine atmospheric field measurements of the CFC replacement compounds, the hydrofluorocarbons and hydrochlorofluorocarbons. *Anal. Chem.* **67**, 717–723.

Stevens, R. K., Drago, R. J., and Mamane, Y. (1993). A long path differential optical absorption spectrometer and EPA-approved fixed-point methods intercomparison. *Atmos. Environ.* **27B**, 231–236.

Stromvall, A.-M. and Petersson, G. (1992). Protection of terpenes against oxidative and acid decomposition on adsorbent cartridges. *J. Chromatogr.* **589**, 385–389.

Sugimoto, A. (1996). GC/GC/C/IRMS system for carbon isotope measurement of low level methane concentration. *Geochem. J.* **30**, 195–200.

Sweet, C. W., and Vermette, S. J. (1992). Toxic volatile organic compounds in urban air in Illinois. *Environ. Sci. Technol.* **26**, 165–173.

Trigg, D. P., Simmonds, P. G. and Nickless, G. (1995). Gas chromatographic determination of volatile alkenes by on-column bromination and electron-capture detection. *J. Chromatogr.* **690**, 197–206.

U.S. EPA (1997a). Compendium of Methods for the Determination of Toxic Organic Compounds in Ambient Air: Method TO-14A, 2nd ed. U.S. Environmental Protection Agency, Research Triangle Park, NC, EPA/625/R-96/010b, January 1997.

U.S. EPA (1997b) Compendium of Methods for the Determination of Toxic Organic Compounds in Ambient Air: Method TO-17, 2nd ed. U.S. Environmental Protection Agency, Research Triangle Park, NC, EPA/625/R-96/010b, January 1997.

Vairavamurthy, A., Roberts, J. M., and Newman, L. (1992). Methods for detection of low molecular weight carbonyl compounds in the atmosphere: a review. *Atmos. Environ.* **26A**, 1965–1993.

Walling, J. F., Bumgarner, J. E., Driscoll, D. J., Morris, C. M., Riley, A. E., and Wright, L. H. (1986). Apparent reaction products desorbed from Tenax used to sample ambient air. *Atmos. Environ.* **20**, 51–57.

Winer, A. M., Arey, J., Atkinson, R., Aschmann, S. M., Long, W. D., Morrison, C. L., and Olszyk, D. M. (1992). Emission rates of organics from vegetation in California's central valley. *Atmos. Environ.* **26A**, 2647–2659.

Woodbury, S. E., Evershed, R. P., Rossell, J. B., Griffith, R. E., and Farnell, P. (1995). Detection of vegetable oil adulteration using gas chromatography combustion/isotope ratio mass spectrometry. *Anal. Chem.* **67**, 2685–2690.

Yokouchi, Y., Bandow, H., and Akimoto, H. (1993). Development of automated gas chromatographic-mass spectrometric analysis for natural volatile organic compounds in the atmosphere. *J. Chromatogr.* **642**, 401–407.

Zeng, Y., Mukai, H., Bandow, H., and Nojiri, Y. (1994). Application of gas chromatography-combustion-isotope ratio mass spectrometry to carbon isotopic analysis of methane and carbon monoxoide in environmental samples. *Anal. Chim. Acta.* **289**, 195–204.

Zielinska, B., Sagebiel, J. C., Harshfield, G., Gertler, A. W., and Pierson, W. R. (1996). Volatile organic compounds up to C_{20} emitted from motor vehicles; measurement methods. *Atmos. Environ.* **30**, 2269–2286.

CHAPTER 5

Reactive Hydrocarbons in the Atmosphere at Urban and Regional Scales

PAOLO CICCIOLI, ENZO BRANCALEONI, AND
MASSIMILIANO FRATTONI

Istituto sull'Inquinamento Atmosferico del CNR, I-00016 Monterotondo Scalo, Italy

The reactivity of hydrocarbons in the atmosphere is discussed
on the basis of aerometric observations made at different tropospheric sites. The rapid and selective removal from the atmosphere of components capable of reacting rapidly with OH radicals, NO_3 radicals, and ozone suggests that photochemical
pollution is the main tropospheric sink for hydrocarbons. Examples showing the selective removal of hydrocarbons during
photochemical smog episodes are provided. Using Rome as a
case study, diurnal changes in reactivity and the effects of dilution and transport in an urban plume are discussed. Differences
between fluxes and emission rates measured in a forest site are
used to assess the reactivity of biogenic hydrocarbons.

I. INTRODUCTION

The achievement of air quality standards in the troposphere is one of the
major issues that modern societies have to face in order to preserve the
existing environment for future generations. The consumption of fossil fuels
necessary to sustain and improve the standard of living of a world population
increasing at an exponential rate is so high that waste materials released for
energy production can affect the secular equilibria established on the earth's
surface.

One important aspect of this changing environment is the continuing
increase in tropospheric ozone concentration observed over the past century.
It has been found (Volz and Kley, 1988; Anfossi *et al.*, 1991) that ground-
level ozone concentrations in remote areas now exceed by a factor of two to
three those existing in preindustrial times. This effect is caused by the increased emission of nitrogen oxides (NO_x) and hydrocarbons released by
human activities. Inextricably linked with the increase in tropospheric ozone
is the formation of oxidized species derived from the transformation of NO_x
and hydrocarbons in air (Finlayson-Pitts and Pitts, 1997).

The complex sequence of chemical reactions, activated when NO_x–
hydrocarbon mixtures are exposed to the actinic region of solar radiation
($\lambda > 290$ nm), is capable of generating a large variety of species, such as
peroxyacetyl nitrate (PAN), nitric acid, and oxygenated hydrocarbons (Finlayson-Pitts and Pitts, 1986), termed as photochemical oxidants. The fate of all
these pollutants is strictly interconnected through an alternate sequence of
day and night cycles (Ciccioli and Cecinato, 1992). Under favorable meteorological conditions both cycles are fully activated. Since these processes are
not linear, rapid accumulation of ozone and photochemical oxidants can take
place in the atmosphere. Within four to five days, the accumulation of pollutants can be so severe that acute episodes of photochemical smog pollution

having adverse effects on human health, vegetation, and materials (U.S. EPA, 1986) can be observed.

The elucidation of the complex cycle leading to ozone and photochemical formation in the troposphere has been one of the major tasks pursued by environmental chemists since the discovery of photochemical pollution in the Los Angeles basin and the first experiments of Haagen-Smit and Fox (1956). Although some aspects of the photochemical cycle still remain obscure, the general picture is clear. Formation of ozone and photochemical oxidants proceeds through a sequence of reactions having the OH radical as driving force during the day and the NO_3 radical, at night (Finlayson-Pitts and Pitts, 1986, 1997; Kley, 1997). OH radicals can oxidize hydrocarbons in the troposphere to species such as peroxyacyl radicals and hydroperoxy radicals that constantly convert NO into NO_2. Photolysis of NO_2 occurring at wavelengths < 420 nm generates ozone, which then accumulates in the atmosphere. Photolysis of ozone, formaldehyde, and nitrous acid together with ozonolysis of alkenes is the main source of OH radicals in air. A detailed description of the reactions leading to photochemical oxidant and ozone formation is reported elsewhere in this book.

The complexity of the chemistry and meteorology occurring makes it difficult to predict the extent to which NO_x and hydrocarbons must be reduced in maintaining the levels of ozone and other photochemical oxidants in the troposphere below targets fixed by the international organization for protecting human health and the environment (e.g. WHO, 1986). This is because the troposphere behaves as a three-dimensional assembly of interconnected photochemical flow reactors in which different boundary conditions and hydrocarbon–NO_x regimes are established in space and time (Finlayson-Pitts and Pitts, 1986, 1997; Ciccioli and Cecinato, 1992; Kley, 1997). In urban areas, for instance, the high emissions of NO limit the buildup of ozone and OH radicals in air. Ozone is constantly removed by reaction with NO to produce NO_2, which, in turn, reduces the availability of OH radicals for hydrocarbon oxidation by forming HNO_3 (Finlayson-Pitts and Pitts, 1997). The removal of nitric acid by deposition on surfaces or conversion into particulate nitrate makes the photochemical regime hydrocarbon-limited. In these conditions, a reduced buildup of ozone is combined with high concentrations in air of NO_2, nitric acid, nitrate, and sulfate particles. Accumulation of fine particles in urban air explains why the term photochemical smog was introduced to describe acute pollution episodes observed in the Los Angeles basin. In hydrocarbon-limited regimes a decrease in NO emission inevitably leads to an increase in ozone, and the only way to control ozone is then to control hydrocarbon emissions.

The same strategy does not work when the plume moves away from the urban area. The reduced supply of NO makes the plume rich in ozone, PAN, nitric acid, and nitrate salts but depleted in NO_x. Formation of ozone proceeds until no more NO is oxidized to NO_2. The plume becomes NO_x-limited and the ozone levels decrease if no further supply of nitrogen oxides is added to the air.

Unfortunately, the distribution, density, and source strength of emissions present today over the earth's surface are such that hydrocarbon/NO_x ratios seldom reach conditions in which little or no production of ozone is achieved. Because of this, national and international programs aimed at reducing ozone and photoxidant levels at local and regional scales concentrate their efforts in reducing the emissions of both NO_x and hydrocarbon (UN-ECE, 1991).

To be effective, however, the control of hydrocarbon emissions should focus on those volatile organic compounds (VOCs) that react faster with OH radicals and generate oxidation products capable of further contributing to ozone formation by photolysis and subsequent reaction with OH radicals. To account for all sources of ozone formation and to derive reactivity criteria for VOCs that are valid in both NO_x and hydrocarbon-limited conditions, the concept of the photochemical ozone creating potential (POCP) has been introduced (Derwent and Jenkin, 1991; Carter, 1994; Derwent et al., 1996). With this approach, the photochemical reactivity of an organic compound is measured by computer simulations carried out with airshed models. The ozone reactivity scale is based on the incremental rate of ozone increase measured when arbitrarily small amounts of an organic compound are added to the VOC emissions. The ranking of VOC reactivity is obtained by averaging the results obtained under different scenarios. Different indices have been proposed in the literature for assessing the reactivity of VOCs with airshed models but consistent results have been obtained for many organic compounds (Derwent and Jenkin, 1991). Although computer simulations provide a simple and flexible means for deriving reactivity scale for VOCs, the results obtained might not always be representative of the real world because they depend on the accuracy with which gas-phase reactions are described by the model, the number and types of organic compounds treated, and the availability of reliable emission data from anthropogenic and biogenic sources present in the domain investigated.

In addition to poor knowledge of the composition and source strength of VOCs and NO_x from biogenic sources, large uncertainties in the use of photochemical models may arise from the limited information available on the degradation pathways followed by certain organic components in air and on the heterogeneous processes responsible for the removal of precursors and products from the atmosphere (Ravishankara, 1997). For instance, the simple inclusion in photochemical models of the nighttime formation of HNO_2 by heterogeneous processes (Febo, 1994; Harrison et al., 1996) can drastically change the diurnal pattern of organic compounds predicted by the model (Neftel and Staffelbach, 1997). This is because hydrocarbon losses by reaction with NO_3 radicals are minimized whereas early morning reactivity is stimulated by a strong injection of OH radicals arising from the photolysis of HNO_2. Ciccioli *et al.* (1996a) found that, under certain conditions, the nighttime formation of HNO_2 determines daytime concentrations of 2-nitrofluoranthene and 2-nitropyrene in air whereas the expected nitration of parent-polycyclic aromatic hydrocarbons (PAH) by reaction with NO_3 radicals is negligible.

Another important limitation of existing models is the way in which degradation of some biogenic components is treated. For instance, many models use α-pinene as a surrogate species for all monoterpene compounds in spite of the rather different reactivities displayed by individual components toward OH radicals and ozone (Atkinson, 1990, 1994; Atkinson *et al.*, 1996). While this assumption may be coherent with aerometric observations, it does not always reflect the real influence of the emission on ozone formation because very reactive terpenes are rapidly eliminated by gas-phase and heterogeneous reactions before they reach the atmospheric boundary layer.

The limitations of existing models in covering some important aspects of the reactivity of VOCs in air explains why aerometric observations are still crucial for a better understanding of atmospheric processes leading to ozone and photooxidant formation in the troposphere. By starting from the content and composition of VOCs in the troposphere, aerometric determinations are used here to gain useful information on the reactivity of VOCs at urban and regional scales.

II. VOLATILE ORGANIC COMPOUNDS IN THE TROPOSPHERE

A. Volatile Organic Compound Composition as a Tool for Understanding Atmospheric Processes

To reveal some unknown aspects of atmospheric chemistry and to improve our capability of predicting photochemical oxidant formation, a detailed knowledge of VOC composition is necessary. Only by looking simultaneously at the precursors and products of photochemical reactions in air is it possible to detect previously unidentified emission sources, to assess the effects of long-range transport, or to identify and quantify degradation and removal processes occurring in the troposphere. The importance of studying in detail VOC composition at tropospheric sites characterized by different emissions and reactivity became evident when the various programs undertaken to reduce levels of tropospheric ozone failed to achieve their task. Concentrations as high as 150 ppbv ozone are still recorded in many areas of the United States (Chock and Heuss, 1987) and Europe (Grennfelt *et al.*, 1988).

In the middle of the 1980s it was already clear that research on VOCs should have been extended to biogenic components because they are capable of generating ozone with the same efficiency as the anthropogenic hydrocarbons (Chameides *et al.*, 1988). However, at the beginning of the 1990s, few aerometric data were available on the levels of isoprene and monoterpene compounds present in forest and rural areas, and most of those that were available were collected in the United States (Singh and Zimmerman, 1992). Moreover, very little was known about polar compounds although data collected in the Appalachian Mountains (Seila, 1984) had clearly shown that more than 50% of the organic fraction was composed of unidentified components.

Only the decisive improvements in the collection, separation, and detection of VOCs achieved in the last few years have made possible the identification of polar and nonpolar organic components in air samples (Ciccioli *et al.*, 1992). By using high resolution gas chromatographic–mass spectroscopic (GC-MS) techniques, the simultaneous detection and quantification of more than 300 components in a single sample were reported (Ciccioli *et al.*, 1993a). This approach is so selective and sensitive that automated VOC monitors based in capillary GC are now available for the semicontinuous monitoring of VOCs at remote sites (Helmig and Greenberg, 1994).

By a systematic collection of data over several years, it has been possible to identify and classify more than 300 different VOCs emitted or formed in the atmosphere (Ciccioli *et al.*, 1994a). For each component, the frequency of observation and relative abundance were reported as a function of the type of site (urban, suburban, forest–rural, and remote) investigated. Indication on the relative retention measured on a DB-5 capillary chromatography column was also provided for making possible the identification by selective ion detection. Data were obtained from field campaigns carried out in central and southern Europe and from three remote sites located in the Himalayan plateau and in the Arctic (Svalbard Islands) and Antarctic regions (Ross Bay). This database has been constantly updated and expanded by adding data from forest ecosystems located in Italy, France, and Spain, many of them collected as part of the Biogenic Emission in the Mediterranean Area (BEMA) project sponsored by the Commission of the European Community (Seufert, 1997).

The database that now includes about 350 components resulting from the analysis of more than 1000 samples collected worldwide is reported in Table 1. Frequency of observation and relative abundance of VOCs are provided only for compounds for which quantification is possible. For the others, only the frequency of observation is reported. Identified compounds are grouped

TABLE 1   Frequency of Observation (FO)[a] and Relative Abundance (RA)[b] of VOCs Identified by GC-MS in More Than 1000 Samples[c] Collected in Urban, Suburban, Forest–Rural, and Remote Areas

| | Type of Site | | | | | | | |
| | Urban | | Suburban | | Forest | | Remote | |
Compound	FO	RA	FO	RA	FO	RA	FO	RA
Alkanes								
Butane, 2-methyl	H	A	H	A	M	M	O	O
n-C$_5$	H	A	H	A	H	M	H	M
Butane, 2,2-dimethyl	H	M	H	M	L	T	L	T
Butane, 2,3-dimethyl	H	T	M	T	L	T	L	T
Pentane, 2-methyl	H	M	H	M	M	M	H	M
Pentane, 3-methyl	H	M	H	M	M	M	M	M
n-C$_6$	H	M	H	M	H	M	H	M
Pentane, 2,4-dimethyl	H	T	M	T	O	O	L	T
Butane, 2,2,3-trimethyl	O	O	O	O	O	O	L	T
Pentane, 3,3-dimethyl	H	T	O	O	O	O	O	O
Hexane, 2-methyl	H	M	H	M	M	T	L	M
Hexane, 3-methyl	H	M	M	T	L	T	M	M
Pentane, 3-ethyl	H	T	O	O	O	O	O	O
Pentane, 2,2,4-trimethyl	H	M	H	M	M	M	L	M
n-C$_7$	H	M	H	M	H	M	H	M
Hexane, 2,5-dimethyl	H	T	M	T	L	T	O	O
Hexane, 2,4-dimethyl	H	T	M	T	L	T	O	O
Pentane, 2,3,4-trimethyl	H	T	M	T	L	T	O	O
Pentane, 2,3,3-trimethyl	H	T	O	O	O	O	O	O
Hexane, 2,3-dimethyl	H	M	M	T	L	T	O	O
Heptane, 2-methyl	H	M	H	T	M	T	O	O
Heptane, 4-methyl	H	T	L	T	L	T	O	O
Heptane, 3-methyl	H	T	L	T	L	T	O	O
n-C$_8$	H	M	H	T	H	M	H	M
Heptane, 2,4-dimethyl	H	T	O	O	O	O	O	O
Hexane, trimethyl	H	T	O	O	O	O	O	O
Heptane, 2,5-dimethyl	H	T	O	O	O	O	O	O
Heptane, 2,3-dimethyl	H	T	O	O	O	O	O	O
Octane, 2-methyl	H	T	O	O	L	T	O	O
Octane, 4-methyl	H	T	O	O	L	T	O	O
Octane, 3-methyl	H	T	O	O	L	T	O	O
n-C$_9$	H	T	H	T	H	M	H	M
Octane, dimethyl	H	T	O	O	O	O	O	O
Nonane, methyl	H	T	O	O	O	O	O	O
Isodecane	H	T	O	O	O	O	O	O
Isodecane	H	T	O	O	O	O	O	O
Octane, dimethyl	H	T	O	O	O	O	O	O
Nonane, methyl	H	T	O	O	O	O	O	O

(continues)

TABLE 1   *(continued)*

Compound	Urban FO	Urban RA	Suburban FO	Suburban RA	Forest FO	Forest RA	Remote FO	Remote RA
Nonane, methyl	H	T	O	O	O	O	O	O
Nonane, 3-methyl	H	T	O	O	O	O	O	O
n-C_{10}	H	T	H	T	H	T	H	M
Isoundecane	H	T	O	O	O	O	O	O
Isoundecane	H	T	O	O	O	O	O	O
Nonane, dimethyl	H	T	O	O	O	O	O	O
Isoundecane	H	T	O	O	O	O	O	O
Decane, methyl	H	T	L	T	O	O	O	O
Isoundecane	H	T	O	O	O	O	O	O
Decane, methyl	H	T	L	T	O	O	O	O
Nonane, dimethyl	H	T	O	O	O	O	O	O
Isoundecane	H	T	O	O	O	O	O	O
n-C_{11}	H	T	H	T	M	T	H	M
Isododecane	H	T	O	O	O	O	O	O
Isododecane	H	T	O	O	O	O	O	O
Isododecane	H	T	O	O	O	O	O	O
Isododecane	H	T	O	O	O	O	O	O
Isododecane	H	T	O	O	O	O	O	O
Isododecane	H	T	O	O	O	O	O	O
Isododecane	H	T	O	O	O	O	O	O
n-C_{12}	H	T	H	T	M	T	H	M
n-C_{13}	H	T	M	T	M	T	M	M
Cycloalkanes								
Cyclopentane	M	T	O	O	L	T	O	O
Cyclopentane, methyl	M	T	H	T	M	M	L	M
Cyclopropane, 1-methyl, 3,3-dimethyl	O	O	O	O	O	O	L	T
Cyclohexane	H	T	M	T	M	M	L	A
Cyclopentane, dimethyl	M	T	O	O	O	O	L	T
trans-Cyclopentane, 1,3-dimethyl	M	T	O	O	O	O	L	T
cis-Cyclopentane, 1,3-dimethyl	M	T	O	O	O	O	L	T
Cyclopentane, 1,2-dimethyl	M	T	O	O	O	O	L	T
Cyclohexane, methyl	H	T	M	T	L	T	L	T
Cyclopentane, ethyl	M	T	O	O	L	T	L	T
Butylcyclopropane	O	O	O	O	L	T	O	O
Cyclohexane, ethyl	H	T	O	O	O	O	L	T
Cyclohexane, trimethyl	H	T	O	O	O	O	O	O
Cyclohexane, 1-ethyl, 4-methyl	H	T	O	O	O	O	O	O
Cyclohexane, 1,2,3-trimethyl	H	T	O	O	O	O	O	O
Butylcyclopentane	H	T	O	O	O	O	O	O
Alkenes								
1-Pentene	H	T	H	T	L	T	L	T
1-Butene, 2-methyl	H	M	M	T	L	M	O	O
trans-2-Pentene	H	M	M	T	L	T	O	O

(continues)

TABLE 1   *(continued)*

Compound	Urban		Suburban		Forest		Remote	
	FO	RA	FO	RA	FO	RA	FO	RA
cis-2-Pentene	H	T	L	T	L	T	O	O
2-Butene, 2-methyl	H	M	M	T	L	T	O	O
1-Pentene, 3-methyl	L	T	O	O	O	O	O	O
1-Hexene	H	T	H	T	H	T	L	T
trans-2-Hexene	H	T	L	T	O	O	O	O
cis-2-Hexene	H	T	L	T	L	T	O	O
Butene, 2,3-dimethyl	H	T	O	O	O	O	O	O
Pentane, 3-methylene	H	T	O	O	O	O	O	O
3-Hexene	H	T	O	O	O	O	O	O
2-Pentene, 3-methyl	H	T	O	O	O	O	O	O
Isoheptene	H	T	O	O	O	O	O	O
Isoheptene	H	T	O	O	O	O	O	O
2-Hexene, 4-methyl	L	T	O	O	O	O	O	O
1-Heptene	M	T	O	O	O	O	L	T
2-Pentene, 3-ethyl	H	T	O	O	O	O	O	O
3-Heptene	M	T	O	O	O	O	O	O
3-Hexene, 3-methyl	H	T	O	O	O	O	O	O
2-Hexene, 2-methyl	H	T	O	O	O	O	O	O
2-Heptene	H	T	O	O	O	O	O	O
1-Pentene, 2,2,4-trimethyl	H	T	O	O	O	O	L	T
2-Hexene, 3-methyl	H	T	O	O	O	O	O	O
Isoheptene	H	T	O	O	O	O	O	O
Isooctene	H	T	O	O	O	O	O	O
Isooctene	H	T	O	O	O	O	O	O
4-Octene	H	T	O	O	O	O	O	O
1-Octene	H	T	L	T	L	T	L	T
Isooctene	H	T	O	O	O	O	O	O
Isooctene	H	T	O	O	O	O	O	O
Isooctene	H	T	O	O	O	O	O	O
Isooctene	M	T	O	O	O	O	O	O
3-Octene	H	T	O	O	O	O	O	O
Isooctene	H	T	O	O	O	O	O	O
2-Heptene, 3-methyl	H	T	O	O	O	O	O	O
Heptene, dimethyl	H	T	O	O	O	O	O	O
Isononene	H	T	O	O	O	O	O	O
4-Nonene	H	T	O	O	O	O	O	O
1-Nonene + 3-nonene	H	T	O	O	O	O	O	O
1-Octene, 2-methyl	H	T	O	O	O	O	O	O
1-Octene, 3-methyl	M	T	O	O	O	O	O	O
Cycloalkenes								
Cyclopentene	H	T	M	T	L	T	O	O
Cyclopentene, 4-methyl	H	T	O	O	O	O	O	O

(continues)

TABLE 1   (continued)

Compound	Urban		Suburban		Forest		Remote	
	FO	RA	FO	RA	FO	RA	FO	RA
Cyclopentene, 3-methyl	H	T	O	O	O	O	O	O
Cyclopentene, 1-methyl	H	T	O	O	O	O	O	O
Cyclohexene	H	T	O	O	O	O	O	O
Cyclopentene, 4,4-dimethyl	H	T	O	O	O	O	O	O
Cyclopentene, 1,5-dimethyl	H	T	O	O	O	O	O	O
Cyclohexene, 4-methyl	H	T	O	O	O	O	O	O
Cyclobutane, isopropyldiene	H	T	O	O	O	O	O	O
Cyclohexene, methyl	H	T	O	O	O	O	O	O
Cyclopentane, 1,2-dimethyl, 3-methylene	H	T	O	O	O	O	O	O
Dienes								
Isoprene	H	T	H	M	H	M	H	T
1,3-Pentadiene	H	T	O	O	O	O	O	O
2,4-Hexadiene	H	T	O	O	O	O	L	T
1,3-Pentadiene, 2-methyl	H	T	O	O	O	O	O	O
2,4-Hexadiene, 2-methyl	H	T	O	O	O	O	O	O
1,4-Heptadien, 3-methyl	O	O	O	O	O	O	L	T
Hexadiene, 2,3-dimethyl	H	T	O	O	O	O	O	O
Diene, iso-C_8	H	T	O	O	O	O	O	O
Diene, iso-C_8	M	T	O	O	O	O	O	O
1,4-Pentadiene-2,3,3-trimethyl	H	T	O	O	O	O	O	O
1,5-Hexadiene,2,5-dimethyl	H	T	O	O	O	O	O	O
Diene, iso-C_8	H	T	O	O	O	O	O	O
Alkynes								
3-Ethyne, 5-methyl	H	T	O	O	O	O	O	O
Monocyclic Arenes								
Benzene	H	A	H	A	H	M	H	M
Toluene	H	A	H	A	H	M	H	A
Ethylbenzene	H	M	H	M	H	T	H	M
$(m + p)$-Xylene	H	A	H	M	H	M	H	M
Styrene	H	T	M	T	L	T	L	T
o-Xylene	H	M	H	M	M	M	H	M
Isopropylbenzene	H	T	M	T	L	T	L	T
Allylbenzene	M	T	O	O	O	O	O	O
n-Propylbenzene	H	M	H	T	M	T	L	T
Benzene, 1-methyl, 3-ethyl + benzene, 1-methyl, 4-ethyl	H	M	H	T	M	T	L	T
Benzene, 1,3,5-trimethyl	H	M	M	T	M	T	L	T
Benzene, 1-methyl, 2-ethyl	H	M	H	T	M	T	L	T
Benzene, tert-butyl	H	T	L	T	O	O	O	O
Benzene, 1,2,4-trimethyl	H	M	H	T	L	T	L	T
Benzene, isobutyl	H	T	L	T	L	T	O	O
Benzene, sec-butyl	H	T	L	T	O	O	O	O

(continues)

TABLE 1   *(continued)*

	Type of Site							
	Urban		Suburban		Forest		Remote	
Compound	FO	RA	FO	RA	FO	RA	FO	RA
Benzene, 1,2,3-trimethyl	H	T	H	T	L	T	L	T
Benzene, 1-methyl, 3-isopropyl	H	T	L	T	M	T	O	O
Benzene, 1-methyl, 4-isopropyl	H	T	L	T	O	O	O	O
Benzene, diethyl +								
Benzene, 1-methyl, 2-isopropyl	H	T	L	T	O	O	O	O
Benzene, methyl, *n*-propyl	H	T	L	T	O	O	O	O
Benzene, *n*-butyl	H	T	L	T	O	O	O	O
Benzene, methyl, *n*-propyl	H	T	L	T	O	O	O	O
Benzene, dimethyl, ethyl	H	T	L	T	O	O	O	O
Benzene, methyl, *n*-propyl	H	T	L	T	O	O	O	O
Benzene, dimethyl, ethyl	H	T	L	T	O	O	O	O
Benzene, dimethyl, ethyl	H	T	L	T	O	O	O	O
Benzene, dimethyl, ethyl	H	T	L	T	O	O	O	O
Benzene, dimethyl, ethyl	H	T	L	T	O	O	O	O
Benzene, dimethyl, ethyl	H	T	L	T	O	O	O	O
Benzene, 1,2,4,5-tetramethyl	H	T	L	T	O	O	O	O
Benzene, 1,2,3,5-tetramethyl	H	T	L	T	O	O	O	O
Benzene, iso-C_5	H	T	O	O	O	O	O	O
Benzene, iso-C_5	H	T	O	O	O	O	O	O
Benzene, 1,2,3,4-tetramethyl	H	T	L	T	O	O	O	O
Benzene, iso-C_5	H	T	O	O	O	O	O	O
Benzene, iso-C_5	H	T	O	O	O	O	O	O
Benzene, iso-C_5	H	T	O	O	O	O	O	O
Benzene, iso-C_5	H	T	O	O	O	O	O	O
Benzene, iso-C_5	H	T	O	O	O	O	O	O
Bicyclic arenes								
Indan	H	M	L	T	L	T	O	O
Indene	H	T	M	T	L	T	O	O
Methylindan	H	T	O	O	O	O	O	O
Methylindan	H	T	O	O	O	O	O	O
Methylindan	H	T	O	O	O	O	O	O
Methylindan	H	T	O	O	O	O	O	O
Methylindan	H	T	O	O	O	O	O	O
Methylindan	H	T	O	O	O	O	O	O
Naphthalene	H	T	L	T	L	T	O	O
Naphthalene, 1-methyl	H	T	O	O	O	O	O	O
Naphthalene, 2-methyl	H	T	O	O	O	O	O	O
Monoterpenes								
Tricyclene	O	O	O	O	L	T	O	O
Thujene	O	O	O	O	L	T	O	O
α-Pinene	M	T	H	T	H	A	L	T
Camphene	O	O	L	T	H	M	L	T

(continues)

TABLE 1 *(continued)*

	Type of Site							
	Urban		Suburban		Forest		Remote	
Compound	FO	RA	FO	RA	FO	RA	FO	RA
Sabinene	O	O	O	O	L	T	O	O
β-Pinene	M	T	L	T	H	M	O	O
Myrcene	O	O	O	O	L	T	L	T
α-Phellandrene	O	O	O	O	L	T	L	T
Δ-3-Carene	O	O	M	T	M	A	L	T
α-Terpinene	O	O	O	O	L	T	O	O
para-Cymene	L	T	L	T	M	M	O	O
β-Phellandrene	O	O	O	O	H	T	O	O
1,8-Cineol	O	O	O	O	L	T	O	O
D-Limonene	M	M	M	T	H	M	L	T
cis-β-Ocimene	O	O	O	O	L	T	O	O
trans-β-Ocimene	O	O	O	O	L	T	O	O
γ-Terpinene	O	O	O	O	L	T	O	O
cis-Linalool oxide	O	O	O	O	L	T	O	O
trans-Linalool oxide	O	O	O	O	L	T	O	O
α-Terpinolene	O	O	O	O	L	T	O	O
Linalool	O	O	O	O	L	T	O	O
Camphor	O	O	O	O	O	O	L	T
β-Caryophyllene	O	O	O	O	L	T	O	O
Halogen-containing compounds								
CFC 11	H	—	H	—	L	—	M	—
Methane, dichloro	M	—	O	—	O	—	L	—
CFC 113	H	T	H	T	H	T	H	M
Chloroform	H	T	H	T	H	T	M	T
1,1,1-Trichloroethane	H	M	H	T	H	M	H	M
Carbon tetrachloride	H	T	H	T	H	T	H	M
Propane, 1,2-dichloro	H	T	H	T	H	T	L	T
Ethene, trichloro	H	M	H	T	H	T	L	T
Ethane, tetrachloro	H	T	H	T	H	T	O	O
Ethene, tetrachloro	H	M	H	T	H	T	L	T
Benzene-chloro	H	T	O	O	O	O	O	O
Benzene, (m + p)-dichloro	H	T	O	O	O	O	O	O
Benzene, o-dichloro	H	T	O	O	O	O	O	O
Sulfur-Containing Compounds								
Carbon disulfide	H	—	H	—	M	—	L	—
Methane isothiocyanate	O	O	O	O	O	O	L	T
Alcohols								
Methanol	H	—	H	—	H	—	H	—
Ethanol	H	—	H	—	H	—	H	—
2-Propanol	H	T	H	T	H	T	H	T
2-Propanol, 2-methyl	H	T	H	T	H	T	H	M
1-Propanol	H	T	H	T	H	T	H	T

(continues)

TABLE 1   *(continued)*

	Type of Site							
	Urban		Suburban		Forest		Remote	
Compound	FO	RA	FO	RA	FO	RA	FO	RA
3-Buten-2-ol, 2-methyl	O	O	O	O	L	T	L	T
1-Propanol, 2-methyl	H	T	H	T	H	T	M	T
1-Butanol	H	T	H	T	H	T	H	A
1-Butanol, 3-methyl	L	T	O	O	M	T	L	T
3-Buten-1-ol, 3-methyl	O	O	O	O	L	T	O	O
1-Pentanol	L	T	O	O	H	T	L	T
1-Hexanol	L	T	O	O	H	T	L	T
3-Buten-1-ol, 3-methyl,acetate	O	O	O	O	L	T	O	O
2-Propanol, 1,3-dichloro	O	O	O	O	O	O	L	T
1-Heptanol	L	T	O	O	M	T	L	T
1-Hexanol, 2-ethyl	L	T	O	O	H	T	L	M
1-Octanol	L	T	O	O	M	T	L	M
1-Nonanol	L	T	O	O	M	T	L	M
1-Decanol	L	T	O	O	M	T	O	O
Alkyloxy alcohols								
2-Butoxy, ethanol	H	T	H	T	H	T	L	M
Etanol, 2-(ethoxy ethoxy)	M	M	O	O	O	O	O	O
2-Propanol, 1-(2-methoxy, 1-methylethoxy)	L	T	O	O	O	O	O	O
2-Propanol, 1-[1-methyl, 2-(propeniloxy) ethoxy]	L	M	O	O	O	O	O	O
2-Propanol, 1-(methoxy propoxy)	L	M	O	O	O	O	O	O
Ethanol, 2-phenoxy	L	T	O	O	O	O	O	O
Diols								
2,4-Pentanediol, 2-methyl	O	O	M	T	H	A	O	O
Aldehydes								
Propanal, 2-methyl	O	O	L	T	M	T	M	T
Butanal	H	T	H	M	H	T	H	M
Butanal, 3-methyl	O	O	L	T	M	T	M	T
Pentanal	H	T	H	M	H	M	M	M
Pentanal, 2-methyl	O	O	O	O	O	O	L	T
Hexanal	H	M	H	M	H	M	H	M
Heptanal	H	T	H	M	H	M	H	M
Octanal	H	M	H	M	H	M	H	M
Nonanal	H	M	H	A	H	M	H	M
Decanal	H	T	H	A	H	M	H	M
Undecanal	O	O	O	O	L	T	O	O
Olefinic aldehydes								
2-Propenal	O	—	O	—	M	—	M	—
2-Propenal, 2-methyl	H	T	H	T	H	T	M	M
2-Pentenal	O	O	O	O	O	O	L	T
2,4-Hexadienal	O	O	O	O	O	O	L	T
1-Formylcyclopentene	O	O	O	O	O	O	L	T

(continues)

TABLE 1   *(continued)*

	Type of Site							
	Urban		Suburban		Forest		Remote	
Compound	FO	RA	FO	RA	FO	RA	FO	RA
Cyclic and aromatic aldehydes								
2-Furanaldehyde	L	T	O	O	O	O	L	T
Benzaldehyde	H	M	H	M	H	M	H	M
2-Furanaldehyde, 5-methyl	O	O	L	T	O	O	O	O
Benzaldehyde, methyl	H	T	M	T	M	T	O	O
Benzaldehyde, methyl	H	T	M	T	M	T	O	O
Benzaldehyde, dimethyl	H	T	O	O	O	O	O	O
Benzaldehyde, dimethyl	H	T	O	O	O	O	O	O
Oxo-aldehydes								
4-Oxopentanal	O	O	O	O	L	M	L	M
Ketones								
2-Propanone	H	—	H	—	H	—	M	—
2-Butanone	M	T	H	M	H	T	M	M
2-Butanone, 3-methyl	O	O	O	O	O	O	L	T
2-Pentanone	O	O	O	O	H	T	L	T
2-Pentanone, 4-methyl	O	O	O	O	O	O	L	T
3-Pentanone, 2-methyl	O	O	O	O	O	O	L	T
2-Hexanone	O	O	O	O	O	O	L	T
Iso-C_7-ketones	O	O	L	T	O	O	O	O
2-Pentanone, 4-hydroxy, 4-methyl	O	O	O	O	L	T	O	O
2-Heptanone	O	O	O	O	O	O	L	T
Iso-C_8-ketones	O	O	L	T	O	O	O	O
Olefinic and cyclic ketones								
3-Buten, 2-one	O	O	H	T	H	T	H	M
Cyclopentanone	H	T	H	T	L	T	O	O
5-Hexen, 2-one, 5-methyl,								
3-methylene	O	O	O	O	O	O	L	M
2(5H)-Furanone, 5,5-dimethyl	O	O	O	O	O	O	L	T
5-Hepten, 2-one, 6-methyl	H	M	H	A	H	A	H	M
2H-Pyran-2-one, tetrahydro	H	T	H	T	H	T	L	T
3,5-Heptadien, 2-one, 6-methyl	O	O	O	O	O	O	L	A
5,9-Undecadien, 2-one, 6,10								
dimethyl(geranyl acetone)	O	O	O	O	L	T	O	O
Furans								
Furan	H	T	H	T	M	T	L	T
Furan, 2-methyl	H	T	H	T	H	T	L	M
Furan, 3-methyl	H	T	M	T	H	T	L	T
Furan, 2,5-dimethyl	O	O	O	O	O	O	L	T
Furan, 2,3-dihydro, 5-methyl	O	O	O	O	L	T	O	O
Acids								
Acetic acid	H	T	H	M	H	A	H	A
Propanoic acid	H	T	H	T	H	T	M	T

(continues)

TABLE 1   (*continued*)

Compound	Urban		Suburban		Forest		Remote	
	FO	RA	FO	RA	FO	RA	FO	RA
Butanoic acid	H	T	H	T	H	T	H	M
Butanoic acid, 3-methyl	H	T	H	T	H	T	M	T
Pentanoic acid	H	T	H	T	H	T	M	T
Hexanoic acid	H	T	H	T	H	T	H	M
Heptanoic acid	H	T	H	T	H	T	H	M
Octanoic acid	H	T	H	T	H	T	H	M
Nonanoic acid	H	T	H	T	H	T	H	M
Esters								
Formic acid, ethylester	O	O	O	O	O	O	L	T
Acetic acid, ethenylester	O	O	O	O	O	O	L	T
Acetic acid, ethylester	M	M	L	T	O	O	L	T
Formic acid, butylester	O	O	O	O	O	O	L	T
Acetic acid, 2-methylpropylester	O	O	O	O	O	O	L	T
Acetic acid, butylester	M	M	L	T	L	T	L	T
Ethers								
Propane, 2-metoxy, 2-methyl	H	M	H	M	H	T	O	O
Butane, 2-metoxy, 2-methyl	H	T	M	T	M	T	O	O
Benzene, 1-methoxy, 4-methyl	O	O	O	O	O	O	L	T
Nitrogen-containing compounds								
Benzonitrile	H	T	O	O	O	O	O	O
Benzothiazole	H	T	M	T	M	T	O	O
Phenols								
Phenol	H	T	H	T	H	T	L	T
o-Cresol	H	T	H	T	H	T	O	O
(*m* + *p*)-Cresol	H	T	H	T	H	T	O	O
Silanols								
Trimethylsilanol	H	T	H	T	H	T	L	T

[a]Ranking: H, high frequency (from 70 to 100% of total samples analyzed); M, medium frequency (from 30 to 70% of total samples analyzed); L, low frequency (up to 30 of total samples analyzed); and O, never observed.

[b]Ranking: A, abundant (more than 5% w/w of total concentration); M, moderately abundant (between 5 and 0.5% w/w of total concentration); T, trace (less than 0.5% w/w of total concentration); O, below the detection limits; and (−), compound for which quantification was not possible.

[c]Typical range of the total VOC concentration measured in: urban, 500–1500 μg/m^3; suburban, 100–250 μg/m^3; forest–rural, 30–200 μg/m^3; and remote, 30–150 μg/m^3.

into 25 different classes. About 13 of them are characterized by functional groups containing oxygen. Although the list is far from exhaustive (some compounds are still unidentified), it provides a rather comprehensive picture of the VOCs that are commonly present in the atmosphere. It is worth noting that most of the components identified at the Mauna Loa Observatory (Helmig *et al.*, 1996) were already present in the early version of our database whereas others (such as geranyl acetone and 4-oxo-pentanal) were only recently observed in vegetated areas located near the Mediterranean coast.

B. DISTRIBUTION AND COMPOSITION OF VOLATILE ORGANIC COMPOUNDS IN THE TROPOSPHERE

Data reported in Table 1 are crucial for understanding the reactivity of VOCs in the atmosphere. The disappearance in a given environment of compounds known by laboratory experiments to be reactive with OH radicals, NO_3 radicals, and ozone must somehow be mirrored by the presence of their degradation products in the ambient atmosphere. However, the occurrence of oxygenated products in the atmosphere does not necessarily imply that they originate from the photochemical oxidation of organic emissions. A critical analysis of these aspects, merely based on statistical observations, is of great help for introducing the topics that will be discussed in the following sections of this chapter. It is also important because it provides an idea of the unresolved questions that still make difficult the modeling of photochemical processes.

Data reported in Table 1 show that the high complexity of the alkane, alkene, and arene fractions rapidly simplifies with distance from urban areas. Only a limited number of components belonging to these classes survive in forest and remote areas. This effect is particularly evident for very reactive components, such as alkenes and cycloalkenes, dienes, and arenes larger than trimethylbenzenes. With few exceptions, these compounds originate from anthropogenic processes related to vehicular emissions, stationary combustion, and solvent evaporation (Singh and Zimmerman, 1992; Ciccioli, 1993; PORG Report, 1993; Derwent *et al.*, 1996). The photochemical decay of alkylbenzenes (Carlier *et al.*, 1986) explains the ubiquitous presence of benzaldehyde in the troposphere and the widespread occurrence of its upper homologues in urban areas.

In forest and rural areas, the decrease in the number of anthropogenic components is mirrored by an increased complexity in the monoterpene fraction. However, only a few of the compounds emitted by vegetation are sufficiently abundant to influence the air composition. Some monoterpenes (namely, α-pinene, β-pinene, and D-limonene) are also present in urban and remote areas. In the former case the contribution comes mainly from anthropogenic sources (such as evaporation of housecleaning products) (Knoeppel and Schauenburg, 1989) while in the latter it is due to long-range transport from vegetated areas. The ubiquitous occurrence of isoprene in the samples investigated confirms that sources other than vegetation emissions are active in releasing this component into the atmosphere. While marine sources (McKay *et al.,* 1996) can play a crucial role in determining isoprene concentrations in areas free from vegetation (Laurila and Hakola, 1996), fossil fuel combustion is a significant source in urban environments (Zielinska *et al.,* 1996).

The distribution of alcohols and metoxy alcohols follows a rather complicated pathway. Components from methanol to butanol are ubiquitous components of the atmosphere whereas upper homologues are observed mostly in forest–rural and remote areas. The ubiquitous occurrence at trace levels of alcohols from methanol (Singh *et al.,* 1995) to butanol is consistent with their photochemical origin (Carlier *et al.,* 1986; Calvert and Madronich, 1987). However, anthropogenic sources can also be responsible for their presence in urban and suburban airsheds (PORG Report, 1993), whereas vegetation emissions contribute to the levels of methanol in forest–rural areas (McDonald and Fall, 1993). The fact that heavier components are not detected in suburban airsheds but are present in forest–rural and some remote areas suggests that biogenic sources exist for these alcohols. Indeed, 3-methyl-3-buten-1-ol has been identified in the emissions of a monoterpene-emitting oak growing in Europe (Loreto *et al.,* 1996) and 2-methyl-3-buten-2-ol seems to be a quite widespread product of vegetation emissions (Goldan *et al.,* 1993; Guenther *et al.,* 1995).

Since the occurrence of many oxyalkyl alcohols is restricted to urban areas, anthropogenic emissions are the most likely source of these components. The ubiquitous presence of 2-butoxy-ethanol is suggestive, instead, of photochemical origin and, very likely, the same source also produces 2-methyl-2,4-pentanediol. Indeed, no presence of these two components was found in emissions from Mediterranean vegetation (BEMA Report 1994, 1995, 1996, 1997a, 1997b).

Degradation of reactive hydrocarbons (Calvert and Madronich, 1987), emissions from biogenic sources, and biomass burning (Bode *et al.,* 1997) might explain the widespread distribution of free acids in the troposphere. It is quite surprising, however, to see that compounds up to nonanoic acid can be still detected in the gas phase (Ciccioli *et al.,* 1993b). Similar considerations can be applied to the aldehyde fraction, from hexanal to undecanal. The fact that these components often account for the largest portion of organic material present in air (Yokouchi *et al.,* 1990; Ciccioli *et al.,* 1993b; Ciccioli *et al.,*

1994a; Helmig *et al.*, 1996) still remains difficult to explain. Emissions from anthropogenic sources (Rogge *et al.*, 1991) and vegetation (Ciccioli *et al.*, 1993b) can account for their presence in urban, suburban, and forest areas. In addition, degradation of 1-octene (Grosjean and Grosjean, 1996), a compound of biogenic and anthropogenic origin quite common in tropospheric samples, certainly contributes to the ubiquitous presence of hexanal in air. All these processes are not sufficient, however, to explain the high levels of semivolatile aldehydes in remote regions free from vegetation (Ciccioli *et al.*, 1994a, 1996b; Helmig *et al.*, 1996). Flux measurements carried out in a pseudo-steppe ecosystem located in proximity to the seashore in Italy showed strong deposition of semivolatile aldehydes, although emissions from soil vegetation were detected with enclosure techniques (BEMA Report, 1994). Very likely, marine aerosol or degradation of organic products associated with this ecosystem can be a possible source for these carbonyl components in air.

It is also difficult to explain the ubiquitous presence in the troposphere of a reactive component, such as 6-methyl-5-hepten-2-one, which often accounts for a substantial portion of the total ketone fraction detected by GC-MS. This compound can be released by vegetation (Ciccioli *et al.*, 1993b; Konig *et al.*, 1995), formed by photochemical degradation of linalool (Shu *et al.*, 1997) or generated by ozonolysis of epicuticular waxes of plants (Fruekilde *et al.*, 1997). The compound has been detected in the emissions of some Mediterranean species, such as rosemary and pine (BEMA Report, 1994), but always as a minor component with respect to terpenes. A normalized emission of 5 ng/m^2 (leaf area)/s^2 was measured from orange trees during recent experiments carried out near Valencia (Spain), where the major component was linalool (BEMA Report, 1997b). So far, substantial emissions of 6-methyl-5-hepten-2-one have only been measured from birch trees (Koenig *et al.*, 1995). Slight fluxes of the compound were detected by gradient techniques in a Mediterranean pseudo-steppe ecosystem (BEMA Report, 1994) and strong fluxes were observed from orange orchards (BEMA Report, 1997a).

While photochemical degradation of linalool can be important during the flowering season, ozonolysis of epicuticular waxes covering tree leaves might be relevant during the whole summer season. The covariance of 6-methyl-5-hepten-2-one with geranyl acetone detected at the Mauna Loa Observatory (Helmig *et al.*, 1996) strongly supports the existence of the heterogeneous pathway because geranyl acetone is one of the major products formed by ozonolysis of epicuticular waxes. These reactions do not justify, however, the presence of 6-methyl-5-hepten-2-one in Arctic (Ciccioli *et al.*, 1994a) and Antarctic (Ciccioli *et al.*, 1996b) air. An interesting possibility is that 6-methyl-5-hepten-2-one comes also from the ultraviolet (UV) irradiation of seawater in which squalene is present. Indeed, Yeo and Shibamoto (1992) demonstrated the production of this component from aqueous micelle solutions. In seawater containing phytoplankton, squalene could be released in sufficient amounts to generate 6-methyl-hepten-2-one in areas where the flux of UV radiation is particularly high.

Some additional comments should be made about volatile carbonyl compounds. 3-Buten-2-one (also called methyl vinyl ketone), 2-methylpropenal (also known as methacrolein), furan, and 2- and 3-methylfurans are formed by photochemical degradation of dienes (Carlier *et al.*, 1986). In particular, 3-buten-2-one, 2-methylpropenal, and 3-methylfuran are oxidation products of isoprene (Atkinson *et al.*, 1989; Atkinson, 1990). A substantial number of the linear, branched, and cyclic carbonyl compounds listed in Table 1 have been identified as ozonolysis products of linear and branched alkenes (Grosjean and Grosjean, 1996, 1997). For instance, 2-butanone and 2-methylpropanal are formed by reaction of ozone with 3-methyl-1-pentene and 4-methyl-1-pentene, respectively. This latter component also produces 3-methyl-butanal.

C. The Importance of Photochemical Reactivity in Determining the Tropospheric Composition of Volatile Organic Compounds

A better view of the atmospheric reactivity of VOCs can be obtained by looking at the data shown in Table 2, which summarizes the percentage of composition of the main organic classes listed in Table 1. Data refer to selected tropospheric sites and all samples were collected during daytime hours and in mild seasons. In the same table the total amount of the organic fraction that was quantified by GC-MS is also reported. Although the contribution of some very volatile species is missing in these calculations, the picture is sufficiently clear for this discussion.

In the proximity of anthropogenic emission sources, alkanes, alkenes, and arenes account for ca. 70% of the organic fraction identified. In forest–rural areas their contribution drops to ca. 20%. Alkenes disappear from the atmosphere and the arene fraction is composed mostly of benzene and toluene because trimethylbenzenes are always detected at trace levels. The decay of anthropogenic components is mirrored by a drastic increase in the concentration of free acids, alcohols, and carbonyl compounds that often exceeds 60% of the total organic compounds detected. Isoprenoids never reach values larger than 18.5% of the total composition.

Data recorded in temperate regions clearly indicate that the transition from urban to forest–rural airsheds leads to a substantial increase in oxygenated

Ciccioli, Brancaleoni, and Frattoni

TABLE 2   Percentage (%) of Composition and Total Amount of VOCs Detected by GC-MS in Different Sites

Site Country Type of site	Milan Italy Urban	Rome Italy Urban	Taranto Italy Urban	Montelibretti Italy Suburban– rural	Madrid Spain Suburban
Alkanes	29.7	34.9	47.0	9.2	33.6
Alkenes	4.5	6.4	11.9	0.5	8.4
Arenes	43.7	32.9	34.1	6.0	34.5
Alcohols	6.7	3.1	1.3	3.0	3.8
Aldehydes	4.0	9.0	3.0	38.6	13
Ketones	0.7	2.4	0.7	6.2	3.5
Free acids	3.4	8.8	1.4	35.2	0.4
Mono- terpenes	0.5	0.7	0.0	0.2	0.6
Isoprene	0.1	0.2	0.2	0.2	0.8
Others	6.7	1.5	0.3	1.0	0.1
Total amount ($\mu g/m^3$)	490.8	583.5	586	360.5	104.3

Site Country Type of site Ecosystem	Castelporziano Italy Forest Pseudo-steppe	Castelporziano Italy Forest Pine–oak forest	Viols en Laval France Forest Maquis	Monti Cimini Italy Forest Pine forest	Storkow Germany Forest Pine forest
Alkanes	11.0	8.6	3.0	12.6	12.6
Alkenes	0.0	0.0	0.0	8.4	0.2
Arenes	2.6	3.9	6.0	8.7	25.8
Alcohols	4.9	14.2	13.0	1.1	0.0
Aldehydes	20.1	28.1	31.0	47.3	34.5
Ketones	7.7	15.0	21.0	10.2	7.3
Free acids	48.8	4.5	14.0	0.2	0.0
Mono- terpenes	1.3	8.9	4.0	10.0	18.5
Isoprene	2.3	0.6	8.0	1.4	0.6
Others	1.3	16.2	0.0	0.1	0.4
Total amount ($\mu g/m^3$)	88.7	95.5	36.0	96.2	91.2

(continues)

products in air. However, less than 50% of these definitely arise from the photochemical degradation of anthropogenic and biogenic emissions. A large fraction of oxygenated VOCs is still composed of products, such as semivolatile aldehydes or 6-methyl-5-hepten-2-one, whose origin is uncertain. In many of the forest areas investigated, true NO_x-limited conditions are never reached

TABLE 2   (continued)

Site	Burriana	Appennini Mountains	Svalbard Islands	Ross Bay	Himalaya
Country	Spain	Italy	Norway	Antarctica	Nepal
Type of site	Rural	Forest-rural	Remote	Remote	Remote
Ecosystem	Orange field	Mixed	Arctic	Antarctic	High elevation
Alkanes	5.0	1.6	14.9	3.3	18.2
Alkenes	0.0	0.0	0.0	1.1	2.2
Arenes	5.0	1.6	5.2	0.5	9.9
Alcohols	3.3	6.4	7.0	9.6	10.5
Aldehydes	24.9	35.2	16.6	30.3	24.5
Ketones	15.8	38.0	6.0	43.1	15.7
Free acids	26.6	12.8	45.3	9.6	14.0
Mono- terpenes	18.3	0.0	.0	0.0	1.2
Isoprene	0.0	0.2	0.3	0.2	0.2
Others	1.2	2.4	4.8	1.5	1.8
Total amount ($\mu g/m^3$)	25.5	34.0	48.3	93.4	126.1

during daytime hours because the VOC/NO$_x$ ratios measured in the afternoon hours are close to 9 and substantial production of ozone takes place until sunset; and ozone levels range from 50 to 80 ppbv all the time. The high-oxidizing capacity of the atmosphere in forest and rural areas is demonstrated by the substantial levels (9 ppbv) of very volatile carbonyl compounds coming from the degradation of hydrocarbons (BEMA Report, 1994, 1997b). Efficient conversion into formaldehyde, acetaldehyde, and acetone, not quantified by GC-MS, is one of the reasons explaining the rather low concentrations of VOCs that were recorded in the forest and rural sites reported in Table 2. Reactivity effects were particularly evident in Castelporziano and Burriana where the typical land–sea breeze circulation occurring in the Mediterranean Basin (Millan et al., 1996) brought substantial levels of ozone (up to 80 ppbv) from the sea after 11 AM.

D. THE REACTIVITY OF VOLATILE ORGANIC COMPOUNDS IN TEMPERATE AND COLD REGIONS

The importance of photochemical reactivity in determining the VOC composition in the troposphere can be better seen by comparing data collected in forest–rural areas with those recorded in remote sites. The photochemical reactivity of cold and remote sites is so low that some semivolatile alkenes and isoprene can be still detected in the atmosphere in spite of their extremely low emission rates from biogenic sources in such hostile environments. The slowdown of photochemical processes at low temperatures allows some reactive biogenic and anthropogenic VOCs to survive longer. For example, reactive compounds released in Nepalese valleys can be still detected at the bottom of Mount Everest (Ciccioli et al., 1993c). Although substantial differences are observed in the percentage of composition of the various classes in which VOCs are grouped, aldehydes and ketones capable of producing radical species by photolysis and reaction with OH radicals still dominate the air composition in remote sites.

Particularly interesting are the data collected in Antarctica where the influence of long-range transport is definitely lower than at other sites. The analysis of the organic fraction extracted from airborne particles has shown that PAH coming from fossil fuel combustion reach the lowest level observed on the earth and no presence of nitrofluoranthenes and nitropyrenes of photochemical origin is detected (Ciccioli et al., 1996a). The analysis of the polar fractions also suggested that the major source of suspended particles in Antarctica is marine aerosol (Ciccioli et al., 1994b). Due to the low mixing height (about 10 m above the ground) and the limited reactivity occurring at this remote site, the VOC composition reflects better that of the emission. Concentrations comparable to or higher than those recorded in temperate regions, where daytime turbulence dilutes organic compounds into a boundary layer typically 1000 m deep, can be thus measured in Antarctica.

The facts that more than 73% of the organic material analyzed by GC-MS was composed of carbonyl compounds and that most of them (Ciccioli et al., 1996b) were semi-volatile components indicate how important these oxygenated species can be in reactive environments. Since data collected in Mauna Loa perfectly fit with these observations, the nature and source strength of potentially reactive oxygenated components in the troposphere deserve more detailed investigations. Whatever the source is, it must be extremely strong to sustain the high atmospheric levels observed in reactive environments.

III. REACTIVE VOLATILE ORGANIC COMPOUNDS AT THE URBAN SCALE

Since ozone production during photochemical smog episodes definitely starts in urban and industrial areas, the identification of reactive VOCs in urban environments is of key importance for designing proper strategies for tropospheric ozone reduction. A significant body of data is available on the composition of VOCs present in urban areas and extensive lists of identified components can be found in the literature (e.g., Singh and Zimmerman, 1992; Ciccioli, 1993; PORG Report, 1993). Although they are not as detailed as Table 1, they include very volatile components not quantified with the method used here. In general, the pattern closely approaches that of gasoline and diesel oil emissions and their combustion products. Alkanes and aromatics largely dominate the hydrocarbon composition, followed by alkenes and acetylene. The most reactive components are some alkylbenzenes and alkenes, listed in Table 1. Their degradation pathway is quite well described by photochemical models. Examples of the ozone production resulting from urban emissions are reported in other chapters of this book.

Emission inventories used in photochemical models often do not include in their list some minor components that can be extremely important for ozone production. Although present at trace levels, some cycloalkenes and dienes reported in Table 1 display a POCP not much different from those of cyclopentene and 1,3-butadiene. Their emissions are so small and their photochemical reactivities are so high that their presence in air cannot be quantified with conventional instruments. Their detection is even difficult in car emissions. In spite of the sophisticated sampling and analytical techniques used in investigations (e.g. Zielinska *et al.*, 1996), none of the pentadienes and hexadienes listed in Table 1 have been identified in tunnel samples. Because of this, no real evidence of their photochemical removal is available except from the observations made in smog chamber experiments.

A. ROME: A CASE STUDY

By using the same methodology developed for generating the data listed in Table 1, daily variations of very reactive components were followed in the urban area of Rome where high levels of photochemical pollution have been observed since 1977 (Cecinato *et al.*, 1979). As a case study, data are used from September 16 and 17, 1991. During these days, limited advection during daytime hours was combined with strong nighttime stability. The removal of precursors and products during the day was so small that levels of PAN and ozone measured in the urban area largely exceeded those recorded downwind (Ciccioli *et al.*, 1993b). Ozone and PAN concentrations observed during this episode are displayed in Fig. 1. In these conditions, the reactivity developed in the urban area was approaching that occurring in a closed chamber in which hydrocarbon-limited conditions occurred. The average VOC/NO$_x$ ratios observed during this episode ranged between 3 and 4, which is a value

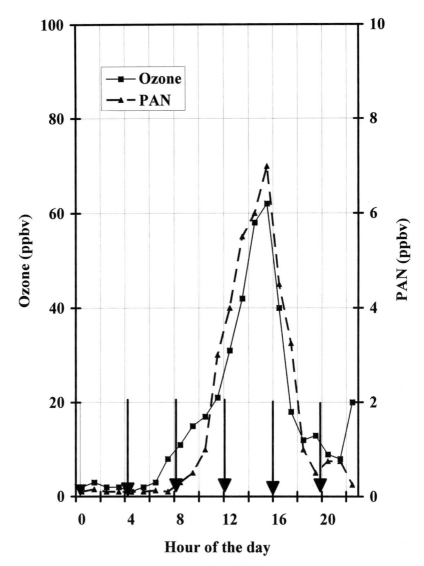

FIGURE 1 Diurnal profiles of ozone and PAN recorded in the center of Rome when stable meteorological conditions were established in the Tiber Valley. Data were collected on September 16 and 17 1991. Arrows indicate the time when collection of VOCs was performed. Normalized concentrations obtained during this episode are reported in Table 3.

quite typical for the city of Rome under stable conditions (Brocco *et al.*, 1994). The only difference between these ambient data and those from smog chamber experiments was the diurnal change in the mixing volume and in the source strength (Singh and Zimmerman, 1992; Ciccioli, 1993). It is possible to account for these effects if concentrations normalized to a less reactive component present in the emission are used for evaluating the reactivity of the VOCs (Roberts *et al.*, 1984). The applicability of this procedure is supported by the fact that diurnal variations of less-reactive components correlate well with the change of mixing height calculated through the concentrations of naturally occurring radioactive components (Allegrini *et al.*, 1995).

The change in VOC reactivity during the episode displayed in Fig. 1 was studied by measuring the concentrations of organic compounds every 4 h. Data obtained were normalized with respect to benzene, which has a rather long atmospheric lifetime and is quite representative of urban emissions. In this way, all compounds that were more reactive than benzene but were released by the same source should have displayed a decrease in the normalized concentrations whenever reactive conditions were established in the atmosphere. The decrease was expected to be proportional to the reactivity of the compounds toward ozone or the radical species formed in air during the day or at night (Roberts, 1984; Grosjean and Fung, 1984).

Since this simple scheme can be complicated by possible changes in the diurnal pattern of emission and in the efficacy of heterogeneous processes in removing different hydrocarbons, the interpretation of the results was based on the largest number of components possible (Satsumabayashi *et al.*, 1992). These were selected to cover a wide range of reactivity toward OH and NO_3 radicals as well as toward ozone.

The observations are summarized in Table 3. They indicate that rather low reactive conditions were observed during nighttime hours from 8 PM to midnight on September 16. Low reactivity was also reached at 8 PM on the following day. Samples collected in these periods were characterized by the highest normalized concentrations for many of the compounds listed in Table 3. In particular, normalized concentrations of alkenes and arenes measured at 8 PM on September 16 closely approached the profiles of vehicular emissions reported by Zielinska *et al.* (1996). Very reactive conditions were observed at 4 AM and at noon on September 17. The transition from these two situations was characterized by a sharp decrease in reactivity observed in the early morning (8 AM). The same conclusions were reached by using the (*m* + *p*)-xylene ethylbenzene ratio as an indicator of the atmospheric reactivity of anthropogenic VOCs (Nelson and Quigley, 1983).

Although some deviations from this general trend are observed, the picture is clear in suggesting the nighttime removal of cycloalkenes and dienes by reaction with NO_3 radicals and their midday consumption by reaction with

TABLE 3 Diurnal Variations of the Normalized Concentrations of Selected VOCs Measured in the Urban area of Rome during Stable Meteorologic Conditions[a]

Date (September 1991)	16	17	17	17	17	17	17
Sampling time	20:00	00:00	04:00	08:00	12:00	16:00	20:00
Benzene concentration (μg/m³)	70.1	24.6	14.3	13.7	9.8	15.4	20.1
Compound	Concentrations of VOCs Normalized with Respect to Benzene Whose Concentration was Taken as 100 Units						
Arenes							
Toluene	255.4	336.2	224.4	274.3	235.5	232.2	260.7
Ethylbenzene	55.1	71.7	44.4	56.0	49.2	47.0	54.7
($m + p$)-Xylene	170.3	220.6	146.2	173.2	149.1	141.7	168.5
Styrene	2.5	2.7	2.9	1.9	1.5	1.7	1.7
o-Xylene	73.2	82.4	58.5	67.5	62.0	60.6	72.5
Benzene, isopropyl	5.1	6.0	4.9	4.2	4.7	3.9	4.6
Benzene, n-propyl	22.9	22.3	19.2	18.9	19.8	17.7	22.2
Benzene, 1-methyl, 3-ethyl + Benzene, 1-methyl, 4-ethyl	91.8	80.3	49.4	73.2	63.8	60.7	92.1
Benzene, 1,3,5-trimethyl	29.5	27.8	18.5	24.1	18.1	19.4	30.2
Benzene, 1-methyl, 2-ethyl	24.1	22.2	16.5	19.5	18.0	17.9	24.4
Benzene, 1,2,4-trimethyl	89.2	84.6	49.2	70.3	64.2	60.0	89.9
Benzene, 1,2,3-trimethyl	19.3	19.2	11.3	15.9	18.2	13.9	20.3
Cycloalkanes							
Cyclopentane	2.3	4.4	3.0	3.2	1.9	2.1	6.2
Cyclopentane, methyl	16.3	21.4	15.8	14.9	15.4	12.3	27.8
Cyclohexane	8.0	10.0	6.7	5.0	7.3	5.4	13.2
Alkenes							
1-Pentene	5.8	5.5	5.5	5.8	4.7	3.6	8.7
1-Butene, 2-methyl	9.3	9.1	6.0	8.4	5.8	5.7	15.8
2-Pentene-trans	11.8	12.3	9.6	11.8	6.1	6.5	19.7
2-Pentene-cis	6.3	6.9	5.0	6.4	3.6	3.6	9.9
2-butene, 2-methyl	18.3	18.1	12.7	18.0	7.9	9.3	26.9
1-Pentene, 3-methyl	2.5	5.2	8.3	5.6	1.3	5.5	3.1
1-Hexene	10.4	12.4	15.5	7.3	6.5	4.0	12.9
2-Hexene-trans	3.7	5.4	3.5	5.9	1.7	2.7	6.1
2-Hexene-cis	9.0	8.1	8.4	6.9	1.8	4.6	12.6
Butene, 2,3-dimethyl	7.7	7.5	5.2	5.5	3.6	3.3	8.4
Pentane, 3-methylene	5.4	5.7	0.3	3.7	0.1	1.8	6.4
3-Hexene	3.3	7.7	1.8	4.0	0.3	1.9	4.5
2-Pentene, 3-methyl	10.6	7.0	3.6	5.6	1.7	3.4	7.1
Cycloalkenes							
Cyclopentene	4.8	3.7	4.1	4.2	1.2	4.4	4.2
Cyclopentene, 4-methyl	1.4	1.1	—	1.3	0.5	1.0	1.0
Cyclopentene, 3-methyl	0.6	0.4	—	0.6	0.2	0.5	0.4
Cyclopentene, 1-methyl	3.5	3.0	2.3	3.2	1.3	2.7	2.8
Cyclohexene	0.8	1.1	1.5	1.2	0.5	0.8	0.7
Cyclopentene, 4,4-dimethyl	1.0	0.8	—	1.1	—	0.8	0.7

(continues)

TABLE 3   *(continued)*

Date (September 1991)	16	17	17	17	17	17	17
Sampling time	20:00	00:00	04:00	08:00	12:00	16:00	20:00
Benzene concentration ($\mu g/m^3$)	70.1	24.6	14.3	13.7	9.8	15.4	20.1
Compound	Concentrations of VOCs Normalized with Respect to Benzene Whose Concentration was Taken as 100 Units						
Cyclopentene, 1,5-dimethyl	0.8	0.6	—	0.5	—	0.6	0.7
Cyclohexene, 4-methyl	0.6	0.2	—	—	—	0.5	—
Cyclobutane, isopropylidene	1.4	1.1	—	1.8	—	1.6	0.9
Cyclohexene, -methyl	0.8	0.6	—	0.4	—	0.6	0.5
Cyclopentane, 1,2-dimethyl, 3-methylene	0.9	—	—	—	—	—	0.7
Dienes							
1,3-Pentadiene	1.9	1.7	1.5	1.6	—	1.0	1.2
1,3-Pentadiene, 2-methyl	0.5	0.8	—	1.0	—	1.2	—
2,4-Hexadiene	0.6	0.5	—	0.7	0.9	0.5	0.6
2,4-Hexadiene, 2-methyl	1.2	0.5	—	0.4	—	0.3	0.7
Iso-C_8-diene	0.6	—	—	—	—	—	—
Iso-C_8-diene	0.3	—	—	—	—	—	—
1,4-Pentadiene, 2,3,3,trimethyl	0.1	—	—	—	—	—	—
Alkynes							
3-Ethyne, 5-methyl	0.1	—	—	—	—	—	—

[a]For ozone and PAN concentrations see Fig. 1.

OH radicals and ozone. Daytime reactivity rapidly leads to ozone and PAN formation, reaching maximum concentrations around 1 PM. It is possible to roughly estimate a 12-h daytime concentration of OH radicals for the episode reported in Fig. 1 by using the data reported by Kley (1997), who derived isolines of OH radicals for the mix of hydrocarbons existing in Rome (Brocco *et al.,* 1994; Brocco *et al.,* 1997) with a box model. This method gives 12-h mean concentrations around 2×10^6 molecules per cubic centimeter for September 17, a value two to three times lower than that which can be produced by the reactant mixture under optimal conditions. From these data, maximum daytime concentrations of OH radicals in the range of 1.5×10^7 molecules per cubic centimeter (ca. 0.5 pptv) were estimated.

Nighttime formation of NO_3 radicals was possible due to the large excess of NO_2 (ca. 150 ppbv) accumulated in the urban airshed. It was clearly indicated by the minimum value reached by the reactivity function ($NO_2 + O_3$) introduced by Febo *et al.* (1997) for describing atmospheric processes. The efficient removal of VOCs observed at 4 AM indicates that NO_3 concentrations larger than 20 pptv were probably present in the atmosphere. This value is consistent with the observations made by D. Perner and P. Ciccioli (unpublished results) who measured an average value of 20 pptv of NO_3 in a suburban forest (Monti Cimini Park) located not far from the city of Rome.

Nighttime removal of reactive components was so efficient that only 5% of 3-methylene-pentane survived the reaction with NO_3 and many cycloalkenes and dienes completely disappeared from the atmosphere. The reactivity of these last components remained so high during the whole day that few of them were able to reach the same concentrations at 8 PM September 17 as those measured the night before. Although alkenes, cycloakenes, and dienes reported in Table 3 accounted for a rather small fraction of the whole emission (3 to 4% at most), they were mainly responsible for the production of ozone, together with tri- and tetramethylbenzenes.

IV. REACTIVE VOLATILE ORGANIC COMPOUNDS AT THE REGIONAL SCALE

A. THE SUBURBAN CASE

The photochemical pollution resulting from primary reactants emitted in urban areas is able to express its whole ozone production potential only when the plume reaches the transition point between hydrocarbon- and NO_x-limited conditions (VOC/NO_x ratios close to 8). This happens when advection brings pollutants to areas where reduced or no emissions of NO take place. This is why largest ozone concentrations are always measured 30 to 50 km downwind of urban source areas (Finlayson-Pitts and Pitts, 1986, 1997; Kley, 1997). However, different reactivity conditions can be reached by the urban plume when it is mixed and diluted with clean background air. Since the oxidation of NO_x by OH radicals proceeds at a faster rate than that of many VOCs, the production of OH radicals in the plume will strongly depend on the rate of mixing and dilution to which the urban plume is subjected (Kley, 1997); the more rapid this process is, the lower the OH radical concentrations that are produced. However, these simple scenarios seldom occur in the real world where advection processes may inject substantial levels of ozone into the urban area and transport precursors and products into a suburban airshed in which photochemical reactivity has already developed. Because of this, the formation of ozone and photochemical oxidants in suburban areas is a rather complex issue and sudden changes in the reactivity of the VOC mixture can be observed.

This aspect has been studied at the suburban site of Montelibretti, which is exposed to the urban plume of Rome whenever the sea–land breeze system drives the air mass circulation in the Tiber Valley (Ciccioli and Cecinato,

1992). The site is located ca. 20 km from the city center and approximately 40 km from the Mediterranean coast. Photochemical smog episodes can be detected from March to late October when cyclonic weather is established in central Italy or when southeastern flow advects warm air from the African continent. Records of ozone and PAN collected in the last decade indicate that levels of photochemical pollution in Montelibretti approach those recorded in the Los Angeles basin (Ciccioli and Cecinato, 1992; Cantuti *et al.*, 1994).

The progressive accumulation of photochemical pollutants in the Tiber Valley is strictly related to the type of local circulation activated under hot and sunny weather conditions. In the morning hours, the emissions in the urban area are mixed into a rather thin layer (ca. 400-m depth), which is virtually isolated from the external air by the heat island generated over the city. Pollutants accumulate and react until the sea breeze pushes them inside the Tiber Valley. The start of the sea breeze and its duration and intensity depend on the temperature gradient established between the sea surface and the urban area and that existing between the urban area and the countryside. Usually the sea breeze becomes effective at 11 to 12 AM.

The amount of photochemical pollutants that the hydrocarbon-limited mixture of Rome can generate downwind is thus a function of the dynamics of the air masses and the accumulation processes occurring in the previous days. If the wind speed exceeds 9 m/sec, the heat island over the city is disrupted. Chemical compounds are well mixed up to an altitude of ca. 800 m and advected more than 50 km inside the Tiber Valley. They reach the suburban site of Montelibretti in about 1 h. Because of the presence of high mountains at the end of the valley, the sea breeze circulation system generates a return flow above 1000 m. Depending on the intensity of the breeze, pollutants transported from the urban area can be channeled into the return layer and moved back toward the sea.

The circular motion lasts until the sea breeze ceases. This occurs after sunset when the land starts to cool radiatively. Stratification of pollutants in separate layers can thus take place because of the low turbulence established at night. Millan *et al.* (1996) have shown that up to three layers containing different amounts of ozone can be generated at night in the Mediterranean coastal zone. They can extend from the mountains to 40 km over the sea. The disruption of these layers in the morning hours followed by the start of the sea breeze generates a constant input of 50 to 80 ppbv of ozone into the Mediterranean coastal zone.

Determinations carried out during the BEMA campaigns have shown that marine air masses are rich in ozone but poor in VOCs and NO_x. The PAN/ozone ratio seldom exceeds a value of 0.05 to 0.06 (BEMA Report, 1997a). The low concentration of PAN relatively to the situation reported in Fig. 1

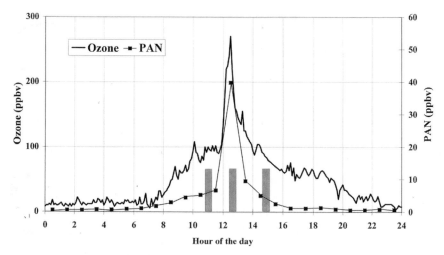

FIGURE 2   Diurnal profile of ozone and PAN recorded in the suburban site of Montelibretti when strong advection from the city of Rome occurred. Data refer to a photochemical smog event observed in July 16, 1992. Bars indicate the collection period of VOCs. Absolute and normalized concentrations of selected VOCs are displayed in Tables 4 and 5 respectively.

suggests that marine air masses approach NO_x-limiting conditions in which PAN dissociation becomes a major source of NO_x.

If photochemical processes are activated for a period of time longer than 4 to 5 days, substantial amounts of photochemical oxidants are accumulated in the Tiber Valley at ground level and in the upper layers. If the sea breeze exceeds 9 m/sec, daily profiles such as those reported in Fig. 2 can then be detected at Montelibretti. Ozone and PAN reach concentrations as high as 275 and 40 ppbv, respectively. Profiles of both pollutants are characterized by a sharp peak at ca. 1 PM that overlaps with local production. It clearly indicates the arrival of polluted air masses from the city of Rome.

During the acute episode reported in Fig. 2, the nitric acid concentration was on the order of 10 ppbv and PAN accounted for the largest portion of gaseous NO_y. The amount of total NO_y (PAN + HNO_3 + NO_3^-) present in air was more than eight times higher than the concentration of NO_x, composed mostly of NO_2 (15 ppbv). A definite transition from hydrocarbon-limited to NO_x-limited conditions occurred in the air masses during the transport from the city of Rome to Montelibretti.

Figure 2 also shows that marked differences in the PAN/ozone ratios were observed before (0.08), during (1.47), and after (0.05) the passage of the urban front. The decoupling of these pollutants with respect to the stationary situation of Fig. 1 (constant PAN/ozone concentrations equal to ca. 1.1) is useful for separating reactivity from transport effects. In situations such as those illustrated in Fig. 2, PAN is definitely a better indicator of atmospheric reactivity than ozone. Its content in marine air masses is negligible and the only important removal process is thermal decomposition. This process is substantially reduced by the cooling of the urban air caused by the disruption of the heat island.

The high intensity and the short duration (3 to 4 h) of the smog episode shown in Fig. 2 provided ideal conditions for studying the mixing and reactivity of VOCs. Samples were collected before, during, and after the plume from Rome reached the site. Bars in Fig. 2 indicate the collection period of VOCs. Absolute concentrations of the main organic components identified during the episode are reported in Table 4 and normalized concentrations with respect to benzene of selected anthropogenic components are listed in Table 5.

Absolute concentrations indicate that the urban plume advected substantial amounts of unreacted organic components into the suburban airshed. The impact was so great that an increase of ca. 3.5 was observed in the concentration of low reactive and volatile components (alkanes, benzene, toluene, and chlorinated compounds). A decay in concentration was detected only for a few components known to be very reactive with both ozone and OH radicals, for example, some methylpropyl- and tetramethylbenzenes and some alkenes and cycloalkenes. In particular, complete removal from the suburban atmosphere was observed for 2-methyl-1-butene, cyclopentene, and *trans*-2-pentene. The slight decrease in isoprene and α-pinene occurring at noon was also indicative of photochemical degradation, since biogenic sources are mainly localized around the site and maximal emission rates were concurrent with the arrival of the urban plume. The advection to Montelibretti of camphene, not present in local emissions, indicates how anthropogenic sources located in the city center can be important in providing a supply of monoterpenes in semi-rural areas.

A large number of oxygenated compounds were also associated with the urban front and substantially contributed to the VOC levels at the suburban site. The inputs of some volatile alcohols, ketones, acids, and ethers were important. Not all these were produced, however, by the oxidation of anthropogenically emitted compounds. For instance, 2-metoxy-2-methyl-propane (also called MTBE) was clearly identified as an anthropogenic component because it is widely used in Italy as an additive for unleaded gasoline. This explains why its increase closely matches that of benzene and toluene in air. The photochemical depletion of anthropogenic compounds occurring during the travel time between the urban area and the suburban site can be roughly estimated by comparing the data reported in Table 5 with those displayed in Table 3. If we look at the values of the normalized concentrations measured

TABLE 4   Change in the Air Concentrations of Selected VOCs Measured in the Suburban Area of Montelibretti during a Photochemical Smog Event Caused by Transport of Pollutants from the Urban Area of Rome[a,b]

Sampling hour	11:00	12:30	15:00
Compound	Air concentrations ($\mu g/m^3$)		
Alkanes and cycloalkanes			
Butane, 2-methyl	0.54	1.56	0.24
n-C_5	2.05	7.02	2.14
Butane, 2,2-dimethyl	1.04	3.65	1.51
Butane, 2,3-dimethyl	0.37	1.30	0.40
Pentane, 2-methyl	0.23	0.84	0.21
Pentane, 3-methyl	1.24	5.08	1.15
n-C_6	1.46	4.66	1.24
Cyclopentane, methyl	0.39	1.61	0.44
Cyclohexane	0.26	0.53	0.20
Hexane, 2-methyl	0.73	3.42	0.70
Hexane, 3-methyl	1.15	3.74	1.18
Pentane, 2,2,4-trimethyl	0.43	4.45	0.63
n-C_7	0.55	1.97	0.71
n-C_8	0.45	1.00	0.57
n-C_9	0.55	0.81	0.55
n-C_{10}	1.19	0.73	0.41
n-C_{11}	1.44	0.40	0.52
n-C_{12}	2.68	1.20	0.96
n-C_{13}	2.86	2.30	0.39
Alkenes and Cycloalkenes			
1-Pentene	0.19	0.38	0.11
1-Butene, 2-methyl	0.08	0.22	0.06
Isoprene	0.71	0.66	0.34
trans-2-Pentene	0.11	0.00	0.00
cis-2-Pentene	0.07	0.46	0.17
2-Butene, 2-methyl	0.25	0.00	0.00
Cyclopentene	0.02	0.00	0.00
1-Octene	0.21	0.42	0.34
Mono and bicyclic arenes			
Benzene	2.70	10.36	1.81
Toluene	5.52	21.88	3.43
Ethylbenzene	0.96	3.80	0.57
($m + p$)-Xylene	1.68	5.42	1.26
Styrene	0.07	0.56	0.00
o-Xylene	0.82	3.17	0.58
Isopropylbenzene	0.08	0.30	0.02
n-Propylbenzene	0.28	1.19	0.20
Benzene, 1-methyl, 3-ethyl + benzene, 1-methyl, 4-ethyl	0.62	2.06	0.47
Benzene, 1,3,5-trimethyl	0.08	0.13	0.07
Benzene, 1-methyl, 2-ethyl	0.23	0.75	0.14

(continues)

TABLE 4   *(continued)*

Sampling hour	11:00	12:30	15:00
Compound	Air concentrations ($\mu g/m^3$)		
Benzene, 1,2,4-trimethyl	0.43	0.89	0.35
Benzene, 1,2,3-trimethyl	0.11	0.20	0.06
Benzene, 1-methyl, 3-isopropyl	0.00	0.04	0.00
Benzene, 1-methyl, 4-isopropyl	0.08	0.10	0.06
Benzene, diethyl	0.14	0.39	0.07
Benzene, methyl, *n*-propyl	0.11	0.23	0.07
Benzene, *n*-butyl	0.07	0.14	0.02
Benzene, methyl, *n*-propyl	0.05	0.18	0.03
Benzene, methyl, *n*-propyl	0.08	0.03	0.02
Benzene, dimethyl, ethyl	0.08	0.11	0.04
Benzene, dimethyl, ethyl	0.08	0.11	0.05
Benzene, dimethyl, ethyl	0.02	0.02	0.00
Benzene, 1,2,4,5-tetramethyl	0.03	0.03	0.01
Benzene, 1,2,3,5-tetramethyl	0.05	0.02	0.02
Benzene, 1,2,3,4-tetramethyl	0.01	0.04	0.00
Indan	0.06	0.13	0.01
Naphthalene	0.04	0.09	0.03
Monoterpenes			
α-Pinene	0.60	0.47	0.54
Camphene	0.00	0.33	0.26
Sabinene	0.00	0.13	0.00
β-Pinene	0.00	0.15	0.15
Halogen containing organic compounds			
Ethene, trichloro	0.27	1.23	0.14
Ethene, tetrachloro	1.26	4.91	0.56
($m + p$)-Dichlorobenzene	0.03	0.20	0.01
Alcohols and oxy-alcohols			
Ethanol	0.41	1.32	0.24
2-Propanol	0.22	0.47	0.17
2-Propanol, 2-methyl	0.16	0.41	0.25
1-Propanol	0.15	0.46	0.06
3-Buten, 2-ol, 2-methyl	0.41	0.58	0.39
1-Butanol	0.60	1.36	0.38
2-Butoxy, ethanol	0.31	3.46	0.40
Aldehydes			
Propanal, 2-methyl	0.18	0.60	0.32
2-Propenal, 2-methyl	0.22	0.18	.17
Butanal	1.11	2.25	0.72
Butanal, 3-methyl	0.00	0.75	0.37
Pentanal	1.01	2.09	0.95
Hexanal	1.16	2.21	1.22
Heptanal	0.94	1.52	1.25

(continues)

TABLE 4   *(continued)*

Sampling hour	11:00	12:30	15:00
Compound	Air concentrations ($\mu g/m^3$)		
Benzaldehyde	0.59	1.21	0.46
Octanal	2.03	2.55	3.07
Nonanal	2.42	2.85	2.62
Decanal	2.81	3.10	6.56
Ketones			
2-Propanone	2.80	4.11	2.74
3-Buten, 2-one	2.40	7.26	2.90
2-Butanone	3.74	6.06	1.97
Iso-C_7-ketones	1.78	11.63	10.23
Iso-C_8-ketones	2.62	5.72	3.26
5-Hepten-2-one, 6-methyl	3.16	4.11	10.33
Iso-C_8-ketones	0.49	2.04	0.78
Acids			
Acetic acid	10.43	21.92	4.88
Esters			
Acetic acid, ethyl ester	0.36	0.55	0.17
Acetic acid, butyl ester	0.35	0.55	0.08
Ethers			
Propane, 2-methoxy, 2-methyl	0.71	2.52	0.40
Silane compounds			
Silanol, trimethyl	0.39	0.90	0.10

[a]Sampling was performed before, during, and after the arrival of the urban plume.
[b]For the levels of ozone and PAN, and for the sampling periods of VOCs see Fig. 2.

at noon in downtown Rome and Montelibretti, we can see how effective a rapid reduction of NO_x emission is in speeding up photochemical reactions. A general decline in normalized concentrations is observed by moving away from hydrocarbon-limited conditions and the effect produced is proportional to the reactivity of the organic compound toward OH and ozone. While only 10% of toluene is lost during transport, losses close to 70 and 90% can be estimated for some xylenes and trimethylbenzenes, respectively. The reactivity of 2-methyl-2-butene, *trans*-2-pentene, and cyclopentene is so high that even local production is completely removed from the atmosphere. The effect is emphasized by the fact that ozonolysis reactions leading to HO_2 radical formation (Paulson and Orlando, 1996) are also highly effective in removing olefin components from the atmosphere.

Data reported in Tables 4 and 5 indicate that after the pollutant front has passed over Montelibretti, the absolute and normalized concentrations of many organic components approach those observed before the arrival of the urban front. Precursors and products of photochemical reactions are effi-

TABLE 5   Change in the Normalized Concentrations of Selected VOCs[a] Recorded during the Photochemical Smog Event Reported in Fig. 2

Sampling Hour	11:00	12:30	15:00
Compound	Normalized concentration		
Monocyclic Arenes			
Toluene	204.4	211.2	189.8
Ethylbenzene	35.6	36.7	31.4
(m + p)-Xylene	62	52.3	69.7
Styrene	2.6	5.4	0
o-Xylene	30.3	30.6	32.1
Isopropylbenzene	2.9	2.9	1.3
n-Propylbenzene	10.6	11.5	11.2
Benzene, 1-methyl, 3-ethyl + benzene, 1-methyl, 4-ethyl	23	19.8	26.2
Benzene, 1,3,5-trimethyl	3.1	1.3	3.7
Benzene, 1-methyl, 2-ethyl	8.5	7.2	7.9
Benzene, 1,2,4-trimethyl	15.9	8.6	19.2
Benzene, 1,2,3-trimethyl	3.9	1.9	3.3
Alkenes and cycloalkanes			
1-Pentene	7.1	3.7	6.2
1-Butene, 2-methyl	2.9	2.1	3.4
2-Pentene-trans	4	—	—
2-Pentene-cis	2.7	4.5	9.2
2-Butene, 2-methyl	9.2	—	—
1-Octene	7.7	4	18.9
Dienes and cycloalkenes			
Cyclopentene	0.8	—	—
Isoprene	26.2	6.4	18.9

[a]Benzene = 100.

ciently removed by advection and replaced by air in which the urban emission is much less consumed by photochemical reactions.

The low PAN/ozone ratio (ca. 0.05) measured in the afternoon hours clearly indicates that the largest portion of ozone present at Montelibretti at 3 PM was advected from the sea. The marine origin of the air masses was suggested by the concurrent increase of water and HCl concentrations after the passage of the urban front. HCl is a rather specific indicator because it is formed by reaction of sea salt with HNO_3 (Allegrini et al., 1993). It has been shown that maximum values of hydrogen chloride in Montelibretti are reached 1 to 2 h later than those of nitric acid, ozone, and PAN.

Although the PAN content measured at 3 PM is rather low, the ozone levels are still remarkable and the oxidizing capacity of the urban plume remains sufficiently high to selectively remove highly reactive components. This explains why cis-2-pentene follows the general increase of other pollutants whereas *trans*-2-pentene, 2-methyl-2-butene, and cyclopentene remain unde-tected. Also indicative of photochemical reactivity is the selective removal of sabinene with respect to camphene and β-pinene and that of styrene with respect to camphene and β-pinene and that of styrene with respect to tri- and tetramethylbenzenes. The high content of 6-methyl-2-hepten-2-one, octanal, and decanal in air masses advected from the sea strongly supports the idea that these components originated upwind of the city of Rome where enough vegetation surface or marine aerosols exist for their formation. Photochemical degradation of semivolatile carbonyl compounds may be responsible for the levels of ozone and PAN measured at Montelibretti at 6 PM because repeated observations have shown that these components are removed from the atmo-sphere before sunset.

The chief difficulty in modeling photochemical processes in the Tiber Valley is the quantification of the reactivity conditions existing in the marine air masses and the ozone content associated with them. This is because the input of pollutants is determined by the return flow from areas located far north of the city of Rome. According to Millan *et al.* (1997), the earth's rotation deforms the air mass trajectories of the sea breeze circulation and moves them down along the Italian coast. As far as the reactivity is concerned, the chemistry of marine masses is also complicated by the possible involve-ment of chlorine radicals in the photochemical cycle. According to some authors (e.g., Finyalson-Pitts and Pitts, 1997) Cl· can accelerate ozone forma-tion in polluted coastal regions because it is capable of oxidizing VOCs in a manner analogous to that of OH radicals. Although chlorine-containing spe-cies and chlorine molecules have been identified in marine air masses, the sources of such halogen atom precursors remain elusive.

B. THE FOREST–RURAL CASE

At long distances from polluted areas the plume is so depleted in reactive VOCs of anthropogenic origin that they cannot further sustain ozone produc-tion even if sufficient amounts of NO_y are injected into the atmosphere. In these conditions, emissions of isoprene and terpenes from vegetation become crucial in determining the levels of photochemical pollution. This is possible because of the substantial emissions of these components in forest–rural areas and their high reactivity toward OH radicals, NO_3 radicals, and ozone.

The effect of emissions from vegetation is particularly relevant at the regional scale where biogenic sources may account for a large portion of total VOC emissions (Fehsenfeld *et al.*, 1992). In the North American continent, for instance, biogenic emissions of VOCs from vegetation exceed those from anthropogenic activities (Singh and Zimmerman, 1992; Guenther *et al.,* 1995). The same is true for some northern European countries like Norway, Sweden, and Finland. In Italy, biogenic emissions account for approximately 50% of the total VOC emissions (Simpson *et al.,* 1995). The contribution made by biogenic emissions is fundamental for a global-scale assessment of ozone formation because emissions from tropical and equatorial vegetation represent the major sources of VOCs in the global atmosphere (Guenther *et al.,* 1995).

One important aspect, often neglected in discussing the role played by vegetation emissions in promoting photochemical smog formation, is the wide range of reactivity covered by emitted compounds. This aspect is summarized by plotting the rate constants for the reaction of isoprene and selected mono-terpenes with ozone and OH radicals in Figs. 3 and 4, respectively. For the sake of comparison, reaction rates of some anthropogenic VOCs are also reported in these same figures. Data are based on the extensive experimental work of Atkinson and co-workers (Atkinson *et al.,* 1985, 1990a, 1990b, 1992), who have also reported a very high reactivity for some sesquiterpene compounds (Shu and Atkinson, 1995). Particularly relevant for regional scale studies are the high ozonolysis rates of some biogenic compounds. Since these reactions lead to the formation of HO_y radicals (Atkinson *et al.,* 1992), they can be important in forest areas to which large quantities of ozone are advected.

The atmospheric lifetime of certain mono- and sesquiterpenes can be short (2 to 3 min), making difficult the accurate determination of their emission rates with enclosure techniques using ambient air for the gas exchange. If the exchange rate of the enclosure is longer than the lifetime of the emitted compounds, extensive removal occurs in the enclosure and little or no emissions are measured. Since the enclosure behaves as a photochemical flow reactor, emissions of oxygenated products coming from degradation of mono- and sesquiterpenes can be also observed. All these reasons, combined with the limited analytical capabilities available a few years ago, might explain why in the past mono- and sesquiterpenes were not detected in plant emissions and the VOC composition reported in enclosure studies was restricted to α-pinene, β-pinene, Δ-3-carene, *p*-cymene, camphene, myrcene, sabinene, and D-limonene.

It is now known that emissions from plants can be extremely complex and can contain highly reactive components. Data collected in the Californian Central Valley (Winer *et al.,* 1992) have clearly shown that a large variety of mono- and sesquiterpenes are emitted by local vegetation together with esters and carbonyl compounds. Similar results have been obtained by Owen *et al.* (1997) who screened 18 Mediterranean species in the frame of the BEMA project. Seventeen different monoterpene compounds were identified (Kessel-

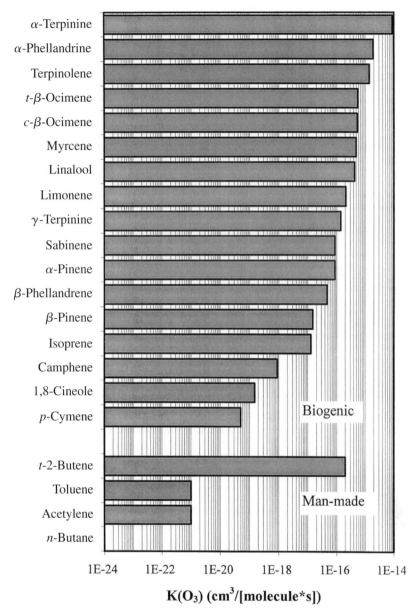

FIGURE 3 Rate constants for the reaction of some biogenic and anthropogenic VOCs with ozone. For the sources used for preparing the figure see the text.

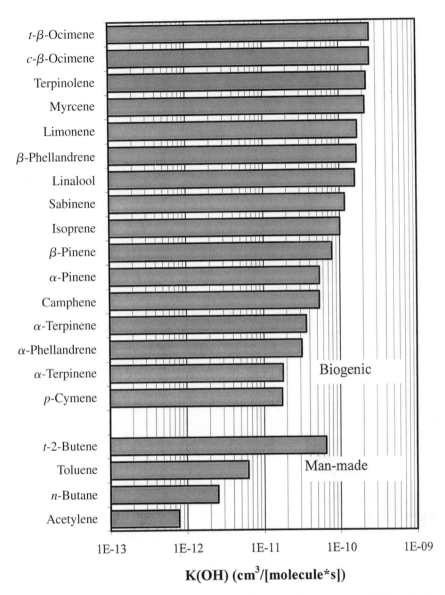

FIGURE 4   Rate constants for the reaction of some biogenic and anthropogenic VOCs with OH radicals. For the sources used for preparing the figure see the text.

meier *et al.*, 1996) in the emissions of *Quercus ilex* L., an isoprene-like monoterpene emitter (Loreto *et al.*, 1996; Bertin *et al.*, 1997) growing in southern Europe. In this plant species, α-pinene accounts for only a third of total emissions. High complexity and strong seasonal variability in monoterpene emissions have also been reported for a coniferous tree (Staudt *et al.*, 1997) and some shrubs growing in the Mediterranean area (Hansen *et al.*, 1997). Heiden *et al.* (1997) reported the emissions of mono- and sesquiterpene and oxygenated VOCs from sunflowers.

If the removal rates by gas-phase reactions of an emitted compound are so high as to approach its diffusion rate through the canopy, only a fraction of the biogenic emission reaches the atmospheric boundary layer unaffected. Evidence of this effect has been provided by the simultaneous determination of emission rates and fluxes carried out in the pine–oak forest of Castelporziano (Valentini *et al.*, 1997). While emission rates were calculated by integrating data obtained by enclosure techniques applied to tree branches and the soil, fluxes were measured by relaxed eddy accumulation (REA). Results for selected components are shown in Fig. 5. During the experiment up to 100 ppbv of ozone were advected from the sea during daytime hours. Although large uncertainties in the upscaling processes caused some evident inconsistencies between the expected and the actual observations (according to the data presented in Figs. 3 and 4, myrcene should have been substantially depleted in the atmosphere), the picture is clear in indicating the selective removal of reactive monoterpenes from the atmosphere.

Similar observations were made in an orange orchard at Burriana, Spain (Seufert, 1997). In this case, evidence for the fast removal of a high reactive sesquiterpene (β-caryophyllene) was collected. Lifetimes on the order of minutes were estimated for this component on the basis of the reaction rate with ozone and OH radicals. This value is comparable with the diffusion time from the canopy to the atmospheric boundary layer.

Unfortunately, the evaluation of atmospheric losses does not tell very much about the real contribution made by reactive mono- and sesquiterpenes to ozone formation because the final fate in the air of the oxidation products is not completely understood. While for isoprene it is possible to track the progressive steps leading to ozone and PAN formation by looking at the atmospheric fate of the main decomposition products (methyl vinyl ketone, methacrolein, and 3-methylfuran) (Yokouchi, 1994) and some reaction chain terminators (O'Brien *et al.*, 1997; Roberts *et al.*, 1997), the same approach fails with terpene compounds because very few of the products observed in smog chamber experiments have actually been detected in ambient air (BEMA Report, 1997a).

For instance, gas-phase chemistry predicts pinonaldehyde as the major ozonolysis product of α-pinene but this compound has not been found in the

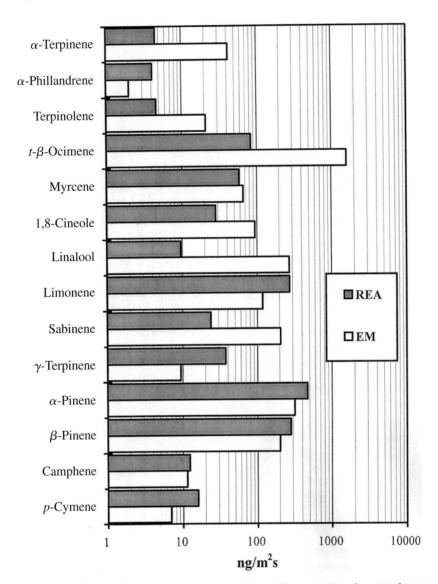

FIGURE 5   Differences between emission rates and fluxes of biogenic VOCs determined over a pine–oak forest in Castelporziano (Italy) August 3, 1994. Emission rates were calculated from soil and branch enclosure data (EM) whereas fluxes were measured by relaxed eddy accumulation (REA), data redrawn and adapted from Valentini *et al.* (1997).

amounts predicted by model calculations. When detected, it was present in the particle phase (Yokouchi and Ambe, 1985). Lahaniati *et al.* (1997) reported that 10 to 40% of α-pinene can be converted into the aerosol phase as a function of the initial α-pinene/ozone ratio and only 50% of the initial concentration goes to pinonaldehyde. Lower yields were reported by Hoffmann *et al.* (1997) who showed, however, that gas to particle conversion is a common degradation route for mono- and sesquiterpenes. It certainly accounts for the large removal of β-caryophyllene from the atmosphere.

Aerosol formation is not, however, the only route blocking the propagating chain leading to ozone formation. Data reported in Figs. 3 and 4 suggest that linalool and limonene should display a similar behavior if photochemical reactions are responsible for their removal. Results reported in Fig. 5 and field observations made in Burriana (Seufert, 1997) consistently indicate that the removal of linalool (ca. 90%) is always much higher than that of limonene (ca. 50%). This difference can be explained only if heterogeneous processes (i.e., partition into water droplets, adsorption on particles) become effective in the selective removal of terpenes from the atmosphere. The presence of the hydroxyl group in the molecule does, indeed, make linalool more soluble in water and more strongly adsorbed onto solid surfaces than limonene. In the real world these processes may play an important role in controlling ozone production because they remove from gas-phase oxidation heavy polar substances that are emitted by vegetation or formed by the photochemical degradation of large hydrocarbons.

The case of linalool cited earlier is not merely academic because this monoterpene is synthesized by many plants during blossoming. It is particularly abundant in the various compartments of the flower in which a specific enzyme for its production (linalool synthase) has been isolated (Pichersky *et al.*, 1994). Together with 6-methyl-5-hepten-2-one, linalool is responsible for the intense smell of orange and jasmine flowers. Its emission during the flowering season largely exceeds that of any other mono- and sesquiterpene (Arey *et al.*, 1991a; BEMA Report, 1997b). Finally, its photochemical degradation leads to volatile carbonyls and 6-methyl-5-hepten-2-one.

Reactive components other than isoprene and monoterpenes are emitted by plants. Among them are 2-methyl-3-buten-2-ol and 2-methyl-3-buten-1-ol. Reactivity data are available only for the former component (Fantechi *et al.*, 1996). Results suggest that 2-methyl-3-buten-2-ol should act as a good ozone producer because it reacts rapidly with OH radicals and ozone to produce acetaldehyde and acetone. *cis*-3-Hexen-1-ol and *cis*-3-hexenyl acetate (Arey *et al.*, 1991b) are also reactive products but they are only released when the plant is injured.

V. CONCLUSIONS

In urban and suburban areas where vehicular emissions are the dominant sources of hydrocarbons, olefins, and arenes are the most reactive organic components. Together, they may account for about 80% of the ozone production (Sagebiel *et al.*, 1996). Particularly important are the C_4 to C_8 branched and cyclic olefins and dienes, and aromatics from xylenes to tetramethylbenzenes. Based on the emission data reported by Zielinska *et al.* (1996) and the results reported in Table 1, the semivolatile fraction from C_8 to C_{20} can be important in producing ozone in areas (such as many European cities) where diesel emissions account for a significant portion (20 to 40%) of the total hydrocarbon emissions.

Isoprene, terpenes, and some oxygenated VOCs are the most reactive components present in forest–rural and remote areas. However, an accurate assessment of their contribution to ozone production is prevented by the present uncertainties in their emission, reactivity, and removal rates. The largest uncertainties exist for oxygenated VOCs for which sources are still undefined. Investigations should focus on 6-methyl-5-hepten-2-one whose reactivity toward OH radicals and ozone is comparable with that of the most reactive monoterpenes displayed in Figs. 3 and 4. Laboratory studies have clearly shown that photochemical decomposition of 6-methyl-5-hepten-2-one leads to substantial production of acetone, pentanal-2-one, and OH radicals (Smith *et al.*, 1996), and its oxidation could be partially responsible for the substantial amounts of acetone present in the atmospheric boundary layer (Singh *et al.*, 1995).

REFERENCES

Allegrini, I., Masia, P., and Sparapani, R. (1993). Measurement of hydrogen chloride in gas phase and particulate chlorides by means of a combination of diffusion denuders and filter packs. *J. Aerosol. Sci.* **24**, 5581–5582.

Allegrini, I., Febo, A., Giusto, M., and Giliberti, C. (1995). Monitoring of benzene in urban areas: the experience of Venice. "Proceedings of the World-Wide Symposium on the Pollution in Large Cities" Padova Fiere, Padova, Italy, pp. 231–240.

Anfossi D., Sandroni, S., and Viarengo, S. (1991). Tropospheric ozone in the nineteenth century: the Moncalieri series. *J. Geophys. Res.* **96**, 17349–17352.

Arey, J., Corchnoy, S. B., and Atkinson, R. (1991a). Emission of linalool from Valencia orange blossoms and its observation in ambient air. *Atmos. Environ.* **25A**, 1377–1381.

Arey, J., Winer, A. M., Atkinson, R., Aschmann, M. S., Long, W. D., and Morrison, C. L. (1991b). The emission of (Z)-3-hexen-1-ol, (Z)-3-hexenylacetate and other oxygenated hydrocarbons from agricultural plant species. *Atmos. Environ.* **25A**, 1063–1075.

Atkinson, R. (1990). Gas-phase tropospheric chemistry of organic compounds: a review. *Atmos. Environ.* **24A**, 1–41.

Atkinson, R. (1994). Gas-phase tropospheric chemistry of organic compounds. *J. Phys. Chem. Ref. Data.* Monograph No. **2**, 1–216.

Atkinson, R., Aschmann, S. M., Winer, A. M., and Pitts, J. N., Jr. (1985). Kinetics and atmospheric implications of the gas-phase reactions of NO_3 radicals with a series of monoterpenes and related organics at 294 ± 2K. *Environ. Sci. Technol.* **19**, 159–163.

Atkinson, R., Aschmann, S. M., Tuazon, E. C., Arey, J., and Zielinska, B. (1989). Formation of 3-methylfuran from the gas-phase reaction of OH radicals with isoprene and the rate constant for its reaction with the OH radical. *Int. J. Chem. Kinet.* **21**, 593–604.

Atkinson, R., Hasegaua, D., and Aschmann, S. M. (1990a). Rate constants for the gas-phase reactions of O_3 with a series of monoterpenes and related compounds at 296 ± 2°K. *Int. J. Chem Kinet.* **22**, 871–887.

Atkinson, R., Aschmann, S. M., and Arey, J. (1990b). Rate constants for the gas-phase reactions of OH and NO_3 radicals and O_3 with sabinene and camphene at 296 ± 2K. *Atmos. Environ.* **24A**, 2647–2654.

Atkinson, R., Aschmann, S. M., Arey, J., and Shorees, B. (1992). Formation of hydroxyl radicals in the gas phase reactions of ozone with a series of monoterpenes. *J. Geophys. Res.* **97**, 6065–6073.

Atkinson, R., Baulch, D. L., Cox, R. A., Hampson, R. F., Jr., Kerr, J. A., Rossi, M. J., and Troe, J., (1996). Evaluated kinetic and photochemical data for atmospheric chemistry: supplement 5. *Atmos. Environ.* **30**, 3903–3904 (summary table on diskette).

Bertin, N., Staudt, M., Hansen, U., Seufert, G., Ciccioli, P., Foster, P., Fugit, J. L., and Torres, L. (1997). Diurnal and seasonal coarse of monoterpene emissions from *Quercus ilex* (L.) under natural conditions. Application of the light and temperature algorithms. *Atmos. Environ.* **31**(S1), 135–144.

Biogenic Emissions in the Mediterranean Area (BEMA) Report (1994). (G. Enders, ed.), Brussels, Belgium, EUR Report 15955 EN.

Biogenic Emissions in the Mediterranean Area (BEMA) Report (1995). (D. Kotzias, M. Staudt, and S. Cieslik, eds.), Brussels, Belgium, EUR Report 16293 EN.

Biogenic Emissions in the Mediterranean Area (BEMA) Report (1996). (C. Coer, V. Jacob, P. Foster, L. Torres, D. Kotzias, S. Cieslik, and B. Versino, eds.), Brussels, Belgium, EUR Report 16449 EN.

Biogenic Emissions in the Mediterranean Area (BEMA) Report (1997a). (G. Seufert, M. Sanz, and M. Millan, eds.), Brussels, Belgium, EUR Report 17305 EN.

Biogenic Emissions in the Mediterranean Area (BEMA) Report (1997b). (G. Seufert, ed.), Brussels, Belgium, EUR Report 17336 EN.

Bode, K., Helas, G., and Kesselmeier, J. (1997). Biogenic contribution to atmospheric organic acids. In "Biogenic Volatile Organic Compounds in the Atmosphere." (G. Helas, J. Slanina, and R. Steinbrecher, eds.), SPB Academic Publishing: Amsterdam, Netherlands, pp. 157–170.

Brocco, D., Petricca, M., and Ventrone, I. (1994). Formazione dello smog fotochimico nell'area urbana di Roma. *Acqua Aria.* **3**, 217–224 (in Italian).

Brocco, D., Fratarcangeli, R., Lepore, L., Petricca, M., and Ventrone, I. (1997). Determination of aromatic hydrocarbons in urban air of Rome. *Atmos. Environ.* **31**, 557–566.

Calvert, J. G., and Madronich, S. (1987). Theoretical study of the initial products of the atmospheric oxidation of hydrocarbons. *J. Geophys. Res.* **92**, 2211–2220.

Cantuti, V., Ciccioli, P., Cecinato, A., Branacleoni, E., Brachetti, A., Frattoni, M., and Di Palo, V. (1994). PAN nella Valle del Tevere. In "Proceedings of the 1st Italian Symposium on the Strategies and Techniques for the Monitoring of the Atmosphere," (P. Ciccioli, ed.) Società Chimica Italiana: Rome, pp. 137–145 (in Italian).

Carlier, P., Hannachi, H., and Mouvier, G. (1986). The chemistry of carbonyl compounds in the atmosphere—a review. *Atmos. Environ.* **20**, 2079–2099.

Carter, W. P. (1994). Development of ozone reactivity scales for volatile organic compounds, *J. Air Waste Manage. Assoc.* **44**, 881–899.

Cecinato, A., Possanzini, M., Liberti, A., and Brocco, D. (1979). Photochemical oxidants in the Rome metropolitan area. *Sci. Total Environ.* **13**, 1–8.

Chameides, W., Lindsay, R. W., Richardson, J., and Kiang, C. S. (1988). The role of biogenic hydrocarbons in urban photochemical smog: Atlanta a case study. *Science.* **241**, 1473–1475.

Chock, D. P., and Heuss, J. M. (1987). Urban ozone and its precursors. *Environ. Sci. Technol.* **21**, 1146–1153.

Ciccioli, P. (1993). VOCs and Air Pollution. *In* "Chemistry and Analysis of Volatile Organic Compounds in the Environment," (H. J. Th. Bloemen, and J. Burn, eds.), Blackie Academic & Professional: Glasgow, UK, pp. 92–174.

Ciccioli, P., and Cecinato A. (1992). Advanced methods for the evaluation of atmospheric pollutants relevant to photochemical smog pollution and acid deposition. *In* "Gaseous Pollutants: Characterization and Cycling." J. O. Nriagu, ed.,John Wiley & Sons: New York.

Ciccioli, P., Cecinato, A., Brancaleoni, E., and Frattoni, M. (1992). Use of carbon adsorption traps combined with high-resolution gas chromatography-mass spectrometry for the analysis of polar and non-polar C_4–C_{14} volatile organic compounds involved in photochemical smog formation. *J. High Resolut. Chromatogr. Chromatogr. Commun.* **15**, 75–84.

Ciccioli, P., Brancaleoni E., Cecinato, A., and Frattoni, M. (1993a). A method for the selective identification of volatile organic compounds (VOC) in air by GC-MS. *In* "Proceedings of the 15th Symposium on Capillary Chromatography," (P. Sandra and G. Devos, eds.), Vol. 2, Dr. Huetig and Verlag: Heidelberg, Germany, pp. 1029–1042.

Ciccioli, P., Brancaleoni, E., Frattoni, M., Cecinato, A., and Brachetti, A. (1993b). Ubiquitous occurrence of semi-volatile carbonyl components in tropospheric samples and their possible sources. *Atmos. Environ.* **27A**, 1891–1901.

Ciccioli, P., Brancaleoni, E., Cecinato, A., Sparapani, R., and Frattoni, M. (1993c). Identification and determination of biogenic and anthropogenic volatile organic compounds in forest areas of Northern and Southern Europe and a remote site of the Himalaya region by high-resolution gas chromatography-mass spectrometry. *J. Chromatogr.* **643**, 55–69.

Ciccioli, P., Cecinato, A., Brancaleoni, E., Brachetti, A., Frattoni, M., and Sparapani, R. (1994a). Composition and distribution of polar and non-polar VOCs in urban, rural, forest and remote areas. *In* "Physico-Chemical (G. Angeletti and G. Restelli, eds.), Behaviour of Atmospheric Pollutants. Proceedings of the 6th European Symposium." EUR Report 15609/1. European Commission-DG XII. Brussels, Belgium, Vol. 1, pp. 549–568.

Ciccioli, P., Cecinato, A., Brancaleoni, E., Montagnoli M., and Allegrini, A. (1994b). Chemical composition of particulate organic matter (POM) collected at Terra Nova Bay in Antarctica. *Int. J. Environ. Anal. Chem.* **55**, 47–59.

Ciccioli, P., Cecinato, A., Brancaleoni, E., Frattoni, M., Zacchei, P., Miguel, A. H., and de Castro Vasconcellos, P. (1996a). Formation and transport of 2-nitrofluoranthene and 2-nitropyrene of photochemical origin in the troposphere. *J. Geophys. Res.* **101**, 19567–19581.

Ciccioli, P., Cecinato, A., Brancaleoni, E., Frattoni, M., Bruner, F., and Maione, M. (1996b). Occurrence of oxygenated volatile organic compounds (VOC) in Antarctica. *Int. J. Environ. Anal. Chem.* **62**, 245–253.

Derwent, R. G., and Jenkin, M. E. (1991). Hydrocarbons and the long-range transport of ozone and PAN across Europe. *Atmos. Environ.* **25A**, 1661–1678.

Derwent, R. G., and Jenkin, M. E., and Saunders, S. M., (1996). Photochemical ozone creation potentials for a large number of reactive hydrocarbons under European conditions. *Atmos. Environ.* **30**, 181–199.

Fantechi, G., Jensen, N. R., Hjorth, J. J., and Peeters, J. (1996). Mechanistic study of the atmospheric degradation of isoprene and MBO. In "Proceedings from the EUROTRAC Symposium '96," (P. M. Borrell, P. Borrell, T. Cvitas, K. Kelly, and W. Seiler, eds.), Vol. 2, Computational Mechanics Publications: Southhampton, UK, pp. 605–609.

Febo, A. (1994). Nitrous acid in the oxidation processes in urban areas. In "Physico-Chemical Behaviour of Atmospheric Pollutants. Proceedings of the 6th European Symposium." (G. Angeletti and G. Restelli, eds.), EUR Report 15609/1. European Commission-DG XII. Brussels, Belgium, pp. 277–292.

Febo, A., Perrino, C., and Giliberti, C. (1997). An interpretation of formaldehyde time evolution in urban areas. In "The oxidizing capacity of the atmosphere. Proceedings of the 7th European Symposium on the Physico-Chemical Behaviour of Atmospheric Pollutants." (B. Larsen, B. Versino, and G. Angeletti, eds.), EUR Report EUR 17482. European Commission-DGXII, Brussels, Belgium, pp. 467–471.

Fehsenfeld, F., Calvert, J., Fall, R., Goldan, P., Guenther, A. B., Hewitt, C. N., Lamb, B., Liu, S., Trainer, M., Westberg, H., and Zimmerman, P. (1992). Emissions of volatile organic compounds from vegetation and the implications for atmospheric chemistry. Global Geophys. Cycles. 6, 389–430.

Finlayson-Pitts, B. J., and Pitts, J. N., Jr. (1986). "Atmospheric Chemistry: Fundamentals and Techniques," John Wiley & Sons: New York.

Finlayson-Pitts, B. J., and Pitts, J. N., Jr. (1997). Tropospheric air pollution: ozone, airborne toxics, polycyclic aromatic hydrocarbons, and particles. Science. 276, 1045–1052.

Fruekilde, P., Hjorth, J., Jensen, N. R., Kotzias, D., and Larsen, B. (1997). Ozonolysis at vegetation surfaces: a source of acetone, 4-oxopentanal, 6-methyl-5-hepten-2-one and geranyl acetone in the troposphere. Atmos. Environ. 32, 1893–1902.

Goldan, P. D., Kuster, W. C., Fehsenfeld, F. C., and Montzka, S. A. (1993). The observation of a C_5 alcohol emission in a North American pine forest. Geophys. Res. Lett. 20, 1039–1042.

Grennfelt, P., Saltbones, J., and Schjoldager, J. (1988). Oxidant data collection in OECD Europe 1985–1987 (OXIDATE). Report on ozone, nitrogen dioxide and peroxyacetyl nitrate. October 1985–March, April, September 1986. Norwegian Institute for Air Research, Lillestrom, Norway, NILU Report OR 31/88.

Grosjean, D., and Fung, K. (1984). Hydrocarbons and carbonyls in the Los Angeles air. J. Air Pollut. Control. Assoc. 34, 537–543.

Grosjean, E., and Grosjean, D. (1996). Carbonyl products of the gas-phase reaction of ozone with 1-alkenes. Atmos. Environ. 24, 4107–4113.

Grosjean, E., and Grosjean, D. (1997). Gas-phase reactions of alkenes with ozone: formation yelds of primary carbonyls and biradicals. Environ. Sci. Technol. 31, 2421–2427.

Guenther, A., Hewitt, C. N., Erickson, D., Fall, R., Geron, C., Graedel, T., Harley, P., Klinger, L., Lerdau, M., McKay, W. A., Pierce, T., Scholes, B., Steinbrecher, R., Tallamraju, R., Taylor, J., and Zimmerman, P. (1995). A global model of natural volatile compound emission. J. Geophys. Res. 100, 8873–8892.

Guenther, A., Zimmerman, P., Klinger, L., Greenberg, J., Ennis, C., Davis, K., Pollock, W., Westberg, H., Allwine, G., and Geron, C. (1996). Estimates of regional natural volatile organic compounds fluxes from enclosure and ambient measurements. J. Geophys. Res. 101, 1345–1359.

Hansen, U., Van Eijk, J., Bertin, N., Staudt, M., Kotzias, D., Seufert, G., Fujit, J. L., Torres, L., Cecinato, A., Brancaleoni, E., Ciccioli, P., and Bomboi, M. M. T. (1997). Biogenic emissions and CO_2 gas exchange investigated on four Mediterranean shrubs. Atmos. Environ. 31(SI), 157–166.

Haagen-Smit, A. J., and Fox M. M. (1956). Ozone formation in photochemical oxidation of organic substances. Ind. Eng. Chem. 48, 1484–1487.

Harrison, R. M., Peak, J. D., and Collins, G. M. (1996). Tropospheric cycle of nitrous acid. J. Geophys. Res. 101, 14429–14439.

Heiden, A. C., Büscher, St., Gabler R., Hoffmann Th., Kahl J., Kobel K., Kolahgar B., Kühnemann, F., Langebartels, Ch., Rockel, P., Rudolph, J., Schraudner, H., Schuh, G., Wedel, A., Wildt, J., and Wolters, H. (1997). Emissions of volatile organic compounds from Tobacco Plants (B. Larsen, B. Versino and G. Angeletti, eds.). The oxidizing capacity of the atmosphere. Proceedings of the 7th European Symposium on the physico-chemical behaviour of Atmospheric Pollutants. EUR Report EUR 17482. European Commission-DGXII, Brussels, Belgium, pp. 462–466.

Helmig, D., and Greenberg, J. P. (1994). Automated *in situ* gas chromatographic-mass spectrometric analysis of ppt level volatile organic trace gases using multistage solid-adsorbent trapping. *J. Chromatogr.* 677, 123–132.

Helmig, D., Pollock, W., Greenberg, J., and Zimmerman, P. (1996). Gas chromatographic mass spectrometry analysis of volatile organic trace gases at Mauna Loa Observatory, Hawaii. *J. Geophys. Res.* 101, 14697–14710.

Hoffmann, T., Odum, J., Bowman, F., Collins, D., Klockow, D., Flagan, R. C., and Seinfeld, J. H. (1997). Organic aerosol formation from biogenic hydrocarbon precursors. In "The oxidizing capacity of the atmosphere. Proceedings of the 7th European Symposium on the physico-chemical behaviour of atmospheric pollutants." (B. Larsen, B. Versino, and G. Angeletti, eds.) EUR Report EUR 17482. European Commission-DGXII, Brussels, Belgium, pp. 537–541.

Isidorov, V. A. (1992). Non-methane hydrocarbons in the atmosphere of boreal forests: composition, emission rates, estimation of regional emission and photo catalytic transformation. *Ecol. Bull.* 42, 71–75.

Kesselmeier, J., Schafer, L., Ciccioli, P., Brancaleoni, E., Cecinato, A., Frattoni, M., Foster, P., Jacob, V., Denis, J., Fugit, J. L., Dutaur, L., and Torres, L. (1996). Emission of monoterpenes and isoprene from a Mediterranean oak species *Quercus ilex L.* measured within BEMA (Biogenic Emissions in the Mediterranean area) project. *Atmos. Environ.* 30A, 1841–1850.

Kley, D. (1997). Tropospheric chemistry and transport, *Science.* 276, 1043–1045.

Konig, G., Brunda, M., Puxbaum, H., Hewitt, C. N., Duckam, S. G., and Rudolph, J. (1995). Relative contribution of oxygenated hydrocarbons to the total biogenic VOC emissions of selected mid-European agricultural and natural plants. *Atmos. Environ.* 29, 861–874.

Knoeppel, H., and Schauenburg, H. (1989). Screening of house hold products for the emission of volatile organic compounds. *Environ. Int.* 15, 413–418.

Lahaniati, M., Nanos, Ch., Kodinari, C., Niccolin, B., Petrucci, G. A., and Kotzias, D. (1997). Formation of aerosols in the gas phase ozonolysis of α-pinene. In "The oxidizing capacity of the atmosphere. Proceedings of the 7th European Symposium on the Physico-Chemical Behaviour of Atmospheric Pollutants." (B. Larsen, B. Versino, and G. Angeletti, eds.), EUR Report EUR 17482. European Commission-DGXII, Brussels, Belgium, pp. 542–546.

Laurila, T., and Hakola, H. (1996). Seasonal cycle of C_2–C_5 hydrocarbons over the Baltic Sea and Northern Finland. *Atmos. Environ.* 30, 1597–1607.

Loreto, F., Ciccioli, P., Brancaleoni, E., Cecinato, A., Frattoni, M., and Sharkey, T. D. (1996). Different sources of reduced carbon contribute to form three classes of terpenoid emitted by *Quercus ilex* L. leaves. *Proc. Natl. Acad. Sci. U.S.A.* 93, 9966–9969.

McDonald, R. C., and Fall, R. (1993). Detection of substantial emissions of methanol from plants to the atmosphere. *Atmos. Environ.* 27A, 1709–1713.

McKay, W. A., Turner, M. F., Jones, M. R., and Halliwell, C. M. (1996). Emissions of hydrocarbons from marine phytoplankton-some results from controlled laboratory experiments. *Atmos. Environ.* 30, 2583–2593.

Millan, M. M., Salvador, R., Mantilla, E., and Artinanao, A. (1996). Meteorology of photochemical air pollution in Southern Europe: experimental results from EC research projects. *Atmos. Environ.* 30, 1909–1924.

Millan, M. M., Salvador, R., and Mantilla, E. (1997). Tropospheric processes in the Mediterranean basin relevant to Global Change. In "The oxidizing capacity of the atmosphere. Proceedings of the 7th European Symposium on the physico-chemical behaviour of atmospheric pollutants." (B. Larsen, B. Versino, and G. Angeletti, eds.), EUR Report EUR 17482. European Commission-DGXII, Brussels, Belgium, pp. 406–410.

Neftel, A., and Staffelbach, T. (1997). Ozone in Europe: VOC or NOX limited? In "The oxidizing capacity of the atmosphere. Proceedings of the 7th European Symposium on the physico-chemical behaviour of atmospheric pollutants." (B. Larsen, B. Versino and G. Angeletti, eds.), EUR Report EUR 17482. European Commission-DGXII, Brussels, Belgium, pp. 391–395.

Nelson, P. F., and Quigley, S. M. (1983). The m,p-xylenes:ethylbenzene ratio. A technique for estimating hydrocarbon age in ambient atmosphere. Atmos. Environ. 17, 659–662.

O'Brien, J. M., Shepson, P. B., Wu, Q., Biesenthal, T., Bottenheim, J. W., Wiebe, H. A., Anlauf, K. G., and Brickell, P. (1997). Production and distribution of organic nitrates and their relationship to carbonyl compounds in an urban environment. Atmos. Environ. 31, 2059–2069.

Owen, S., Boissard, C., Street, R. A., Duckham, C. S., Csiky, O., and Hewitt, C. N. (1997). Screening of 18 Mediterranean plant species for volatile organic compound emissions. Atmos. Environ. 31(SI), 101–117.

Paulson, S. E., and Orlando, J. J. (1996). The reactions of ozone with alkenes: an important source of HO_x in the boundary layer. Geophys. Res. Lett. 23, 3727–3730.

Photochemical Oxidants Review Group (PORG) Third Report (1993). Ozone in the United Kingdom 1993. Department of the Environment, Air Quality Division. London, UK.

Pichersky, E., Raguso, R. A., Lewinsohn, E., and Croteau, R. (1994). Floral scent production in Clarkia (Onagraceae). I. Localization and developmental modulation of monoterpene emission and linalool synthase activity. Plant Physiol. 106, 1533–1540.

Ravishankara, A. R., (1997). Heterogeneous and multiphase chemistry in the troposphere. Science. 276, 1058–1065.

Rogge, W. F., Hildemann, M. L., Mazurek, M. A., and Cass, G. R. (1991). Sources of fine aerosols. I. Char broilers and meat cooking operations. Environ. Sci. Technol. 22, 1112–1125.

Roberts, J. M., Fehsenfeld, C. F., Liu, S. C., Bollinger, M. J., Hahn, C., Albritton, D. L., and Sievers, R. E. (1984). Measurements of aromatic hydrocarbon ratios and NO_x concentrations in the rural troposphere: observation of air mass photochemical aging and NO_x removal. Atmos. Environ. 18, 2421–2432.

Roberts, J. M., Williams, J., Bertman, S. B., Williams, E., Baumann, K., Buhur, M. P., Ryerson, T. B., Trainer, M., Hubler, G., and Fehesenfeld, F. C. (1997). Using MPAN, PPN, and PAN as probes for regional ozone production from biogenic hydrocarbons. Can we reconcile observations, mechanisms and models? In "Workshop on Biogenic Hydrocarbons in the Atmospheric Boundary Layer." (J. D. Fuentes, ed.), AMS, Charlottesville, VA, p. 142.

Sagebiel, J. C., Zielinska, B., Pierson, W. R., and Gertler, A. W. (1996). Real-world emissions and calculated reactivities of organic species from motor vehicles. Atmos. Environ. 12, 2287–2296.

Satsumabayashi, H., Kurita, H., Chang, Y. S., Carmichael, R., and Ueda, H. (1992). Diurnal variation of OH radical and hydrocarbons in polluted air mass during long-range transport in Central Japan. Atmos. Environ. 26A, 2835–2844.

Seila, R. L. (1984). Atmospheric volatile hydrocarbon composition in five remote sites in northwestern North Carolina. In "Environmental Impact of Natural Emissions." (V. Aneja, ed.), Air Pollut. Control Assoc., Pittsburgh, PA, pp. 125–140.

Seufert, G. (1997). The BEMA-Project on biogenic emissions in the Mediterranean area-overview and some aspects of relevance to air chemistry. In "The oxidizing capacity of the atmosphere. Proceedings of the 7th European Symposium on the physico-chemical behaviour of atmospheric pollutants." (B. Larsen, B. Versino, and G. Angeletti, eds), EUR Report EUR 17482. European Commission-DGXII, Brussels, Belgium, pp. 260–269.

Shu, Y., Kwok, E. S. C., Tuazon, E. C., Atkinson, R., and Arey, J. (1997). Products of the gas-phase reactions of linalool with OH radicals, NO_3 radicals and O_3. Environ. Sci. Technol. 31, 896–904.

Shu, Y., and Atkinson, R. (1995). Atmospheric lifetimes and fates of a series of sesquiterpenes. *J. Geophys. Res.* **100**, 7275–7281.

Simpson, D., Guenther, A., Hewitt, C. N., and Steinbrecher, R. (1995). Biogenic emissions in Europe. I. Estimates and uncertainties. *J. Geophys. Res.* **100**, 506–512.

Singh, H. B., and Zimmerman, P. B. (1992). Atmospheric distribution and sources of non-methane hydrocarbons. *In* "Gaseous Pollutants: Characterization and Cycling." (J. O. Nriagu ed.), John Wiley & Sons: New York.

Singh, H. B., Kanakidou, M., Crutzen, P. J., and Jacob, D. J. (1995). High concentrations and photochemical fate of oxygenated hydrocarbons in the global troposphere. *Nature.* **378**, 50–54.

Smith, A. M., Rigler, E., Kwok, E. S. C., and Atkinson, R. (1996). Kinetics and products of the gas-phase reactions of 6-methyl-5-hepten-2-one and *trans*-cinnamaldehyde with OH and NO$_3$ radicals and O$_3$ at 296 + 2°K. *Environ. Sci. Technol.* **30**, 1781–1785.

Staudt, M., Bertin, N., Hansen, U., Seufert, G., Ciccioli, P., Foster, P., Frenzel, B. and Fugit, J. L., (1997). Diurnal and seasonal pattern of monoterpene emissions from *Pinus Pinea* (L.) under field conditions. *Atmos. Environ.* **31**(SI), 145–156.

United Nations—Economic Commission for Europe (UN-ECE). (1991). Protocol for the 1979 convention of long-range transboundary air pollution concerning the control of emission of volatile organic compounds and their transboundary fluxes. EB. AIR/R54, Geneva, Switzerland, November 1991.

U.S. Environmental Protection Agency (U.S. EPA). (1986). Air Quality Criteria for Photochemical Oxidants, Vol. 2 to 5, U.S. Environmental Protection Agency, Environmental Criteria and Assessment Office, EPA Reports No. EPA-600/8-84/020 d, e and f. Research Triangle Park, Raleigh, NC.

Valentini, R., Greco, S., Seufert, G., Bertin, N., Hansen, U., Ciccioli, P., Cecinato, A., Brancaleoni, E., and Frattoni, M. (1997). Fluxes of biogenic VOC from Mediterranean vegetation by trap enrichment relaxed eddy accumulation. *Atmos. Environ.* **31**(SI), 229–238.

Volz, A., and Kley, D. (1988). Evaluation of the Montsouris series of ozone measurements made in the nineteenth century. *Nature.* **332**, 240–242.

Winer, A. M., Arey, J., Atkinson, R., Aschmann, S. M., Long, W. D., Morrison, C. L., and Olszyk, D. M. (1992). Emission rates of organics from vegetation in California's Central Valley. *Atmos. Environ.* **26A**, 2647–2659.

World Health Organization (WHO). (1986). Air quality Guidelines. Review Draft. Regional Office of Europe, Copenhagen.

Yeo, H. C. H., and Shibamoto, T. (1992). Formation of formaldehyde and melonaldehyde by photo oxidation of squalene. *Lipids.* **27**, 50–53.

Yokouchi, Y. (1994). Seasonal and diurnal variation of isoprene and its reaction products in a semi-rural area. *Atmos. Environ.* **28**, 2651–2658.

Yokouchi, Y., and Ambe, Y. (1985). Aerosols formed from the chemical reaction of monoterpenes and ozone. *Atmos. Environ.* **19**, 1271–1276.

Yokouchi, Y., Mukai, H., Nakajima, K., and Ambe, Y. (1990). Semi-volatile aldehydes as predominant organic gases in remote areas. *Atmos. Environ.* **24A**, 439–442.

Zielinska, B., Sagebiel, J. C., Harshfield, G., Gertler, A. W., and Pierson, W. R. (1996). Volatile organic compounds up to C$_2$O emitted from motor vehicles; measurement methods. *Atmos. Environ.* **30**, 2269–2286.

CHAPTER 6

Global Distribution of Reactive Hydrocarbons in the Atmosphere

B. BONSANG * AND C. BOISSARD†

*Laboratoire des Sciences du Climat et de L'Environnement CE Soclay, Gif-Sur-Yvette Cédex, France
†LISA, Laboratoire Interuniversitaire des Systèmes Atmosphériques, Universités Paris 7-12, Créteil, France

This chapter focuses on the global distributions of the C_2 to C_{10} nonmethane hydrocarbons (NMHCs) and derived light oxygenated species. This group of reactive volatile organic compounds (VOCs) has been extensively measured since the early 1970s, mainly in the boundary layer of continental areas, half of these measurements being carried out over North America. A much smaller number of measurements have been made in the marine troposphere or at altitude. Due to relatively large areas where no data are available and to the high reactivity of these compounds, leading to scattered data, global latitudinal and seasonal distributions can be well described only for species, such as ethane, propane, and acetylene, with lifetimes on the order of several weeks or months. These distributions reveal, at ground level as well as in the free troposphere, a clear strong interhemispheric gradient, with ambient concentrations at mid-northern latitudes one to two orders of magnitude higher than in the Southern Hemisphere. The seasonality of the species distributions is due to the variability of both their sources and their major atmospheric sink (OH radicals). This leads, in the Northern Hemisphere, to maximum concentrations during winter. In addition, the species distributions reveal the influence of long-range transport of species from nonlocal sources to Arctic or inhabited area. These species can thus be used as tracers for specific sources. The situation is much more complex for more reactive species, such as higher ($> C_4$) alkanes or alkenes, and biogenic NMHCs (isoprene and terpenes). In the free troposphere high concentrations of even highly reactive NMHCs have been reported. Although the sources of biogenic compounds are believed to be dominant among the reactive VOC budgets, their high reactivity (lifetimes of minutes to hours) and the strong dependence of their sources on photosynthesis processes, lead to a tremendous variability in their ambient concentrations.

I. INTRODUCTION

Since about 1970, nonmethane hydrocarbons (NMHCs) have been extensively measured in different atmospheric environments, including in the background atmosphere. NMHCs consist of a variety of compounds: straight-chain hydrocarbons, such as alkanes, alkenes, and alkynes; polyunsaturated species, such as isoprene and terpenes; and aromatic or polyaromatic species, such as toluene and benzene. Their global distributions are the result of both their emission sources, and their chemical reactivities that vary over a wide range giving lifetimes from several months for the most stable species (such as ethane) to hours or less for the very reactive species (such as terpenes). In addition, transport processes tend to redistribute these species within the troposphere, mainly by the influence of interhemispheric, zonal, and vertical exchanges. The major anthropogenic sources of NMHCs are related to transport and chemical and petrochemical industries including organic solvent production. They are responsible for the production of aromatic and saturated hydrocarbons, which represent the largest fraction, and, to a lesser extent, of unsaturated species. Ethene and propene are also generated during the combustion of organic matter and are produced by gasoline-powered vehicle exhausts.

In rural areas the concentrations of unsaturated hydrocarbons may be increased by the mixing of emissions from soil and vegetation. This is particularly the case for isoprene and terpenes (see e.g., Zimmerman, 1979a; Müller, 1992; Singh and Zimmerman, 1992). In forest areas the contribution of natural sources to the overall amount of hydrocarbons in the atmosphere becomes dominant, leading to a strong and very significant effect on photochemistry on local, regional, or even global scale (Zimmerman, 1979a; Müller, 1992; Lübkert and Shoepp, 1989; Lamb *et al.*, 1987). Far from industrialized, urban, and suburban areas, biomass burning can also be a major source of unsaturated hydrocarbons that can significantly change the relative composition of trace gases (see, e.g., Rasmussen and Khalil, 1988; Bonsang *et al.*, 1995).

Of the large group of reactive hydrocarbons, this chapter will mainly focus on the C_2 to C_{10} nonmethane hydrocarbons that have a strong impact on the photochemistry of the atmosphere, and that have been sufficiently investigated to enable a global picture of their atmospheric distributions to be drawn. A short section will describe some oxygenated compounds. The anthropogenic and biogenic compounds will be separately considered, following the presentation of the other chapters of this book. However, the main interest in the investigation of biogenic hydrocarbons has been the assessment of their fluxes and emission rates, and their measurement is usually made very close to the sources and do not provide a useful set of data for obtaining global distributions. First, the vertical distributions, including the stratospheric compartment, will be presented; then the distribution within the boundary layer and, particularly, the latitudinal and seasonal variations will be described.

II. CAMPAIGNS

The ambient concentrations of hydrocarbons and other volatile organic compounds (VOCs) have been extensively measured around the world since the early 1970s, at ground level, over continents or oceans, and during vertical profile experiments, using balloons or research aircraft flights up to the lower stratosphere. Tables 1A to 1C summarize the more than 100 experiments reported in the literature. They consist of some of the major campaigns carried out since 1975 over a time period of at least one week. However, for some locations or periods where few data were available, some measurements obtained over a period of measurement shorter than one week are presented.

More campaigns have been carried out over continental areas, compared to shipborne measurements made over the oceans. Half of these ground-based measurements have been performed over North America (Table 1A) with the

TABLE 1A List of Campaigns Conducted at Ground Level[a]

Name	Date	Location	Latitude[b]	Longitude[c]	n	Compounds	Comments	Reference
Ground based								
	Jul. 1975	Chackatawbut Hill, MA	42	−72		ISO, MT	Hardwood, oak	Lonneman et al., 1977
	Jul. 16–Aug. 14, 1975	Glasgow, IL	40	−90		C_2–C_{10}, paraffinic Hc	Rural farmland	Rasmussen et al., 1977
	Aug. 21–Sep. 2, 1975	Elkton, MO	38	−90		C_2–C_{10}, paraffinic Hc		Rasmussen et al., 1976
	Apr.–May 1976	Jetmore, KA	38	−98		C_2–C_{10}, paraffinic Hc		Martinez and Singh, 1979
	May 1976	Central–South Florida	26	−82		C_2–C_{10}, paraffinic Hc	Orange groves, everglade	Lonneman et al., 1978
	Summer 1976	McKee, KT	38	−85		ISO, MT	Hardwood, oak	Lonneman et al., 1978
	Summer 1976	Keysville, VA	37	−77		C_2–C_5, alkylbenzene	Hardwood, oak	Lonneman et al., 1978
	Jan. 1978	Jones State Forest, TX	33	−96		C_2–C_{10}, paraffinic Hc	Picnic area, forest interior	Seila, 1984
	Summer 1978	Keysville, VA	37	−77		Paraffinic HC		Ferman, 1981
	Jul. 1978	Rio Blanko County, CO	38	−105		ISO	Sparce vegetation	Arnts and Meeks, 1981
	Jul.–Sep. 1978	Smoky Mountain, CO	38	−105		C_2–C_{10}, paraffinic Hc	Pine and fir	Arnts and Meeks, 1981
	Jul.–Sep. 1978	Moscow Mountain, ID	45	−106		ISO, MT	Coniferous, spruce	Holdren et al., 1979
	Aug. 1978	Gjerdum, Norway	62	10		ISO, MT		Schjoldager and Wathne, 1978
	Sep. 1978	Great Smoky Mountain, CO	38	−105		Paraffinic Hc	Coniferous and deciduous	Cronn and Harsch, 1980
	Sep. 1978	Great Smoky Mountain, CO	35	−87		ISO, MT		Arnts and Meeks, 1981
	Sep. 1978	Great Smoky Mountain, CO	35	−87		ISO, MT	Live oak, slash pine	Cronn and Harsch, 1980
	Sep. Oct. 1978	Houston, TX	30	−96		MT		Zimmerman, 1979
	Aug. 1979	York County, PA	43	−83		ISO	Oak	Quarles et al., 1980
	Jul. 1979	Mt. Kanobili, USA	42	42		ISO, MT	Valley floor above canopy	Stevens et al., 1981
	Sep. 1979, Summer 1980	Brazil	−5	−55	150	C_2–C_6, ISO, MT	Cerrado, selva	Greenberg and Zimmerman, 1984
	Jul. 13–Aug. 3, 1981	Atlanta, GA	34	−84		C_2–C_6	Urban site	Chamiedes et al., 1992
	May 1981–Nov. 1982	Niwot Ridge, CO	40	−105	160	MT	Forest, 3048 m	Roberts et al., 1983
	Spring 1982–1985	Alaska	>60			C_2–C_8	From flights?	Rasmussen et al., 1983
	Mar.–Jun. 1982	Barrow, Arctic	70		67	C_2–C_7	Ground station, 0–4 km	Rasmussen and Khalil, 1983
	1982–1989	Antarctic	−70	−8		Light ALK, C_2H_2		Rudolph et al., 1989, 1992

Campaign	Location	Date	Lat	Long	Number	Compounds	Site	Reference
	Norwegian Arctic	Spring 1983, 1985, 1986	79	12	60		Weather ship	Hove et al., 1989
	Point Arena, CA	Jan. 15–28, 1984	39	−124	237		Real-time analysis	Singh et al., 1988
	Point Arena, CA	Apr. 25–May 3, 1985	39	−124	200		Real-time analysis	Singh et al., 1988
	Amazon basin	Jul. 1985	−2	−60		ISO		Rasmussen and Khalil, 1988
	Amsterdam Island	Jan. 1984–May 1987	−37	77	40	C_2–C_5		Bonsang et al., 1990
	Amsterdam Island	Mar. 1986–May 1987	−37	77	13	C_2H_2		Kanakidou et al., 1988
	Indian Ocean	Mar. 1986–May 1987	20	−155				Kanakidou et al., 1988
	Norway	May 1987–1988	59	8	143	C_2–C_5	Rural coastal site	Hov et al., 1991
	Birkenes, Norway	May 1987–Sep. 1989	58	8	350	C_2–C_5	Rural sites	Hov et al., 1991; Hov, 1992
	Ny-Alesund, Norway	May 1987–Sep. 1989	78	4		C_2–C_5	Rural sites	Hov et al., 1991; Hov, 1992
DECAFE	Congo	Feb. 1988	3	15	50	C_2–C_6	Biomass burning influence	Rudolph et al., 1992a
EUROTRAC-TOR	Europe	Feb. 1989–Oct. 1990	28–59	−16 to 21		C_2–C_7		Lindskog, 1996
	Barrow, Alaska	Mar. 1989	72	−156	10	C_2–C_8		Doskey and Gaffney, 1992
	Jadraas, Sweden	Jun.–Aug. 1989	60	16		MT	Scots pine forest	Janson, 1992
OCEANONOX	Penmarc'h, France	Summer 1989	48	−5	43	C_2–C_6		Bertrand, 1989
EUROTRAC-TOR	Rorvik, Sweden	Spring 1989–1990	58	13		C_2–C_5		Lindskog et al., 1992
FOS/DECAFE	Lamto, Ivory Coast	Jan. 1989, 1991	6	−5	50	C_2–C_6	Savanna burnings	Bonsang et al., 1994, 1995
	Southeastern United States	Summer 1990	30	−82		VOC, ARO, BNHMC	Forested rural site	Fehsenfeld et al., 1992
ROSE	Western central Alabama	Jun. 8–Jul. 19, 1990	32	−88	1000	C_2–C_6, ISO, MT, ARO	Loblolly pine	Goldan et al., 1995
	Fraserdale, Canada	Apr. 1990–Oct. 1992	50	−82		C_2–C_5, ISO	Boreal forest (fir, spruce, birch)	Jobson et al., 1994
	Canada	Nov. 1990–Dec. 1991	40	−65 to −125	400	C_2–C_6	Rural sites	Bottenheim and Shepherd, 1995
	Edmonton, Canada	1991–1993	54	−114	100	C_2–C_{12}	Urban sites	Cheng et al., 1997
	Landes, France	Jun. 1992	44	0		MT	Maritime pine forest	Simon et al., 1994
SAFARI	Southern Africa	Jun.–Dec. 1992	−30	20		Trace gases	From fires and soils	J. Geophys. Res., 101, Oct/1996 special issue
	Rome, Italy	1992–1993	42	12		C_2–C_6, ARO	Urban site	Brocco et al., 1997
	Ontario and Saskatchewan	1992–1995	50–54	−82 to −106	130	C_2–C_6, ISO	Boreal forest	Young et al., 1997
	Alaska	Mar.–May 1993	64	−147	35	Total C_2–C_5	Boreal forest	Beine et al., 1996
	Ontario, Canada	Apr.–Nov. 1993	44	−81		ISO, MT	Deciduous forest	Fuentes et al., 1996

continued

TABLE 1A (continued)

Name	Date	Location	Latitude[b]	Longitude[c]	n	Compounds	Comments	Reference
BEMA	Jun. 1–9, 1993	Castelporziano, Italy	42	12	Hundreds	BVOC		Atmos. Environ. 31, dec 1997, special issue
	Aug. 1993	Raleigh, NC	36	–77	~75	C_2–C_{10}	0–450 m	Lawrimore et al., 1995
	Aug. 26–Oct. 1, 1993	Colorado Mountains	35–40	–110	530	C_3–C_{10}, C_1–C_5	Rural site	Goldan et al., 1997
Tropical OH Photo-chemical Expedition	Aug.–Oct. 1993	Rocky Mountains, CO	40	–105		NMHC, VOC		Mount and Williams, 1997
BEMA	May 19–27, 1994	Castelporziano, Italy	42	12	Hundreds	BVOC		Atmos. Environ. 31, dec 1997, special issue
BEMA	Oct. 21–28, 1994	Castelporziano, Italy	42	12	Hundreds	BVOC		Atmos. Environ. 31, dec 1997, special issue
	Nov. 1994–Feb. 1995	Orinoco basin, Venezuela	9	–65	110	C_2–C_6, ISO	Rural site	Donoso et al., 1996
BEMA	Jun., Sep. 1995	Montpelier, France	43	4	Hundreds	BVOC		Atmos. Environ. 31, dec 1997, special issue
BEMA	Apr., Sep. 1996	Burriana, Spain	41	0	Hundreds	BVOC		Atmos. Environ. 31, dec 1997, special issue
BEMA	Jun. 1997	Burriana, Spain	41	0	Hundreds	BVOC		Atmos. Environ. 31, dec 1997, special issue
Ground network								
UK Monitoring Network	From Jun. 1991	11 Sites in UK	52–56	–5 to 0	8000 per year	C_2–C_6	Urban and rural sites	Dollard et al., 1995
Porspoder station	1992–1995	Porspoder, France	48	–4	Hundreds	C_2–C_6	EUROTRAC-TOR station	Boudries et al., 1994
EMEP	1992–1995	Rural sites in Europe	40–80	10		C_2–C_6, carbonyls	In cooperation with TOR	Solberg et al., 1996

[a]Abbreviations: ALK, alkane; Hc, hydrocarbon; BNMHC, biogenic nonmethane hydrocarbon; ARO, aromatics hydrocarbons; ISO, isoprene; MT, monoterpene; VOC, volatile organic compound; and n, number of samples.

[b]90°–0°N: > 0; 0°–90°S: < 0.

[c]0°–180°W: < 0; 0°–180°E: > 0.

214

TABLE 1B List of Campaigns Conducted over the Oceans Onboard Ship[a]

Name	Date	Location	Latitudinal range[b]	Longitudinal range[c]	n	Compounds	Comments	Reference
Meteor 51, 52	Jan.–Sep. 1979	North Atlantic	48 to −4	4 to −25	40	C_2–C_5		Rudolph and Ehhalt, 1981b
	1979–1981	Point Arena, CA	39	−123	150	C_2–C_4	Coastal site	Singh and Salas, 1982
Polar Sea	Nov. 30–Dec. 21, 1981	Eastern Pacific	33 to −32	−125 to −70	30	C_2–C_5	Shipboard	Singh and Salas, 1982
R/V Conrad	Jul. 1982	Pacific	0	−90		C_2–C_6		Greenberg and Zimmerman, 1984
R/V Melville	Dec. 1982	Pacific	0 to 20	−140		C_2–C_6		Greenberg and Zimmerman, 1984
Polar Star	Nov. 15–Dec. 28, 1984	Eastern Pacific	48 to −48	−125 to −70	406		Real-time analysis	Singh et al., 1988
SINODE	Apr. 1985	Indian Ocean	13 to −25	40 to 55	27	C_2–C_6, ISO		Bonsang et al., 1988
R/V Polarstern	Mar.–Apr. 1987	Atlantic	50 to −40	−40 to −20	200	C_2–C_5		Rudolph and Johnen, 1990
	May 1987	Indian Ocean	20	−155	12	C_2H_2		Kanakidou et al., 1988
RE-87-03-RITS (NOAA)	Apr.–Jul. 1987	Pacific–Indian oceans	45 to −45	−155 to 90	314	C_2–C_4, RCOOH		Arlander et al., 1990
SAGA II	Sep.–Oct. 1988	Atlantic Ocean	45 to −30	−30	92	C_2–C_5		Koppman et al., 1992
SAGA 3	Feb.–Mar. 1990	Equatorial Pacific Ocean	20 to −15	140 to 170	Cont.	C_2–C_5, HALO		Donahue and Prinn, 1993
EREBUS94	Dec. 12–Feb. 25, 1994	Antarctic Ocean	−43 to −77	145 to 172	20	C_2–C_5		Gros et al., 1998

[a]Abbreviations: ISO, isoprene; HALO, halocarbon hydrocarbons; n, number of samples, and RCOOH, carboxylic acids.
[b]90°–0°N: > 0; 0°–90°S: < 0.
[c]0°–180°W: < 0; 0°–180°E: > 0.

TABLE 1C List of Campaigns Conducted Onboard Aircraft or Balloons[a]

Name	Date	Location	Latitude[b]	Longitude[c]	n	Compounds	Comments	Reference
Onboard Aircraft								
	May 1976	Tampa Bay area				ISO, MT		Lonneman et al., 1978
	Apr. 1977	San Francisco Bay,	37	−123	60	C_2H_6, C_2H_2	Up to lower stratosphere	Cronn and Robinson, 1979
	Jul. 1977	Panama Canal	9	−80		C_2H_6, C_2H_2		Cronn and Robinson, 1979
	Jun. 1979	Southern France	44	2	20	C_2H_6, C_3H_8	Lower stratosphere	Rudolph et al., 1981a
	Jun. 1979	Southern France	44	2	30	C_2H_2	Lower stratosphere	Rudolph et al., 1984
STRATOZ II	Apr.–May 1980	Global			110	C_2–C_5		Ehhalt et al., 1985
	Sep. 1981	Texas	32	−96	16	C_2H_2		Rudolph et al., 1984
	Apr. 5, 1984	New Mexico	33			Light Hc	Lower stratosphere	Aikin et al., 1987
STRATOZ III	Jun. 1984	Global			120	C_2–C_5	Lower stratosphere	Rudolph, 1988
	Jul. 18–28, 1984	Boulder, Colorado	40	−105	94		Samples	Singh et al., 1988
	Aug. 9–21, 1984	Moffett Field, CA	36	−122	75		Samples	Singh et al., 1988
	Dec. 1982–Mar. 1986	Northwestern midlatitudes	50 to 55	0	62	C_2–C_6, HALO		Lightman et al., 1990
	Feb.–Apr. 1985	U.S. East Coast	20 to 40	−60 to −90	58	C_2–C_4		Van Valin and Luria, 1988
	Apr. 1985	U.S. West Coast, Alaska	33 to 71	−117 to −157	48	C_2–C_5	0–13 km	Greenberg et al., 1990
ATMOS	Apr.–May 1985		25 to 31	−70 to −345	21	C_2H_6, C_2H_2	Lower stratosphere, upper troposphere	Rinsland et al., 1987

Experiment	Date	Location	Latitude	Longitude	n	ISO	Notes	Reference
GTE/ABLE 2A	Jul. 1985	Amazon basin	−5	−55 to −65	25	C_2–C_6	0–5000 m	Rasmussen and Khalil, 1988
TROPOZI	May 31–Jun. 11, 1987	Hao Atoll, Pacific Ocean	−18	−141				Bonsang et al., 1991
	Dec. 1988	0°–50°N						Boissard, 1992
ABLE 3A	Jul.–Aug. 1988	Sub-Arctic Alaska	40 to 83	−70 to −155	1000	C_2–C_5, ISO		Blake et al., 1992
ABLE 3B	Jul.–Aug. 1990	Eastern Canadian	45 to 60	−70 to −85	900	C_2–C_6, ISO, ARO, HALO	Forest fires	Blake et al., 1994
TROPOZ II	Jan. 1991	Global			100			Boissard et al., 1996
PEM-West A	Sep.–Oct. 1991	Northwestern Pacific	−0.5 to 59.5	−122 to 114	1667	C_2–C_6, ISO, HALO		Blake et al., 1996a
ASTEX/MAGE	Jun. 1992	North Atlantic	27.5 to 33.1	−16.3 to 25	808	C_2–C_7, C_1–C_2 HALO		Blake et al., 1996b
STARE (TRACE A/SAFARI)	Jun.–Dec. 1992	Southern tropical Atlantic (Brazil/southern Africa)	−23 to 90			Trace gases	From fires and soils	*J. Geophys. Res.* **D19**, 1996
AASE-II (Airborne Arctic Stratospheric Experiment)	Jan.–Mar. 1992				260	C_2–C_4	DC-8 (NASA)	Anderson et al., 1993
					1600	CO_2, CO, CH_4	DC-8 (NASA)	Anderson et al., 1993
Balloon/tethered balloon/remote control device								
GTE/ABLE 2A	Jul.–Aug. 1985	Amazon basin	−2	−60	100	C_2–C_{10}	Tethered balloon (0–900 m)	Zimmerman et al., 1988
	Jul. 1986	Auch, France	44	0	15	C_2–C_6, ISO	Rural site/balloon	Kanakidou et al., 1989
	Jun. 1987	Hao, French Polynesia	−18	−141	12	C_2H_2	Remote controlled device	Kanakidou et al., 1988
	Jul.–Aug. 1992	Oak Ridge, TN	36	−84			Oak forest/tethered balloon	Guenther et al., 1996

[a]Abbreviations: Hc, hydrocarbon; ARO, aromatics hydrocarbons; HALO, halocarbon hydrocarbons; ISO, isoprene; MT, monoterpene; and n, number of samples.

[b]90°–0°N: > 0; 0°–90°S: < 0.

[c]0°–180°W: < 0; 0°–180°E: > 0.

others roughly equally distributed between Europe, the tropics, and polar regions. Very few data are available for Asia, with only two references reported in Table 1A for this continent (Stevens et al., 1981; Yokouchi and Ambe, 1988). Half of all these campaigns, and particularly those in the last 10 years, were devoted to biogenic VOC studies. Indeed, since the early 1990s, understanding biogenic emissions, especially over tropical regions, became a new priority, within the frame of the International Geosphere–Biosphere Program (IGBP, for further details see Andreae et al., 1996). Rasmussen et al. (1976) were the first to report extensive biogenic VOC data. Data collected within European ground networks have been reported since the early 1990s (see e.g., Dollard et al., 1995; Boudries et al., 1994; Lindskog, 1996).

About 10 shipborne campaigns are reported in Table 1B. Cruises have been made over large latitudinal gradients, one of the most important being that carried out on the German research vessel Polarsten during spring 1987 (50° N to 40°S, Rudolph and Johnen, 1990) and during the ANT VII/1 cruise in summer 1988 (45° N to 30° S, Koppman et al., 1992). Both of these campaigns were made over the Atlantic Ocean. The American vessels R/V Melville and R/V Conrad also have been used for measurements over the Pacific regions (Greenberg and Zimmerman, 1984), as well as the Soviet R/V Akademik Korolev vessel for the SAGA expeditions (Donahue and Prinn, 1993). The atmosphere of the Indian Ocean was investigated during the SINODE and RE-87-03-RITS missions (Bonsang et al., 1988; Kanakidou et al., 1988, respectively).

Global vertical and latitudinal distributions were obtained for the lighter NMHCs during the extensive aircraft campaigns STRATOZ II and III (Ehhalt et al., 1985; Rudolph, 1988, respectively) and later during the TROPOZ I and II campaigns (Kanakidou et al., 1988; Boissard et al., 1996). Vertical profiles and distributions were also obtained during airborne campaigns over oceanic regions, PEM-West A for the Pacific (Blake et al., 1996b) and ASTEX/MAGE for the Atlantic (Blake et al., 1996a), and ABLE 3A and 3B campaigns over high northern latitudes (Blake et al., 1992; Blake et al., 1994). For more details see Table 1C.

III. VERTICAL DISTRIBUTION

To complete the understanding of the ozone budget and related species on a global scale, vertical measurements of reactive VOCs (mainly C_2 to C_6 hydrocarbons) together with CO and CO_2 have been carried out since the early 1980s, in and above the boundary layer as well as in the lower stratosphere. Such measurements were useful as inputs to global photochemistry models. They were obtained by means of grab samples taken in aircraft and free or tethered balloons, over an area of a few hundreds of square kilometers. In Section III.A local and global tropospheric vertical distributions are presented. Stratospheric VOC concentration data are summarized in Section III.B.

A. TROPOSPHERIC DISTRIBUTIONS

1. Vertical Profiles

Decreasing concentrations are usually observed as a function of increasing altitude for C_2 to C_6 NMHCs in the lower atmosphere (Singh *et al.*, 1988; Rudolph *et al.*, 1992; Boissard *et al.*, 1996), with a less pronounced gradient within the boundary layer (Singh *et al.*, 1988). However, high concentrations of even highly reactive VOCs are surprisingly often encountered in the free troposphere (Ehhalt *et al.*, 1985; Rudolph, 1988; Singh *et al.*, 1988; Bonsang *et al.*, 1991; Boissard *et al.*, 1996), as well as sulfur compounds and radon. These observations highlight the significance of rapid vertical mixing via different processes including cloud pumping (Ehhalt *et al.*, 1985). Such dynamic processes must be very rapid compared to chemical processes, even for short-lived NMHCs, and lead to relatively constant mixing ratios of short-lived species up to 1 km. Early observations of this kind were made by Rasmussen and Khalil (1983) during vertical profile measurements carried out in the Arctic free troposphere. These authors measured acetylene and propane mixing ratios higher in the 1 to 2-km layer than at ground level. During the two experiments PEM-WEST A (1991) and ASTEX-MAGE (1992), Blake *et al.* (1996b, 1996a) reported enhanced hydrocarbon and halocarbon mixing ratios in the continental boundary layer (1 to 2 km) compared to the oceanic boundary layer. Factors of 2, 5, 2, and 6 times enhancement were found, for ethane, propane, acetylene, and benzene, respectively. By using back trajectories, the authors were able to link high VOC mixing ratios above the marine boundary layer with old continental polluted air (Figs. 1 and 2). Thus, they demonstrated the influence, on a medium scale, of anthropogenic sources and dynamic phenomena on the distributions of VOCs in the remote atmosphere. Such influence, on a larger scale, was previously mentioned by Blake *et al.* (1992) during the Arctic Boundary Layer Experiment 3A (ABLE 3A). Their measurements revealed the influence of long-range transport of pollutants, such as C_2H_2 and C_3H_8, from nonlocal sources (wildfires, oil exploitation) from the Arctic or other high latitude inhabited areas. The transport of VOCs from polluted areas or the local biogenic emissions of short-lived compounds can probably therefore lead to significant production of regional-scale tropospheric ozone in remote areas of the globe.

Several attempts to model these vertical distributions of selected NMHCs

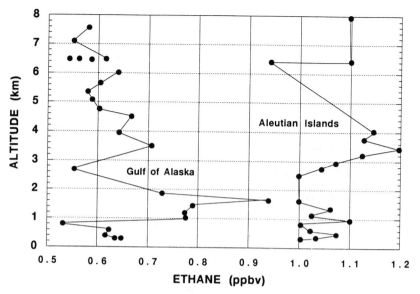

FIGURE 1   Vertical profiles of ethane over the Gulf of Alaska and the Aleutian Islands during the 1991 Pacific Exploratory Mission (PEM-West A); adapted from Blake *et al.* (1996b).

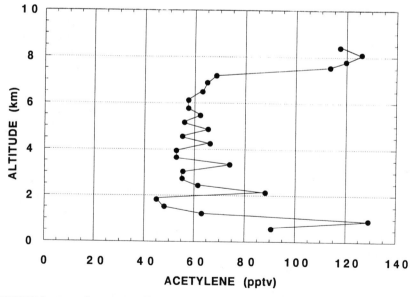

FIGURE 2   Typical vertical profile of acetylene over the Pacific (28° N, 147° E), southeast of Japan (results from PEM-West A experiment), adapted from Blake *et al.* (1996b).

in the troposphere (e.g., Singh *et al.*, 1988) using only eddy diffusion processes have so far only given inconsistent results. Studies by Lawrimore *et al.* (1995) using eddy diffusivity in addition to surface concentrations and species reactivity showed better agreements for compounds, such as isoprene and *n*-butane, but unexpected results for propane and benzene. Thus, it is not possible at present to obtain a reasonable description of the large inhomogeneous tropospheric distributions of selected NMHCs, either from measurements or from models.

As far as biogenic compounds are concerned, very few vertical profiles in the boundary layer have been reported, since these species react rapidly in the troposphere with OH radicals and ozone, leading to a rapid decrease in concentration down to undetectable values. Even close to source areas, isoprene concentrations drop rapidly by an order of magnitude in the first few hundred meters of the boundary layer (Davis *et al.*, 1994). However, as for less reactive hydrocarbons, significant and rapid upward transport can lead in some specific cases to elevated concentrations at altitude: measurements by Blake *et al.* (1992) over Wallop Island (Virginia) during the ABLE 3A expedition indicate significant isoprene concentrations (700 pptv) up to several hundred meters of altitude.

2. Global Distributions

Global distributions of C_2 to C_{10} NMHCs have been obtained using specially equipped research aircraft, in coordinated international research campaigns. One set of campaigns used the French Caravelle and consisted of four major experiments: (1) the STRATOZ II experiment in April–May 1980, (2) the STRATOZ III mission in June 1984, and (3) the TROPOZ I campaign in December 1987 that was used as a precampaign for (4) the TROPOZ II flights undertaken in January 1991. The aims of these campaigns were to study the impact of VOCs on ozone distribution and also the seasonality of the global distribution of major VOCs. Of the reactive VOCs, only light NMHCs and aldehyde were measured, from 70° N to 60° S and between ground level up to the tropopause. Two meridian sections were made from the eastern coast of North America to the western coast of South America (hereafter referred as the Western part), and from the eastern coast of South America to the western coasts of Africa and Europe (hereafter referred as the Eastern part). Around 100 samples were taken during the campaign. Figure 3 shows some distributions obtained for acetylene, propane and ethane, during TROPOZ II (Boissard *et al.*, 1996).

These studies led to these major observations. (1) The existence of a significant latitudinal gradient with higher mixing ratios in the midnorthern latitudes at ground level as well as at different altitudes (Ehhalt *et al.*, 1985;

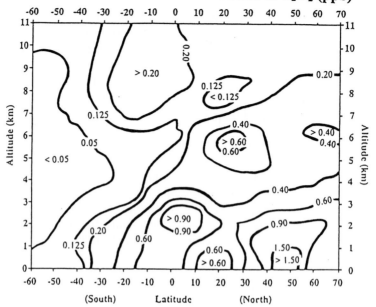

a. TROPOZ II Western distribution of C₂H₂ (ppb)

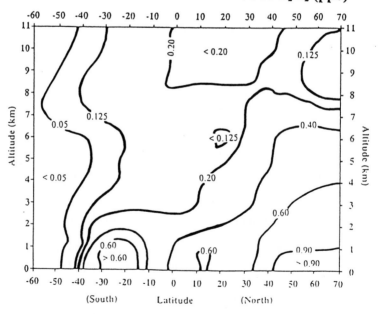

TROPOZ II Eastern distribution of C₂H₂ (ppb)

FIGURE 3 Two-dimensional distributions of (a) acetylene, (b) propane, and (c) ethane obtained from the TROPOZ II experiment (January 1991); from Boissard *et al.* (1996).

b. TROPOZ II Western distribution of C₃H₈ (ppb)

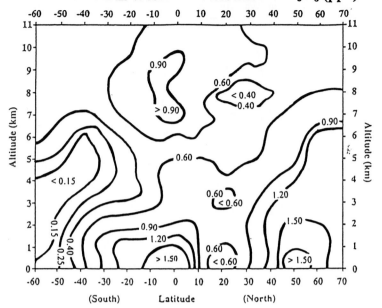

TROPOZ II Eastern distribution of C₃H₈ (ppb)

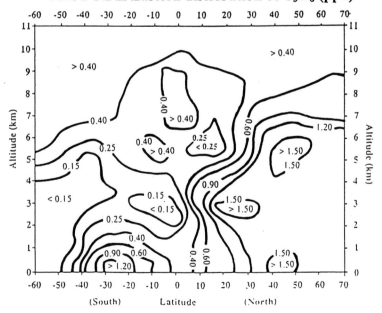

FIGURE 3 *(Continued)*

c. TROPOZ II Western distribution of C₂H₆ (ppb)

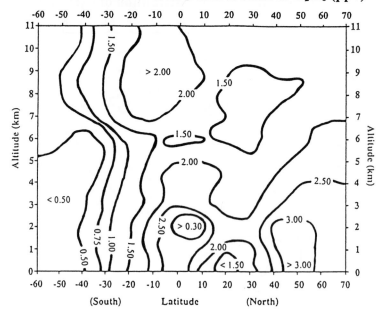

TROPOZ II Eastern distribution of C₂H₆ (ppb)

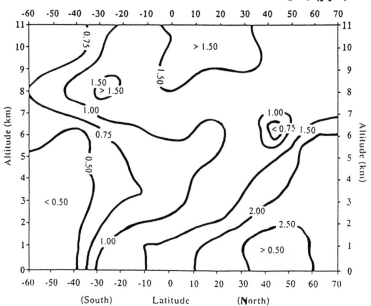

FIGURE 3   (*Continued*)

Rudolph, 1988; Boissard et al., 1996). Boissard et al. (1996) report an inter-hemispheric ratio, at ground level, of up to 70 for acetylene (1890 and 25 pptv in the Northern and Southern Hemisphere, respectively), of 30 for propane (2000 and 70 pptv in the Northern and Southern Hemisphere, respectively), and of 20 for ethane (4000 and 200 pptv in the Northern and Southern Hemisphere, respectively). In June 1984, lower maximum interhemispheric ratios were measured: 30, 10, and 15 for acetylene, propane, and ethane, respectively. (2) Meridian heterogeneity was, however, observed with higher mixing ratios over America than over Europe; during winter 1991 maximum acetylene mixing ratios in the free troposphere were more than twice over the United States (1900 pptv) than over Europe (800 pptv) (Boissard et al., 1996). This meridian discrepancy was lower, for ethane (4000 and 3140 pptv, respectively), and for propane (1890 and 1740 pptv, respectively) but of the same magnitude [for light alkenes (4270 and 1700 pptv ethene over America and Europe, respectively, and 1400 and 680 pptv propene over America and Europe, respectively]. (3) The strong influence of fast vertical transport on NMHC distribution was inferred; this effect led to high mixing ratios of even short-lived species at midaltitude up to the tropopause; air mass back trajectories were used to attribute high-mixing ratios measured during the TROPOZ II experiment in the boundary layer and the upper troposphere to significant dynamic phenomena [convection or vertical transfers (Boissard et al., 1996)]. However, at altitudes higher than 8 km, high alkene mixing ratios remain unexplained (Ehhalt et al., 1985; Boissard et al., 1996). Ehhalt et al. (1985) attributed these values to various meteorological phenomena, such as strong frontal activity and convection in connection with the inter-tropical convergence zone (ITCZ) or in regions of enhanced tropospheric–stratospheric air mass exchanges, such as zonal polar jet streams. (4) Strong local sources such as biomass burning in the tropics were found to significantly influence the latitudinal distributions. (5) Due to the effect of both photochemistry and fast transport phenomena, the global distributions of light NHMCs remain difficult to calculate by global models.

During three months of winter 1992, a NASA DC 8 was used to explore the stratosphere [Arctic Airborne Stratospheric Expedition (AASE II)], from 90° N to 23° S. The results obtained confirmed the latitudinal gradient observed in the previous campaigns, with maximum mixing ratios in subpolar and Arctic latitudes (Anderson et al., 1993). At 40° N, the mixing ratios were found to increase significantly over the course of the three-month measurement period for most light NMHCs measured, due to a reduced efficiency of removal processes. Thus, the concentrations were much higher within the tropics and subtropics. This reflects both the seasonal reduction in the rates of photochemical destruction and the greater abundance of biogenic and anthropogenic sources in the northern latitudes. The NMHC increases during

TABLE 2 Stratospheric Measurements of Nonmethane Hydrocarbons

Date	Location	Altitude (km)	Latitude[a]	Longitude[b]	Concentration[c]			Reference
					min	mean	max	
C$_2$H$_6$								
Apr. 1985	U.S. West Coast, Alaska	13	71	−157		87		Greenberg et al., 1990
Apr. 1985	U.S. West Coast, Alaska	12.4	66	−147		145		Greenberg et al., 1990
Apr. 1985	U.S. West Coast, Alaska	13	56	−141		46		Greenberg et al., 1990
Apr. 1985	U.S. West Coast, Alaska	13	48	−126		534		Greenberg et al., 1990
Jun. 1979	Southern France	11.0–30.0	44	2	10		165	Rudolph et al., 1981a
Jul. 1984	Colorado	>5.5	40	−105		1260		Singh et al., 1988
Apr. 1985	U.S. West Coast, Alaska	11	38	−124		1075		Greenberg et al., 1990
Apr. 1977	S. Francisco Bay, CA	11.5–14.5	37	−123	100		800	Cronn and Robinson, 1979
Feb. 1985	Pacific/California	>5.5	36	−125		1260		Singh et al., 1988
Aug. 1985	Pacific/California	>5.5	36	−125		980		Singh et al., 1988
Apr. 1984	New Mexico	11.6	33	−105		441		Aikin et al., 1987
Apr. 1984	New Mexico	14.1	33	−105		362		Aikin et al., 1987
Apr. 1985	Atlantic Ocean	13.9–16.8	30	−69		210		Rinsland et al., 1987
May 1985	Baja peninsula	6.2–15.7	30	−115		813		Rinsland et al., 1987
Apr. 1985	U.S. West Coast, Alaska	13	30	−117		1157		Greenberg et al., 1990
May 1985	Libya	15.2–18.4	27	15		327		Rinsland et al., 1987
May 1985	Atlantic Ocean	9.6–17.8	26	−31		604		Rinsland et al., 1987
May 1985	Bahama Islands	13.0–17.4	25	−77		378		Rinsland et al., 1987
C$_2$H$_4$								
Apr. 1985	U.S. West Coast, Alaska	13	71	−157		<10		Greenberg et al., 1990
Apr. 1985	U.S. West Coast, Alaska	12.4	66	−147		<10		Greenberg et al., 1990
Apr. 1985	U.S. West Coast, Alaska	13	56	−141		<10		Greenberg et al., 1990
Apr. 1985	U.S. West Coast, Alaska	13	48	−126		<10		Greenberg et al., 1990
Apr. 1985	U.S. West Coast, Alaska	11	38	−124		<10		Greenberg et al., 1990
Apr. 1985	U.S. West Coast, Alaska	13	30	−117		98		Greenberg et al., 1990

C_2H_2

	Date	Location							Reference
	Apr. 1985	U.S. West Coast, Alaska	13	71	−157		<10		Greenberg et al., 1990
	Apr. 1985	U.S. West Coast, Alaska	12.4	66	−147		<10		Greenberg et al., 1990
	Apr. 1985	U.S. West Coast, Alaska	13	56	−141		<10		Greenberg et al., 1990
	Apr. 1985	U.S. West Coast, Alaska	13	48	−126		76		Greenberg et al., 1990
	Jun. 1979	Southern France	11.0–30.0	44	2	20		250	Rudolph et al., 1981
	Jul. 1984	Colorado	>5.5	40	105		90		Singh et al., 1988
	Apr. 1985	U.S. West Coast, Alaska	11	38	−124		173		Greenberg et al., 1990
	Apr. 1977	San Francisco Bay, CA	11.5–14.5	37	−123	<40		250	Cronn and Robinson, 1979
	Feb. 1985	Pacific/California	>5.5	36	−125		190		Singh et al., 1988
	Aug. 1985	Pacific/California	>5.5	36	−125		60		Singh et al., 1988
	Apr. 1984	New Mexico	11.6	33	−105		60		Aikin et al., 1987
	Apr. 1984	New Mexico	14.1	33	−105		50		Aikin et al., 1987
	Sep. 1981	Texas	11.0–35.0	32	−96	<10		40	Rudolph et al., 1984
	Apr. 1985	U.S. West Coast, Alaska	13	30	−117		<10		Greenberg et al., 1990
	May 1985	Baja peninsula	7.7–16.6	30	−115		60		Rinsland et al., 1987

C_3H_8

	Date	Location							Reference
	Apr. 1985	U.S. West Coast, Alaska	13	71	−157		<6		Greenberg et al., 1990
	Apr. 1985	U.S. West Coast, Alaska	12.4	66	−147		<6		Greenberg et al., 1990
	Apr. 1985	U.S. West Coast, Alaska	13	56	−141		<6		Greenberg et al., 1990
	Jul. 1984	Colorado	>5.5	40	−105		400		Singh et al., 1988
	Apr. 1985	U.S. West Coast, Alaska	13	48	126		76		Greenberg et al., 1990
	Jun. 1979	Southern France	11.0–30.0	44	2	2		1000	Rudolph et al., 1981a
	Apr. 1985	U.S. West Coast, Alaska	11	38	−124		84		Greenberg et al., 1990
	Feb. 1985	Pacific/California	>5.5	36	−125		160		Singh et al., 1988
	Aug. 1985	Pacific/California	>5.5	36	−125		450		Singh et al., 1988
	Apr. 1984	New Mexico	11.6	33	−105		84		Aikin et al., 1987
	Apr. 1984	New Mexico	14.1	33	−105		139		Aikin et al., 1987
	Apr. 1985	U.S. West Coast, Alaska	13	30	−117		115		Greenberg et al., 1990

continued

TABLE 2 *(Continued)*

Date	Location	Altitude (km)	Latitude[a]	Longitude[b]	Concentration[c] min	mean	max	Reference
C_3H_6								
Apr. 1985	U.S. West Coast, Alaska	13	71	-157		<6		Greenberg *et al.*, 1990
Apr. 1985	U.S. West Coast, Alaska	12.4	66	-147		<6		Greenberg *et al.*, 1990
Apr. 1985	U.S. West Coast, Alaska	13	56	-141		<6		Greenberg *et al.*, 1990
Apr. 1985	U.S. West Coast, Alaska	13	48	-126		<6		Greenberg *et al.*, 1990
Jul. 1984	Colorado	>5.5	40	-105		40		Singh *et al.*, 1988
Apr. 1985	U.S. West Coast, Alaska	11	38	-124		<6		Greenberg *et al.*, 1990
Feb. 1985	Pacific/California	>5.5	36	-125		50		Singh *et al.*, 1988
Aug. 1985	Pacific/California	>5.5	36	-125		980		Singh *et al.*, 1988
Apr. 1985	U.S. West Coast, Alaska	13	30	-117		<6		Greenberg *et al.*, 1990
$n\text{-}C_4H_{10}$								
Apr. 1985	U.S. West Coast, Alaska	13	71	-157		<5		Greenberg *et al.*, 1990
Apr. 1985	U.S. West Coast, Alaska	12.4	66	-147		<5		Greenberg *et al.*, 1990
Apr. 1985	U.S. West Coast, Alaska	13	56	-141		<5		Greenberg *et al.*, 1990
Apr. 1985	U.S. West Coast, Alaska	13	48	-126		<5		Greenberg *et al.*, 1990
Jul. 1984	Colorado	>5.5	40	-105		60		Singh *et al.*, 1988
Apr. 1985	U.S. West Coast, Alaska	11	38	-124		20		Greenberg *et al.*, 1990
Feb. 1985	Pacific/California	>5.5	36	-125		20–110		Singh *et al.*, 1988
Aug. 1985	Pacific/California	>5.5	36	-125		40		Singh *et al.*, 1988
Apr. 1985	U.S. West Coast, Alaska	13	30	-117		<5		Greenberg *et al.*, 1990
C_6H_6								
Apr. 1985	U.S. West Coast, Alaska	13	71	-157		<3		Greenberg *et al.*, 1990
Apr. 1985	U.S. West Coast, Alaska	12.4	66	-147		<3		Greenberg *et al.*, 1990
Apr. 1985	U.S. West Coast, Alaska	13	56	-141		<3		Greenberg *et al.*, 1990
Apr. 1985	U.S. West Coast, Alaska	13	48	-126		7		Greenberg *et al.*, 1990
Apr. 1985	U.S. West Coast, Alaska	11	38	-124		35		Greenberg *et al.*, 1990
Apr. 1985	U.S. West Coast, Alaska	13	30	-117		37		Greenberg *et al.*, 1990

[a] 90°–0°N : >0; 0°–90°S : <0.

[b] 0°–180°W : <0; 0°–180°E : >0.

[c] Values in pptv (parts per trillion volume of compound.)

the winter were found to parallel CO and CO_2, increases, suggesting that this accumulation may be caused by combustion-related sources.

B. STRATOSPHERIC DISTRIBUTIONS

Few stratospheric VOC data are available and those that are were mainly obtained in the late 1970s and early 1980s. Table 2 summarizes some NMVOC values found in the literature. The first vertical distribution that extends into the lower stratosphere was reported by Cronn and Robinson (1979) for C_2H_6 and C_2H_2. By using a balloon-borne cryosampler, measurements of profiles of C_2 to C_5 hydrocarbons were made by Rudolph *et al.* (1981a) and Rudolph *et al.* (1984) up to 30-km altitude. Among these compounds only C_2H_2 (Rudolph *et al.*, 1984) and C_2H_6 and C_3H_8 (Rudolph *et al.*, 1981a) were found to be above analytical detection limits. These were 60, 3, and 15 ppt for C_2H_2, C_2H_6, C_3H_8, respectively.

As seen in Fig. 4 (from Rudolph *et al.*, 1984) and Fig. 5 (from Rudolph *et al.*, 1981a), a regular and rapid decrease with altitude was observed above the

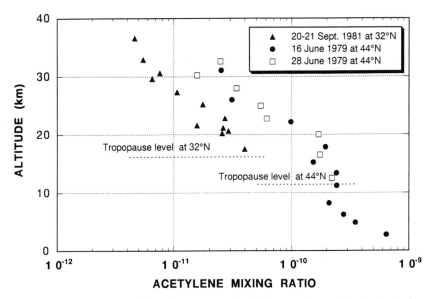

FIGURE 4   Vertical profiles of acetylene mixing ratios in the troposphere and in the stratosphere over southwest France (44° N, 2° E) on June 16 and 28, 1979 (circles and squares) and over Texas on September 20–21, 1981 (triangles). Arrows indicate measurements below the detection limit; adapted from Rudolph *et al.* (1984).

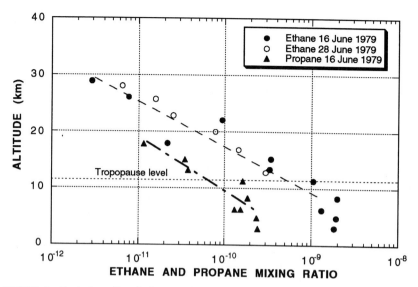

FIGURE 5   Vertical profile of ethane and propane in the troposphere and stratosphere over southern France (44° N, 2° E) June 1979; adapted from Rudolph *et al.* (1981).

tropopause at 15 to 20 km for acetylene and ethane. A similar observation was made by Cronn and Robinson (1979). However, Rudolph *et al.* (1984) reported a variability, for various latitudes (44° to 32° N), of a factor of four in the acetylene mixing ratios at 32 km. The authors showed that upper tropospheric and lower stratospheric measurements of acetylene reveal merely a reflection of the latitudinal gradient found at the surface, which itself reflects the distribution of sources.

IV. GROUND-LEVEL DISTRIBUTIONS

A. ACETYLENE

Acetylene is one of the hydrocarbons whose origins are the most clearly identified, these consisting of urban and industrial sources, mainly related to transportation in urban areas and biomass burning in remote areas (Rudolph *et al.*, 1992; Bonsang *et al.*, 1995). The oceanic contribution is negligible (Kanakidou *et al.*, 1988; Boissard *et al.*, 1996). As mentioned earlier, the average global and tropical lifetimes of acetylene are 24 and 6.7 days, respectively (Singh and Zimmerman, 1992). Data from more than 25 references have been compiled. Acetylene mixing ratios were reported, for every month, in a 10° width grid and averaged regardless of the longitude, that is, regardless of the local sources influencing these measurements (oceanic, continental, urban, or remote). A global latitudinal and seasonal distribution was thus obtained and this is presented in Fig. 6 (white areas symbolize the periods and or locations where no data were available).

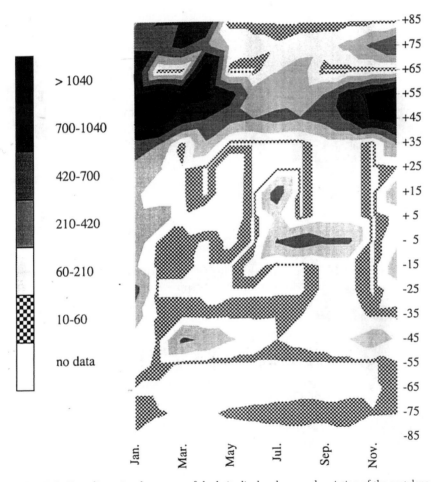

FIGURE 6   Two-dimensional contours of the latitudinal and seasonal variation of the acetylene mixing ratio in the troposphere. The scale on the left side is in pptv of acetylene; the right scale is the latitude.

As expected, the well-defined sources of acetylene and the seasonal and latitudinal variations of OH radical concentrations lead to a two-dimensional distribution showing a clear annual cycle as well as a strong latitudinal gradient, with higher concentrations in the midnorthern latitudes. The well-described winter accumulation of anthropogenic VOCs (e.g. Jonson *et al.*, 1996; Greenberg *et al.*, 1996) in the mid- to high northern latitudes can be observed from January until early May between 40° and 70° N. More generally, highest mixing ratios are observed from mid-September until the beginning of May, reflecting a slower chemical sink (oxidation with OH radicals). The tropical influence of biomass burning can be seen as well during summertime, especially in the Southern Hemisphere where it represents the dominant source. Spring and autumn maxima are observed in the midsouthern latitudes (between 35° and 50° S). In the Indian Ocean, this maximum is generally observed between September and October and coincides with the period of maximal burning and enhanced transport of continental-derived compounds. In more remote locations such as the Antarctic, a seasonal variation is still observed. Rudolph *et al.* (1989) have reported, on the basis of a six-year survey, a very clear seasonal effect with a minimum of 10 pptv in February–March and a maximum of 30 pptv in August and September.

B. ALKANES

1. Ethane

a. Overview and Latitudinal Distribution

Ethane, one of the less reactive NMHCs observed in the troposphere, results mainly (like methane) from the exploitation of natural gas, and it is the second most abundant organic trace gas in the atmosphere. Other minor sources include soils, wetlands, oceans, engine exhaust, and losses from industrial processes. A large uncertainty remains in the assessment of the role of vegetation as a source or sink for ethane (Rudolph, 1995). Due to its long lifetime and the importance of the anthropogenic and continental sources, ethane is used as a test for atmospheric transport and chemical models. Moreover, ethane could be used as a tracer to observe changes in the magnitude and distribution of natural gas and biomass burning, in particular, to understand the seasonal and geographic distribution of biomass burning in the Southern Hemisphere.

The first large-scale distributions of ethane were obtained by Rudolph *et al.* (1981a) over the Atlantic Ocean, by Bonsang and Lambert (1985) for the Atlantic and the Pacific oceans and by Blake and Rowland (1986), along the west coast of North America and in the Pacific Ocean. All these distributions show a similar strong latitudinal gradient, with mixing ratios of 2000 to 3000 pptv in midnorthern latitudes and of 200 to 300 pptv in midsouthern latitudes. Longitudinal gradients were not systematically studied but the lifetime of ethane in the troposphere is long enough to assume that heterogeneity in the distributions of sources is smoothed and that a relative good estimate of the ethane global distribution can be obtained by combining measurements undertaken in different locations.

Measurements undertaken at ground level, and mainly above the ocean, were compiled by several authors, particularly by Kanakidou *et al.* (1991) and Rudolph (1995) who published an extensive review of ethane distributions to define the two-dimensional latitudinal and seasonal distributions of this compound (Figs. 7 and 8). Although they exhibit a strong variability in the mixing ratios reported in the literature, a very clear latitudinal gradient appeared. Maximal values of several parts per billion volume are observed at mid- and northern latitudes, compared with background values on the order of 100 to 500 pptv in the Southern Hemisphere. The steepest gradients in ethane mixing ratio were found for the latitudinal bands 65° to 50° N and 25° to 5° N. No clear latitudinal variation was found in the Southern Hemisphere. A global

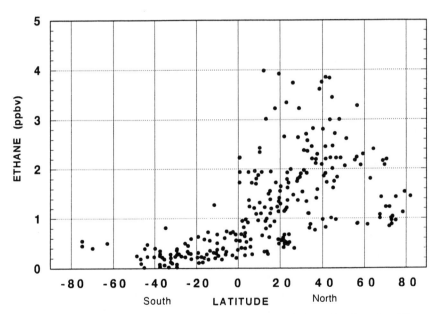

FIGURE 7   Latitudinal variation of ethane mixing ratios in the boundary layer for different seasons. Data are taken from Ehhalt and Rudolph (1984), Singh and Salas (1982), Greenberg and Zimmerman (1984), Bonsang and Lambert (1985), Blake and Rowland (1986), Tille and Bachmann (1986), Singh *et al.* (1988), Bonsang *et al.* (1988, 1990), and Rudolph *et al.* (1989); adapted from Kanakidou *et al.* (1991).

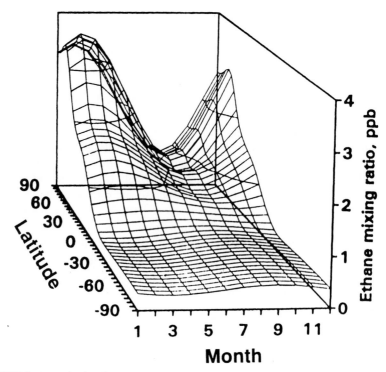

FIGURE 8 Latitudinal and seasonal variation of the average ethane mixing ratio in the troposphere, adapted from Rudolph (1995).

annual mean ethane mixing ratio of 860 pptv (384 and 1330 pptv in the Southern and Northern Hemisphere, respectively) was estimated.

b. Seasonal Variation

The seasonal variation of ethane concentrations has been followed at various sites in the Northern and Southern Hemisphere, for example, Greenberg *et al.* (1996) at Moana Loa. Blake and Rowland (1986) were the first to show evidence for a seasonal variation in the concentrations of ethane for a given latitudinal range. This large seasonal variation can be explained by the variability of the sources combined with the variability of the major atmospheric sink, the OH radical. Thus, in the Northern Hemisphere, maximum ethane concentrations are generally observed during winter (December) and, in accordance with the seasonality of OH concentrations, low concentrations are measured in July and August. Some fluctuation with latitude was noticed in the seasonal amplitudes (maximum at midnorthern latitudes, with a factor of four, and minimum around the equator). As expected, the seasonal variation is reversed in the Southern Hemisphere, but with a much lower magnitude due to the presence of lesser continental sources and the existence of some interhemispheric transfers, from the north to the south (half of the ethane in the Southern Hemisphere results from transport from the Northern Hemisphere, Rudolph, 1995). This effect tends to smooth the seasonal cycle in the Southern Hemisphere. Therefore, the global distribution of ethane shows a stronger interhemispheric gradient in the northern winter when the sink effect is at a minimum, and a relatively low interhemispheric gradient in the northern summer. A similar seasonal effect was also reported by Rudolph *et al.* (1992) for the Antarctic troposphere over the 1982 to 1990 period. The authors reported a stable seasonal amplitude of 300 to 500 pptv, with maximal ethane mixing ratios observed in August and September, coinciding with the maximum in biomass burning activities.

2. Propane

As for acetylene, propane data reported in the literature have been used to draw a two-dimensional latitudinal–seasonal distribution, as presented in Fig. 9. Although more reactive than acetylene or ethane, and with more widespread and complex sources, a clear enough picture emerges with a well-defined latitudinal gradient in concentrations. There is a large uncertainty in the background concentration of propane for remote locations, since propane is a relatively short-lived compound and is influenced by regional-scale advection. Northern mean concentrations of 1000 pptv are 10 times higher than the values found over southern latitudes. The seasonality in the Northern Hemisphere is less pronounced than for acetylene, but above 65° N a maximum is observed during springtime, followed by lower concentrations during summer. This spring accumulation of propane is noticed for most of the Northern Hemisphere. Propane mixing ratios measured at Amsterdam Island (37° S) over a two-year period show a maximum in July and August (90 and 92 pptv, respectively), compared to winter values of less than 40 pptv. Such seasonality in propane has been reported in the marine atmosphere as well as in the continental troposphere (Singh and Salas, 1982; Rudolph *et al.,* 1989). This is mainly due to the strong seasonality in the chemical sink strength, in addition to the seasonality in propane sources (biomass burning) and the seasonality in the advective vertical mixing processes, these being stronger during summer, which leads to a decline in the propane mixing ratios.

A propane/ethane ratio close to 0.5 is generally observed in the midlatitude troposphere as well as in the Arctic (Doskey and Gaffney, 1992). However, far from anthropogenic sources, the propane/ethane ratio is on the order of 0.2. In remote locations in the Southern Hemisphere, the propane/ethane ratio drops down to values between 0.15 and 0.08 at Scott base (Antarctic) and Baring Head (New Zealand) (Clarkson and Martin, 1996). Although the

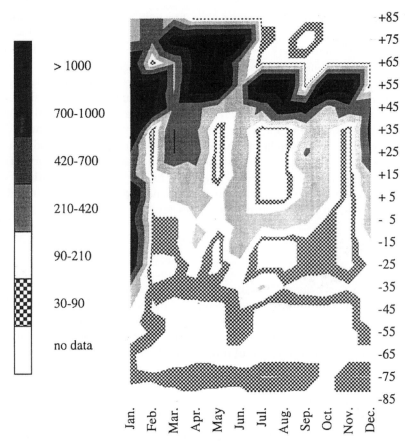

FIGURE 9 Two-dimensional contours of the latitudinal and seasonal variation of the propane mixing ratio in the troposphere. The scale on the left side is in pptv of propane, and the right scale is the latitude.

latitudinal variation of ethane and propane concentrations are similar, the variability in the propane/ethane ratio is in accordance with the faster removal of propane by OH during transport. Although a clear latitudinal gradient can be seen, there is considerable scatter in propane concentrations, due to its relative short lifetime, especially in close proximity to strong local sources.

3. n-Butane and Higher Alkanes

Due to a global lifetime of 80 days (Singh and Zimmerman, 1992) and sources similar to those of propane, the latitudinal and seasonal variation of n-butane concentrations should be close to those of propane. However, too few data are available to enable a two-dimensional global distribution to be drawn. Table 3 gives some examples of n-butane concentrations. Except very close to anthropogenic sources, where the concentrations can reach several parts per billion volume (Hov et al., 1991), n-butane mixing ratios are generally on the order of fractions of parts per billion volume. Typical rural concentrations range between 100 and 500 pptv. n-Butane is not known to be produced in significant amounts by biomass burning or vegetation activity and therefore measurements in the remote continental atmosphere (e.g., the Amazon basin, Venezuela) do not show any clear enhancement due to these sources. Background values in the Atlantic or Pacific oceans seem to be significantly less than 100 pptv. So far, no global distributions can be drawn for higher alkanes, due to a lack of sensitive measurements in the free troposphere. Their higher reactivities lead to direct involvement in tropospheric chemistry after emission.

C. ALKENES

1. Ethene and Propene

Due to their short lifetime in the atmosphere and to the multiplicity of their sources, ethene and propene concentrations show a large variability, from parts per billion volume levels in the urban or rural atmosphere to a few parts per trillion volume in the remote atmosphere. Penkett et al. (1993) report, for industrialized cities in the United States and western Europe, concentrations in the range 12 to 15 ppbv for ethene and 2.6 to 8 ppbv for propene. Similar values are generally found at high northern latitudes. Due to the impact of their photochemical sink, the concentrations are generally strongly dependent on the season of the measurements. For instance, Jobson et al. (1994) found maximal concentrations in the Arctic on the order of 400 to 800 pptv in January, and a decline to lower values in April (50 pptv). At Barrow (Alaska) in March 1989, Doskey and Gaffney (1992) found high concentrations on the order of 330 to 1000 pptv for ethene and 160 to 330 pptv for propene. In rural areas of the Northern Hemisphere, the concentrations are subparts per billion. Greenberg and Zimmerman (1984) reviewed measurements made in midlatitudes. They reported values of 460 pptv for ethene and 160 pptv for propene at Niwot Ridge in the central United States, and of 1820 pptv for ethene and 540 pptv for propene at Pawnee grasslands. In the tropics, higher concentrations are observed, due to the impact of vegetation and biomass burning emissions. Indeed, the same authors report concentrations of 1400 to 2000 pptv for ethene and 500 pptv for propene in Brazil. The same range of

TABLE 3 Latitudinal Distribution of n-Butane Concentration

Date	Location	Latitude[a]	Longitude[b]	Concentration[c] min	mean	max	Reference
Winter 1992	Arctic	90 to 70			149		Anderson et al., 1993
Jan. 1992	Arctic and subpolar	90 to 40			87		Anderson et al., 1993
Mar. 1992	Arctic and subpolar	90 to 40			106		Anderson et al., 1993
Sep. 1979, 11 GMT[d]	Atlantic Ocean	75	25		50		Rudolph et al., 1980
Apr. 1985	U.S. West Coast, Alaska	71	-157		35		Greenberg et al., 1990
Apr. 1985	U.S. West Coast, Alaska	66	-147		254		Greenberg et al., 1990
Apr. 1985	U.S. West Coast, Alaska	61	-150		402		Greenberg et al., 1990
Winter 1992	Northern subpolar	60 to 40			75		Anderson et al., 1993
Summer 1987	Norway	59	8		1670		Hov et al., 1991
Fall 1987	Norway	59	8		1800		Hov et al., 1991
Winter 1988	Norway	59	8		4780		Hov et al., 1991
Spring 1988	Norway	59	8		3570		Hov et al., 1991
Apr. 1979, 12 GMT[d]	West coast of Ireland	53	-10		400		Rudolph et al., 1991
Apr. 1985	U.S. West Coast, Alaska	48	-126		124		Greenberg et al., 1990
Summer, daytime	Central Europe	45	10		500		Rudolph et al., 1980
Winter, daytime	Central Europe	45	10		1000		Rudolph et al., 1980
Mar.-Apr.	Atlantic Ocean	45 to 0	-30		94		Rudolph and Johnen, 1989
Sep.-Oct.	Atlantic Ocean	45 to 0	-30		85		Koppman et al., 1992
Aug.-Sep. 1980	Niwot Ridge, CO	40	-105	270	510	1010	Greenberg and Zimmerman, 1984
Aug.-Sep. 1980	Pawnee grasslands, CO	40	-105	300	610	1050	Greenberg and Zimmerman, 1984
Dec. 1981	Eastern Pacific	40 to 30	-117 to -124		510		Singh and Salas, 1982
Mar. 1985	U.S. East Coast	40 to 20	-60 to -90		171-1060[e]		Van Valin and Luria, 1988
Apr. 1985	U.S. West Coast, Alaska	38	-124		106		Greenberg et al., 1990
Apr. 1985	Bermuda	35	-70		205[e]		Van Valin and Luria, 1988
Apr. 1985	U.S. West Coast, Alaska	30	-117		276		Greenberg et al., 1990
Sep. 1979, 12 GMT[d]	Sahara Desert	30 to 20	10 to -10		400		Rudolph et al., 1980
Dec. 1981	Eastern Pacific	30 to 20	-108 to -117		660		Singh and Salas, 1982

Date	Location	Latitude[a]	Longitude[b]				Reference
Jan. 1979, 9 GMT[d]	Atlantic Ocean	30 to 15			—		Rudolph et al., 1980
Winter 1992	Northern subtropical	30 to 10			<5		Anderson et al., 1993
Feb.–Mar. 1990	Equatorial Pacific Ocean	20 to -15	-140 to -170	11	47	154	Atlas et al., 1993
Dec. 1981	Eastern Pacific	20 to 10	-97 to -108		650		Singh and Salas, 1982
Jul., Dec. 1982	Pacific	17 to -11	-80 to -160		—	180	Greenberg and Zimmerman, 1984
Apr. 1985	Indian Ocean	15 to 10	40 to 55	80	120	150	Bonsang et al., 1988
Apr. 1985	Indian Ocean	10 to 0	40 to 55	70	250	400	Bonsang et al., 1988
Dec. 1981	Eastern Pacific	10 to 0	-89 to -97	120	300		Singh and Salas, 1982
Apr. (11:00–16:00)	Orinoco basin, Venezuela	9	-65		38		Donoso et al., 1996
Sep. (11:00–16:00)	Orinoco basin, Venezuela	9	-65		15		Donoso et al., 1996
May 1979, 14–18 GMT[d]	Equatorial Atlantic	2 to 0	-22		30–180		Rudolph et al., 1980
May 1979, 10 GMT[d]	Atlantic Ocean	2 to -2	-22		100		Rudolph et al., 1980
Apr. 1985	Indian Ocean	0 to -10	40 to 55	90	200	390	Bonsang et al., 1988
Dec. 1981	Eastern Pacific	0 to -10	-79 to -89		190		Singh and Salas, 1982
Winter 1992	Southern tropical	0 to -23			<5		Anderson et al., 1993
Mar.–Apr.	Atlantic Ocean	0 to -30	-30		20		Rudolph and Johnen, 1990
Sep.–Oct.	Atlantic Ocean	0 to -30	-30		<10		Koppman et al., 1992
May 1979, 10 GMT[d]	Equatorial Atlantic	-1	-22 to -26		70–100		Rudolph et al., 1980
Jul.–Aug.	Amazon basin (30–305 m)	-2	-65	60	90	170	Zimmerman et al., 1988
Aug.–Sep. 1980	Brazil	-5	-55		240		Greenberg and Zimmerman, 1984
Dec. 1981	Eastern Pacific	-10 to -20	-75 to -79		130		Singh and Salas, 1982
Jun. 1987	Hao Atoll, Pacific Ocean	-18	-141	51	247	422	Bonsang et al., 1991
Apr. 1985	Indian Ocean	-20 to -10	40 to 55	250	330	410	Bonsang et al., 1988
Apr. 1985	Indian Ocean	-20 to -25	40 to 55	30	145	260	Bonsang et al., 1988
Dec. 1981	Eastern Pacific	-20 to -32	-72 to -75		140		Singh and Salas, 1982

[a] 90°–0°N: > 0; 0°–90°S: < 0.
[b] 0°–180°W: < 0; 0°–180°E: > 0.
[c] Values in pptv (parts per trillion volume of compound).
[d] GMT, Greenwich mean time.
[e] Averaged over 1000 m.

concentrations occurs over tropical Africa: 500 to 1000 pptv and 200 to 400 pptv for ethene and propene, respectively (Rudolph et al., 1992a).

In the marine atmosphere, ethene and propene concentrations are an order of magnitude lower. In the northern Pacific Ocean, Blake et al. (1996b) showed, during the PEM-WEST A experiment, a large variability in concentrations, with averaged boundary layer values of 82 pptv for ethene and 27 pptv for propene.

Tables 4 and 5 summarize the considerable variability in the measurements of ethene and propene reported in the literature. However, the possibility of contamination of light alkenes by the sampling and analytical systems used in these determinations is still an open question, as discussed by Donahue and Prinn (1993).

Since 1981, several cruises have been performed to describe the global latitudinal variation of light hydrocarbons in the low troposphere. Cruises carried out along a large band of latitudes over the Atlantic and Pacific oceans were reported by Rudolph et al. (1981a), Bonsang and Lambert (1985), Rudolph and Johnen (1990), and Koppman et al. (1992), as well as by Donahue and Prinn (1993) for the equatorial Pacific. An interhemispheric ratio of at least 2 was generally observed. Ethene concentrations were in the range of 20 to 150 pptv in the Northern Hemisphere and of 10 to 40 pptv in the Southern Hemisphere. For propene, the same order of magnitude was observed in the Northern Hemisphere, but lower values (0 to 20 pptv) were observed at southern latitudes. Figures 10 and 11 depict the data reported by Rudolph and Johnen (1990), Donahue and Prinn (1993), and Carsey et al. (1997) for the Atlantic ASTEX/MAGE cruise. A large variability in ethene and propene concentrations is usually observed at tropical latitudes ($20°$ to $30°N$), as confirmed by Blake et al. (1996a) during the ASTEX experiment. They suggested the occurrence of advection from the continents to explain this variability. At the same latitudes, but for the Pacific Ocean (several hundred kilometers west of Hawaii) Greenberg et al. (1992) reported mixing ratios in agreement with those previously reported, with median figures for ethene and propene of 140 and 70 pptv, respectively. However, a relatively strong latitudinal gradient emerged, as confirmed by the measurements of Donahue and Prinn (1993), carried out close to the equator, which showed that the largest gradient is close to the equatorial regions.

As pointed out by Rudolph and Johnen (1990), variability in the strength of the marine source of alkenes (see Bonsang et al., 1988; Plass-Dülmer et al., 1995), and the biological activity and wind speed responsible for enhanced emissions at the air–sea interface, need to be taken into account in explaining this large variability. In addition, reaction with the OH radical has a strong influence on the removal rate and consequently on the budget of these species. High southern latitude data are relatively scarce. Some values were

TABLE 4 Measurements of Ethene in the Troposphere

Date	Location	Latitude[a]	Longitude[b]	Concentration[c]			Reference
				min	mean	max	
Winter, oceanic[d]							
Winter 1992	Arctic	90 to 80			45		Anderson et al., 1993
Jan.–Apr. 1992	Alert	82.5	−62.3	10	31	750	Jobson et al., 1994
Jan. 1992	Arctic and subpolar	90 to 80			38		Anderson et al., 1993
Mar. 1992	Arctic and subpolar	90 to 80			45		Anderson et al., 1993
Winter 1992	Arctic	80 to 70			58		Anderson et al., 1993
Jan.–Mar. 1992	Arctic and subpolar	80 to 50		31		38	Anderson et al., 1993
Mar.–Apr.	Atlantic Ocean	45 to 0	−30		58		Rudolph and Johnen, 1989
Jan. 1979, 9 GMT[d]	Atlantic Ocean	30 to 15	−22		1000		Rudolph et al., 1980
Dec. 1981	Eastern Pacific	30 to 20	−117 to −108		100		Singh and Salas, 1982
Feb.–Mar. 1990	Equatorial Pacific Ocean	20 to −15	140 to 170	40	495	1519	Atlas et al., 1993
Dec. 1981	Eastern Pacific	20 to −15	−108 to −97		50		Singh and Salas, 1982
Feb.–Mar. 1990	Equatorial Pacific Ocean	20 to −15	140 to 170	40	426	1519	Atlas et al., 1993
Mar.–Apr.	Atlantic Ocean	0 to −30	−30		25		Rudolph and Johnen, 1989
Feb.–Mar. 1990	Equatorial Pacific Ocean	20 to −15	140 to 170	40	426	1519	Atlas et al., 1993
Mar.–Apr.	Atlantic Ocean	0 to −30	−30		25		Rudolph and Johnen, 1989
Dec. 1981	Eastern Pacific	−20 to −32	−75		80		Singh and Salas, 1982
Jan.–Mar. 1984	Amsterdam Island	−37.47	77.31	147		294	Bonsang et al., 1990
Jan.–Feb. 1985	Amsterdam Island	−37.47	77.31	178		434	Bonsang et al., 1990
Feb. 1987	Amsterdam Island	−37.47	77.31		549		Bonsang et al., 1990
Jan.	Antarctic	−70	−8		250		Rudolph et al., 1989
Feb.	Antarctic	−70	−8		200		Rudolph et al., 1989
Mar.	Antarctic	−70	−8		410		Rudolph et al., 1989
Mar.	Antarctic	−70	−8		470		Rudolph et al., 1989
Winter, continental[f]							
Winter 1988	Norway	59	8		2860		Hov et al., 1991
Winter daytime	Central Europe	48	10		1500		Rudolph et al., 1980
Jan. 1992	Arctic and subpolar	50 to 40			31		Anderson et al., 1993
Mar. 1992	Arctic and subpolar	50 to 40			38		Anderson et al., 1993
Winter 1992	Northern subpolar	60 to 40			44		Anderson et al., 1993
Spring, oceanic[e]							
Apr. 1985	U.S. West Coast, Alaska	71	−157		<10		Greenberg et al., 1990
Apr. 1985	U.S. West Coast, Alaska	66	−147		<10		Greenberg et al., 1990
Apr. 1985	U.S. West Coast, Alaska	61	−150		719		Greenberg et al., 1990

continued

TABLE 4 (*Continued*)

Date	Location	Latitude[a]	Longitude[b]	Concentration[c]			Reference
				min	mean	max	
Apr. 1979, 12 GMT[d]	West coast of Ireland	55	-10		300		Rudolph et al., 1980
Apr. 1985	Bermuda	35	-70		64[e]		Van Valin and Luria, 1988
Mar. 21–Apr. 18, 1987	Atlantic	30 to 0	-30	22		120	Rudolph and Johnen, 1990
Mar.–Apr. 1982	Atlantic Ocean	45 to 0	-30		58		Rudolph and Johnen, 1990
May 1979, 14–18 GMT	Equatorial Atlantic	2 to 0	-22		100–250		Rudolph et al., 1980
Mar. 21–Apr. 18, 1987	Atlantic	0 to -30	-30	10		130	Rudolph and Johnen, 1990
Mar.–Apr.	Atlantic Ocean	0 to -30			25		Rudolph and Johnen, 1990
Feb.–Mar. 1990	Pacific Ocean	20 to -12	-145 to -170	30		150	Donahue and Prinn, 1993
May 1979, 10 GMT[d]	Atlantic Ocean	2 to -2	-22		450		Rudolph et al., 1980
May 1979, 10 GMT[d]	Equatorial Atlantic	-1	-25		450–500		Rudolph et al., 1980
May–Jun. 1984	Amsterdam Island	-37.47	77.31	177		337	Bonsang et al., 1990
Apr.–Jun. 1986	Amsterdam Island	-37.47	77.31	447		624	Bonsang et al., 1990
May 1987	Amsterdam Island	-37.47	77.31		579		Bonsang et al., 1990
Apr.–Jun. 1985	Amsterdam Island	-37.47	77.31	312		660	Bonsang et al., 1990
Apr.	Antarctic	-70	-8		900		Rudolph et al., 1989
May	Antarctic	-70	-8		280		Rudolph et al., 1989
Jun.	Antarctic	-70	-8		400		Rudolph et al., 1989
Spring, continental[f]							
Spring	Arctic (Barrow, Alaska)	70			75		Rasmussen and Khalil, 1983
Spring 1988	Norway	59	8		2000		Hov et al., 1991
Apr.	Orinoco basin, Venezuela	9	-65		600		Donoso et al., 1996
Summer, oceanic[f]							
Jul. 1982	Norwegian Arctic	74	19		255		Hov et al., GRL 11, 1984
Aug.–Sep. 1986	Amsterdam Island	-37.47	77.31	336		446	Bonsang et al., 1990
Aug.–Sep. 1984	Amsterdam Island	-37.47	77.31	563		572	Bonsang et al., 1990
Jul.–Sep. 1985	Amsterdam Island	-37.47	77.31	265		486	Bonsang et al., 1990
Aug.–Sept. 1996	Amsterdam Island	-37.47	77.31	199		467	Bonsang et al., 1990
Jul.	Antarctic	-70	-8		370		Rudolph et al., 1989
Aug.	Antarctic	-70	-8		320		Rudolph et al., 1989
Sep.	Antarctic	-70	-8		190		Rudolph et al., 1989
Summer, continental[f]							
Summer 1987	Norway	-59	8		1670		Hov et al., 1991
Summer, daytime	Central Europe	48	-10		100–300		Rudolph et al., 1980
Aug.–Sep. 1980	Niwot Ridge, CO	40	-105	140	460	880	Greenberg and Zimmerman, 1984

Date	Location	Latitude[a]	Longitude[b]				Reference
Aug.–Sep. 1980	Pawnee grasslands, CO	10	−105	950	1820	3440	Greenberg and Zimmerman, 1984
Sep. 1979, 12 GMT[d]	Sahara Desert	30 to 20	−10 to 10		200		Rudolph et al., 1980
Sep. (11:00–16:00)	Orinoco basin, Venezuela	9	−65		1520		Donoso et al., 1996
Jul.–Aug. (30–305 m)	Amazon basin	−2	−60	630	970	1680	Zimmerman et al., 1988
Aug.–Sep. 1980	Brazil	−5	−55	1150	1890	3440	Greenberg and Zimmerman, 1984
Fall, oceanic[f]							
Sep.–Oct.	Atlantic Ocean	45 to 0	−30		45		Koppman et al., 1992
Dec. 1981	Eastern Pacific	40 to 30	−124 to −117		120		Singh and Salas, 1982
Sep.–Oct. 1982	Atlantic	30 to 0		12		58.5	Bonsang and Lambert, 1985
Sep.–Oct. 1982	Atlantic	0 to −30		10		56	Bonsang and Lambert, 1985
Sep.–Oct.	Atlantic Ocean	0 to −30	−30		22		Koppman et al., 1992
Dec. 1981	Eastern Pacific	0 to −10	−89 to −79		70		Singh and Salas, 1982
Dec. 1981	Eastern Pacific	−10 to −20	−79 to −75		70		Singh and Salas, 1982
Sep.–Oct.	Atlantic Ocean	−10 to −30	−30		22		Koppman et al., 1992
Oct.–Dec. 1984	Amsterdam Island	−37.47	77.31	718		993	Bonsang et al., 1990
Oct.–Dec. 1985	Amsterdam Island	−37.47	77.31	720		855	Bonsang et al., 1990
Oct.–Dec. 1986	Amsterdam Island	−37.47	77.31	238		260	Bonsang et al., 1990
Nov.	Antarctic	−70	−8		120		Rudolph et al., 1989
Dec.	Antarctic	−70	−8		190		Rudolph et al., 1989
Fall, continental							
Fall 1987	Norway	59	8		1600		Hov et al., 1991

[a] 90°–0°N: > 0; 0°–90°S: < 0.
[b] 0°–180°W: < 0; 0°–180°E: > 0.
[c] Values in pptv (parts per trillion volume of compound).
[d] GMT, Greenwich mean time.
[e] Averaged over 1000 m.
[f] Seasons are defined for North Hemisphere.

TABLE 5 Measurements of Propene in the Troposphere

Date	Location	Latitude[a]	Longitude[b]	Concentration[c]			Reference
				min	mean	max	
Mar. 1989	Point Barrow			180	283	310	Doskey and Gaffney, 1992
Spring 1985	Ny-Alesund	78.5	11.6		106		Hov et al., 1989
Spring 1986	Ny-Alesund	78.5	11.6				Hov et al., 1989
Jan.–Apr. 1992	Alert	82.5	−62.3	nd[d]		nd[d]	Jobson et al., 1994
Sep. 1979, 11 GMT[e]	Atlantic Ocean	75	25		50		Rudolph et al., 1980
Apr. 1985	U.S. West Coast, Alaska	71	−157		49		Greenberg et al., 1990
Apr. 1985	U.S. West Coast, Alaska	66	−174		32		Greenberg et al., 1990
Apr. 1985	U.S. West Coast, Alaska	61	−150		89		Greenberg et al., 1990
Summer 1987	Norway	59	8		820		Hov et al., 1991
Fall 1987	Norway	59	8		810		Hov et al., 1991
Winter 1988	Norway	59	8		1000		Hov et al., 1991
Spring 1988	Norway	59	8		610		Hov et al., 1991
Apr. 1979, 12 GMT[e]	West coast of Ireland	53	−10		200		Rudolph et al., 1980
Apr. 1985	U.S. West Coast, Alaska	48	−126		<10		Greenberg et al., 1990
Aug.–Sep. 1980	Niwot Ridge, CO	40	−105	90	160	360	Greenberg and Zimmerman, 1984
Aug.–Sep. 1980	Pawnee grasslands, CO	40	−105	230	540	1760	Greenberg and Zimmerman, 1984
Dec. 1981	Eastern Pacific	40 to 30	−124 to −117		50		Singh and Salas, 1982
Mar. 1985	U.S. East Coast	40 to 20	−60 to −90		51–77		Van Valin and Luria, 1988
Summer, daytime	Central Europe				100–500		Rudolph et al., 1980
Winter, daytime	Central Europe				400		Rudolph et al., 1980
Apr. 1985	U.S. West Coast, Alaska	38	−124		42		Greenberg et al., 1990
Apr. 1985	Bermuda	35	−70		47		Van Valin and Luria, 1988
Apr. 1985	U.S. West Coast, Alaska	30	−117		<60		Greenberg et al., 1990
Dec. 1981	Eastern Pacific	30 to −20	−117 to −108		180		Singh and Salas, 1982

Mar.–Apr. 1987	Atlantic	30 to 0		10		40	Rudolph and Johnen, 1990
Jan. 1979, 9 GMT[e]	Atlantic Ocean	30 to 15	−22		200		Rudolph et al., 1980
Dec. 1981	Eastern Pacific	20 to 10	−108 to −97		300		Singh and Salas, 1982
Feb.–Mar. 1990	Equatorial Pacific Ocean	20 to −15	140 to 170	24	228	1111	Atlas et al., 1993
Jul., Dec. 1982	Pacific	17 to −11	−160 to −80	70		490	Greenberg and Zimmerman, 1984
Dec. 1981	Eastern Pacific	10 to 0	−97 to −89		160		Singh and Salas, 1982
Apr.	Orinoco basin, Venezuela	9	−65		380		Donoso et al., 1996
Sep.	Orinoco basin, Venezuela	9	−65		630		Donoso et al., 1996
May 1979, 14–18 GMT[e]	Equatorial Atlantic	2 to 0	−22		90–140		Rudolph et al., 1980
May 1979, 10 GMT[e]	Atlantic Ocean	2 to −2	−22		100		Rudolph et al., 1980
May 1979, 10 GMT[e]	Equatorial Atlantic	−1	−26 to −22		90–100		Rudolph et al., 1980
Jul.–Aug.	Amazon basin	−2	−60	210	310	490	Zimmerman et al., 1988
Aug.–Sep. 1980	Brazil	−5	−55	130	310	550	Greenberg and Zimmerman, 1984
Dec. 1981	Eastern Pacific	0 to −10	−89 to −79		150		Singh and Salas, 1982
Dec. 1981	Eastern Pacific	−10 to −20	−79 to −75		280		Singh and Salas, 1982
Dec. 1981	Eastern Pacific	−20 to −32	−75 to −72		70		Singh and Salas, 1982
Mar.–Apr. 1987	Atlantic	0 to −30		0		30	Rudolph and Johnen, 1996
Jan.–Mar.	Antarctic	−70	−8	90	310		Rudolph et al., 1989
Apr.–Jun.	Antarctic	−70	−8	190	490		Rudolph et al., 1989
Jul.–Sep.	Antarctic	−70	−8	120	240		Rudolph et al., 1989
Nov.–Dec.	Antarctic	−70	−8	90	100		Rudolph et al., 1989

[a]90°–0°N: >0; 0°–90°S: <0.

[b]0°–180°W: <0; 0°–180°E: >0.

[c]Values in parts per trillion volume of compound.

[d]nd: no data.

[e]GMT, Greenwich mean time.

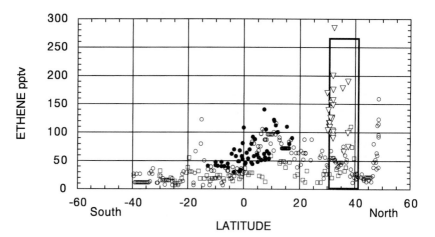

FIGURE 10 Latitudinal variation of the ethene mixing ratio in the marine boundary layer. For the Atlantic Ocean: data are taken from Rudolph and Johnen (1990), open circles; Koppmann *et al.* (1992), squares; and for the ASTEX experiment from Carsey *et al.* (1997), triangles; results from Blake *et al.* (1996) for the same experiment (aircraft data, boundary layer) are within the rectangle. For the equatorial Pacific results are taken from Donahue and Prinn (1993), closed circles.

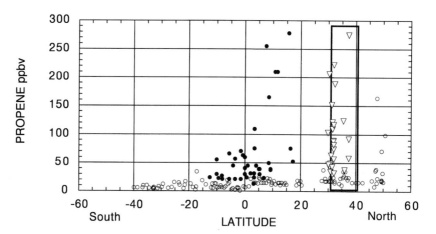

FIGURE 11 Latitudinal variation of the propene mixing ratio in the marine boundary layer. For the Atlantic Ocean: data are taken from Rudolph and Johnen (1990), open circles; for the ASTEX experiment from Carsey *et al.* (1997), triangles; results from Blake *et al.* (1996) for the same experiment (aircraft data, boundary layer) are within the rectangle. For the equatorial Pacific results are taken from Donahue and Prinn (1993), closed circles.

obtained, using flask samples, by Rudolph *et al.* (1989) at 70° S, who calculated yearly averaged mixing ratios of 360 pptv for ethene and of 210 pptv for propene, with minima close to 180 and 130 pptv, respectively. Unfortunately, such relatively high mixing ratios cannot be compared with other measurements.

2. Seasonal Variability of Alkenes

Some seasonal variabilities in light alkene concentrations have been reported in the literature. Seasonal variability was discussed on the basis of an extensive set of measurements by Greenberg *et al.* (1996) for samples collected during the MLOPEX 2 experiment conducted at Mauna Loa, from September 1991 through August 1992. They observed a decreasing trend for ethene from January (40 pptv) to the end of February (10 pptv), similar to those found for other hydrocarbons and CO. However, on the whole, the seasonal variation in alkenes is not as clear as that of the other hydrocarbons; and whatever the season, ethene and propene concentrations are in the range of 0 to 40 and 0 to 20 pptv, respectively.

D. Aromatic and Oxygenated Hydrocarbons

Relatively few data are available in the literature for the aromatic (mainly toluene and benzene, Table 6) and oxygenated hydrocarbons (Table 7), although analytical improvements for the latter group should allow more data to be obtained in the near future. Thus, it is difficult to get a clear picture of the global distribution of this group of reactive VOCs at ground level, as well as of their tropospheric vertical distributions.

A detailed study made by Singh *et al.* (1995) on aromatic hydrocarbons suggested a dramatic reduction in the ambient concentration of benzene and toluene between 1970 and 1980 around the southern coast of the United States, due to the introduction of catalytic converters on motor vehicles. Strong diurnal and seasonal variations were reported for toluene, benzene (the dominant aromatic species), and other aromatic hydrocarbons [ethyl benzene, (*m* + *p*)-xylene] in urban atmospheres in the United States (see, e.g., Singh and Zimmerman, 1992). A clear minimum was observed in the afternoon, due to both the reactivity of aromatic species (chemical loss by OH radicals) and the prevailing meteorology (convective mixing). Benzene, with its relatively long lifetime in the troposphere, is usually found at levels on the order of 100 pptv in the atmosphere. In June 1992, an extensive set of

TABLE 6 Concentration of Aromatic Hydrocarbons

Date	Location	Latitude[a]	Longitude[b]	Concentration[c]			Reference
				min	mean	max	
Benzene							
Apr. 1985	U.S. West Coast, Alaska	71	−157		179		Greenberg et al., 1990
Spring	Arctic (Barrow, Alaska)	70		319			Rasmussen et al., 1983
Apr. 1985	U.S. West Coast, Alaska	66	−147		136		Greenberg et al., 1990
Apr. 1985	U.S. West Coast, Alaska	61	−150		160		Greenberg et al., 1990
Apr. 1985	U.S. West Coast, Alaska	48	−126		122		Greenberg et al., 1990
Aug.–Sep. 1980	Niwot Ridge, CO	40	−105	30	240	900	Greenberg and Zimmerman, 1984
Aug.–Sep. 1980	Pawnee grasslands, CO	40	−105	180	580	1660	Greenberg and Zimmerman, 1984
Apr. 1985	U.S. West Coast, Alaska	38	−124		99		Greenberg et al., 1990
Apr. 1985	U.S. West Coast, Alaska	30	−117		151		Greenberg et al., 1990
Jul., Dec. 1982	Pacific	17 to −11	−80 to −160	40		1860	Greenberg and Zimmerman, 1984
Jul.–Aug.	Amazon basin (0–305 m)	−2	−60	60	90	140	Zimmerman et al., 1988
Aug.–Sep. 1980	Brazil	−5	−55	310	500	720	Greenberg and Zimmerman, 1984
Dec. 1991–Dec. 1992	New Zealand (polluted)	−41	174	0	46	386	Clarkson et al., 1996
Dec. 1991–Dec. 1992	New Zealand (marine)	−41	174	0	6	127	Clarkson et al., 1996
Toluene							
Aug.–Sep. 1980	Niwot Ridge, CO	40	−105	10	140	320	Greenberg and Zimmerman, 1984
Aug.–Sep. 1980	Pawnee grasslands, CO	40	−105	190	360	770	Greenberg and Zimmerman, 1984
Jul., Dec. 1982	Pacific Ocean	17 to −11	−80 to −160	40		1490	Greenberg and Zimmerman, 1984
Jul.–Aug.	Amazon basin (0–305 m)	−2	−60	350	740	2550	Zimmerman et al., 1988
Aug.–Sep. 1980	Brazil	−5	−55	40	120	190	Greenberg and Zimmerman, 1984
Dec. 1991–Dec. 1992	New Zealand (polluted)	−41	174	0	49	604	Clarkson et al., 1996
Dec. 1991–Dec. 1992	New Zealand (marine)	−41	174	0	4	144	Clarkson et al., 1996

[a]90°–0°N: >0; 0°–90°S: <0.
[b]0°–180°W: <0; 0°–180°E: >0.
[c]Values in parts per trillion volume of compound.

TABLE 7 Measurements of Organic Acids, Aldehydes and Ketones in the Troposphere

Date	Location	Latitude[a]	Longitude[b]	Concentration[c]					
				HCOOH			CH$_3$COOH		
				min	mean	max	min	mean	max
Organic acids (from Arlander et al., 1990)									
May–Jun.	Pacific–Indian oceans	45 to 0	160 to 170	70	800	1720	70	780	1920
Jun.–Jul.	Indian Ocean	45 to 0	90	360	860	1100	270	690	1190
Jul.	Pacific–Indian oceans	5	104 to −155	20	400	1920	10	450	1510
May–Jun.	Pacific–Indian oceans	0 to −45	160 to 170	30	220	630	100	290	780
Jun.–Jul.	Indian Ocean	0 to −45	90	20	190	790	50	280	1120
Jun.	Indian Ocean	−40	90 to 175	30	110	260	40	210	680
				HCHO			CH$_3$CHO		
				min	mean	max	min	mean	max
Aldehyde (from Solberg et al., 1996)									
Jan.	Europe, rural sites	70 to 80	0 to 10		—			—	
		50 to 60	0 to 10		229–614			221–357	
		40 to 50	0 to 10		571–2344			319–1344	
Feb.	Europe, rural sites	70 to 80	0 to 10		—			—	
		50 to 60	0 to 10		378–788			207–519	
		40 to 50	0 to 10		706–2036			420–1282	
Mar.	Europe, rural sites	70 to 80	0 to 10		—			—	
		50 to 60	0 to 10		284–533			160–356	
		40 to 50	0 to 10		760–2394			472–1320	
Apr.	Europe, rural sites	70 to 80	0 to 10		153			168	
		50 to 60	0 to 10		442–991			217–446	
		40 to 50	0 to 10		887–1888			477–1055	
May	Europe, rural sites	70 to 80	0 to 10		276			316	
		50 to 60	0 to 10		720–827			334–407	
		40 to 50	0 to 10		1322–2064			577–991	
Jun.	Europe, rural sites	70 to 80	0 to 10		271			98	
		50 to 60	0 to 10		728–788			275–325	
		40 to 50	0 to 10		1289–3439			556–1309	

(continues)

TABLE 7 (continued)

Date	Location	Latitude[a]	Longitude[b]	Concentration[c]						
				min	max	mean	max	min	mean	max
Jul.	Europe, rural sites	70 to 80	0 to 10			—			—	
		50 to 60	0 to 10			899–967			399–483	
		40 to 50	0 to 10			1372–4248			647–1484	
Aug.		70 to 80	0 to 10			197			80	
		50 to 60	0 to 10			641–754			243–439	
		40 to 50	0 to 10			1872–5903			787–1504	
Sep.	Europe, rural sites	70 to 80	0 to 10			144			81	
		50 to 60	0 to 10			468–706			212–391	
		40 to 50	0 to 10			738–2487			454–1027	
Oct.		70 to 80	0 to 10			—			—	
		50 to 60	0 to 10			341–710			240–460	
		40 to 50	0 to 10			894–1737			541–1086	
Nov.	Europe, rural sites	70 to 80	0 to 10			—			—	
		50 to 60	0 to 10			292–868			176–640	
		40 to 50	0 to 10			792–2017			524–1229	
Dec.	Europe, rural sites	70 to 80	0 to 10			—			—	
		50 to 60	0 to 10			230–665			106–433	
		40 to 50	0 to 10			452–2026			288–1276	
Acetone (from Solberg et al., 1996)										
Jan.	Europe, rural sites	70 to 80	0 to 10							
		50 to 60	0 to 10			0.397–1.167				
		40 to 50	0 to 10			0.678–1.875				
Feb.	Europe, rural sites	70 to 80	0 to 10							
		50 to 60	0 to 10			0.333–1.283				
		40 to 50	0 to 10			0.855–1.997				
Mar.	Europe, rural sites	70 to 80	0 to 10							
		50 to 60	0 to 10			0.406–1.119				
		40 to 50	0 to 10			1.108–2.184				

Month	Location	Latitude	Longitude	Value
Apr.	Europe, rural sites	70 to 80	0 to 10	0.538
		50 to 60	0 to 10	1.085–1.325
		40 to 50	0 to 10	1.308–1.970
May	Europe, rural sites	70 to 80	0 to 10	0.624
		50 to 60	0 to 10	0.835–1.481
		40 to 50	0 to 10	1.657–1.793
Jun.	Europe, rural sites	70 to 80	0 to 10	0.548
		50 to 60	0 to 10	0.773–1.248
		40 to 50	0 to 10	1.650–1.847
Jul.	Europe, rural sites	70 to 80	0 to 10	—
		50 to 60	0 to 10	1.029–1.629
		40 to 50	0 to 10	1.724–1.894
Aug.	Europe, rural sites	70 to 80	0 to 10	0.552
		50 to 60	0 to 10	0.823–1.528
		40 to 50	0 to 10	1.566–2.019
Sep.	Europe, rural sites	70 to 80	0 to 10	0.494
		50 to 60	0 to 10	0.803–1.227
		40 to 50	0 to 10	1.125–1.495
Oct.	Europe, rural sites	70 to 80	0 to 10	—
		50 to 60	0 to 10	0.562–1.153
		40 to 50	0 to 10	1.009–1.592
Nov.	Europe, rural sites	70 to 80	0 to 10	—
		50 to 60	0 to 10	0.390–1.112
		40 to 50	0 to 10	0.884–2.025
Dec.	Europe, rural sites	70 to 80	0 to 10	—
		50 to 60	0 to 10	0.237–0.669
		40 to 50	0 to 10	0.507–1.837

[a] 90°–0°N: >0; 0°–90°S: <0.
[b] 0°–180°W: <0; 0°–180°E: >0.
[c] Values in parts per trillion volume of compound.

measurements was performed by Blake *et al.* (1996a) over the Atlantic during the ASTEX/MAGE cruise. In the marine boundary layer, the authors reported benzene mixing ratios from 20 to 120 pptv. Free tropospheric values were much lower (20 to 6 pptv). This large variability in the marine boundary layer was clearly related to the origin of the air masses and, in particular, to the impact of industrialized continents. For instance, the impact of European outflow led to values higher by a factor of four compared to samples collected in clean Atlantic air.

In remote areas of the Southern Hemisphere, benzene is still measurable at levels on the order of 10 to 15 pptv (Rudolph *et al.*, 1984), whereas toluene remains frequently below the detection limit of a few parts per thousand volume. Measurements by Clarkson *et al.* (1996) in New Zealand confirm that the levels generally observed in clean marine air are around 10 pptv and exhibit a seasonal cycle with a winter maximum. Significant increases from 40 to 50 pptv for both compounds are correlated with the advection of polluted air masses. The benzene/toluene ratio remains close to 0.6 and agrees with the observations made in more polluted regions

Latitudinal and seasonal distributions of organic acids (formic and acetic) and formaldehyde over the Pacific and Indian oceans were measured during the Soviet-American Gas and Aerosol (SAGA II) expedition cruise (Arlander *et al.*, 1990). Distributions of aldehydes (formadehyde and acetaldehyde) and acetone over rural European sites are reported in Solberg *et al.* (1996).

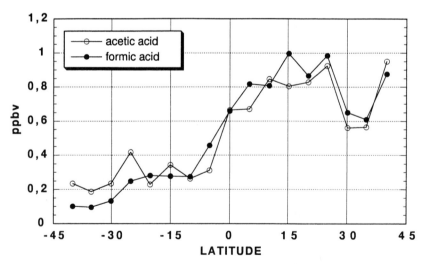

FIGURE 12   Latitudinal variations of acetic and formic acids and acetic/formic acid ratio during the 1987 Korolev cruise in the Pacific Ocean; adapted from Arlander *et al.* (1990).

Arlander *et al.* (1990) observed seasonal and latitudinal dependencies for organic acids and formaldehyde, reflecting the latitudinal gradient of their suspected precursors (nonmethane hydrocarbons). Reported values are shown in Table 7. A clear hemispheric gradient can be observed, with a decreasing trend with latitude for acetic and formic acids mixing ratios from 40° N to 40° S (Fig. 12). The authors reported a maximum for formaldehyde mixing ratios between 20° N and the equator, probably due to CH_4 and O_3 alkene photochemical oxidation, as modeled by Logan *et al.* (1981) (Fig. 13). Mean Northern and Southern Hemisphere values of 860 and 190, and of 690 and 290 pptv, are given for formic and acetic acids, respectively. A longitudinal study clearly suggested anthropogenic influences on the organic acids distributions.

Formaldehyde, acetaldehyde, and acetone were found to represent 85% or more of the total carbonyls measured by Solberg *et al.* (1996). The seasonal distributions of carbonyl concentrations were quite opposite to those of the hydrocarbons (Fig. 14), with a summer maximum, whatever the type of site (remote or rural). Nevertheless, this maximum and the seasonal amplitude can vary, with a more pronounced fluctuation for acetone, reflecting differences in biogenic emissions of VOCs and in the efficiency of formation of these secondary products by the oxidation of their precursors. In summer, the sum of the carbonyl concentrations was found to be of the same order as the sum of the C_2 to C_3 hydrocarbon concentrations. On a global scale, Singh *et*

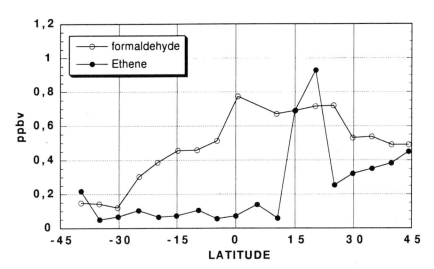

FIGURE 13   Latitudinal variations of formaldehyde and ethene during the 1987 Korolev cruise in the Pacific Ocean; adapted from Arlander *et al.* (1990).

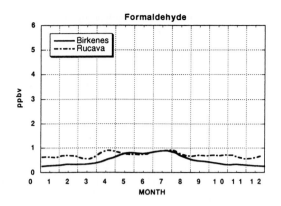

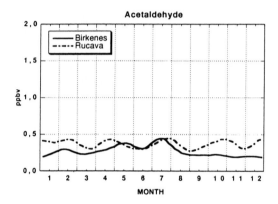

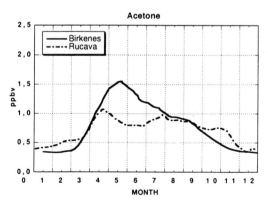

FIGURE 14 Average seasonal cycles of formaldehyde, acetaldehyde, and acetone (all in ppbv) for several locations of the Northern Hemisphere: Birkenes (Norway), Rucava (Lithuania), Ispra (Italy), Kosetice (Czech Republic), Waldhof (Germany); adapted from Solberg *et al.* (1996).

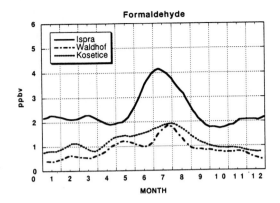

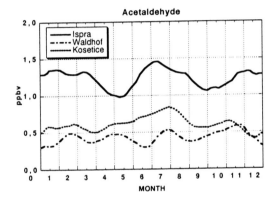

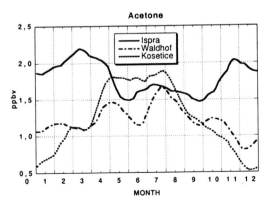

FIGURE 14   *(Continued)*

TABLE 8 Measurements of Biogenic NMHC in the Troposphere[a]

Date	Location	Latitude[b]	Longitude[c]	Concentration[d]			Reference
				min	mean	max	
Isoprene							
Summer 1982	Niwot Ridge, CO	40	−105	0.22	0.63	1.76	Greenberg and Zimmerman, 1984
Nov. 1982	Niwot Ridge, CO	40	−105	0.03	0.11	0.16	Greenberg and Zimmerman, 1984
Jul., daytime	Amazon basin				2.24		Rasmussen and Khalil, 1988
Jul., daytime	Amazon basin				1.93		Rasmussen and Khalil, 1988
Apr.	Orinoco basin, Venezuela	9	−65		1.6		Donoso et al., 1996
Sep.	Orinoco basin, Venezuela	9	−65		3.29		Donoso et al., 1996
	Lamto, Ivory Coast				1.14		Bonsang et al., 1991
Jul.–Aug.	Amazon basin	−2	−60	1.14	2.04	2.72	Zimmerman et al., 1988
Aug.–Sep. 1980	Brazil	−5	−55	1.00	2.40	5.24	Greenberg and Zimmerman, 1984
α-Pinene							
Summer 1982	Niwot Ridge, CO	40	−105	0.01	0.14	0.66	Greenberg and Zimmerman, 1984
Nov. 1982	Niwot Ridge, CO	40	−105	0.03	0.07	0.11	Greenberg and Zimmerman, 1984
Jul.–Aug.	Amazon basin	−2	−60	0.05	0.1	0.15	Zimmerman et al., 1988
β-Pinene							
Summer 1982	Niwot Ridge, CO	40	−105	0.01	0.08	0.39	Greenberg and Zimmerman, 1984
Nov. 1982	Niwot Ridge, CO	40	−105	0.01	0.07	0.11	Greenberg and Zimmerman, 1984
Jul.–Aug.	Amazon basin	−2	−60	0.01	0.03	0.04	Zimmerman et al., 1988
Aug.–Sep. 1980	Brazil	−5	−55	0.07	0.27	0.54	Greenberg and Zimmerman, 1984

Compound / Period	Site	Lat.	Long.				Reference
δ-3-Carene							
Summer 1982	Niwot Ridge, CO	40	−105	0.01	0.05	0.19	Greenberg and Zimmerman, 1984
Nov. 1982	Niwot Ridge, CO	40	−105	0.01	0.03	0.04	Greenberg and Zimmerman, 1984
Jul.–Aug.	Amazon basin	−2	−60	0	0.01	0.04	Zimmerman et al., 1988
Aug.–Sep. 1980	Brazil	−5	−55	0.07	0.27	0.54	Greenberg and Zimmerman, 1984
Myrcene							
Aug.–Sep. 1980	Brazil	−5	−55	0.07	0.27	0.54	Greenberg and Zimmerman, 1984
α-Phellandrene							
Aug.–Sep. 1980	Brazil	−5	−55	0.11	0.18	0.28	Greenberg and Zimmerman, 1984
α-Terpinene							
Summer 1982	Niwot Ridge, CO	40	−105	0.01	0.04	0.05	Greenberg and Zimmerman, 1984
Aug.–Sep. 1980	Brazil	−5	−55	0.12	0.49	0.81	Greenberg and Zimmerman, 1984
γ-Terpinene							
Aug.–Sep. 1980	Brazil	−5	−55	0.03	0.11	0.18	Greenberg and Zimmerman, 1984
α-Terpineol							
Aug.–Sep. 1980	Brazil	−5	−55	0.04	0.76	1.46	Greenberg and Zimmerman, 1984
Linalool							
Aug.–Sep. 1980	Brazil	−5	−55	0.11	0.20	0.30	Greenberg and Zimmerman, 1984

[a] Over continent.
[b] 90°–0°N: >0; 0°–90°S: <0.
[c] 0°–180°W: <0; 0°–180°E: >0.
[d] Values in parts per billion volume of compound.

al. (1995) have reported airborne measurements of acetone and methanol in the free troposphere (5 to 10 km altitude) over a wide range of latitude (40° N to 5° S) in the Pacific Ocean, showing for both compounds a regular decrease: for acetone from 500 to 600 pptv at northern latitudes to 250 pptv in southern areas and for methanol from 600–800 to 400 pptv. These results tend to suggest that oxygenated species are globally ubiquitous at relatively high concentrations in the troposphere.

E. BIOGENIC HYDROCARBONS

Sources of biogenic hydrocarbons, mainly isoprene (C_5H_8) and monoterpenes ($C_{10}H_{16}$), are widely spread over the earth's surface and they are believed to dominate the global flux of reactive VOCs to the atmosphere (Müller, 1992; Guenther *et al.*, 1995). However, the high reactivity (atmospheric lifetimes of minutes to hours) of these hydrocarbons and the very large spatial and temporal variations in their emission rates lead to a tremendous variability in their ambient concentrations. A global distribution is thus impossible to obtain or to calculate. They are mainly found in the lower boundary layer and in forest canopies, with virtually no isoprene or monoterpenes in the free troposphere. However, it is now clear that these compounds strongly influence ozone formation at a local or regional scale (Lübkert and Shoepp, 1989; Lamb *et al.*, 1987). Table 8 presents measurements of isoprene and of the most abundant monoterpenes in different environments. Clear and strong diurnal cycles can be observed for isoprene, with maximal emissions and removal (by reaction with O_3, and OH) during daytime, but with minimal ambient levels during the night. In contrast to isoprene, monoterpenes can be stored in specific reservoirs in the plants and their emissions are more dependent on temperature. Hence, a different diurnal cycle can be observed with maximal ambient concentrations during the nighttime (minimal convective mixing and chemical removal).

Seasonality has also been observed in long-term studies in rural or forested areas (see e.g., Yokouchi and Ambe, 1988; Seila, 1984; Janson, 1992) with higher ambient levels expected during the summer.

V. CONCLUSIONS

The global distributions of reactive hydrocarbons in the atmosphere are relatively well described for a limited number of species, such as ethane, propane, and acetylene. For these compounds it is possible to draw a clear picture of their latitudinal and seasonal variations. A strong interhemispheric gradient exists with a reduction close to one order of magnitude near the equator. For these species it is also possible to estimate their global source strength and even establish a separate budget for the northern and southern hemispheres. Blake and Rowland (1986) estimated the chemical loss of ethane per latitudinal band, and integrated over the globe, giving a value of 13 ± 3 t per year. Rudolph (1995) used a compilation of about 1500 measurements and an evaluation of the chemical loss per interval of 10° latitude and 1-km altitude, to arrive at a global ethane flux of 15.5 ton per year. Hemispheric budgets have also been estimated by different authors: in the Southern Hemisphere Clarkson *et al.* (1997) have established the turnover of ethane and propane and compared it to the north to south flux and the magnitude of the main sources identified for this hemisphere. A reasonable agreement was obtained with figures of 4.9 t per year for ethane and 2.2 t per year for propane. Light NMHC global budgets were also derived by Boissard *et al.* (1996) on the basis of the large scale TROPOZ II measurements. A figure, similar to that reported by the other authors has been obtained for ethane (16 to 17 t per year). For propane a budget of 26 t per year, was estimated, but with a very large uncertainty. However, this value seems relatively high if we take into account the probable small contribution of the Southern Hemisphere. An acetylene budget of 7 to 10 t per year was also assessed, with large contributions from anthropogenic sources and from biomass burning emissions.

The situation is much more complex for higher alkanes ($> C_4$), for alkenes, dienes (isoprene), and terpenes whose lifetimes are shorter (on the order of a day or less). Measurements of these short-lived compounds show a considerable scatter in the boundary layer as well in the free troposphere. This is partly due to analytical difficulties and also to the occurrence of strong source impacts that creates large inhomogeneities in their distribution. Such effects are clearly visible in northern mid-latitudes in the vicinity of industrialized areas, under the influence of fast transport processes, or in the tropical latitudes under the influence of biomass burning emissions. Considerable work remains to be done in arriving at a quantitative description of the global distribution of these and other reactive VOCs in the atmosphere.

REFERENCES

Aikin, A. C., Gallagher, C. C., Spicer, C. W., and Holdren, M. W. (1987). Measurement of methane and other light hydrocarbons in the troposphere and lower stratosphere. *J. Geophys. Res.* **92**, 3135–3138.

Anderson, B. E., Collins, J. E., Sachse, G. W., Whiting, G. W., Blake, D. R., and Rowland, F. S. (1993). AASE-II observations of trace carbon species distributions in the mid to upper troposphere. *Geophys. Res. Let.* **20**, 2539–2542.

Andreae, M. O., Fishman J., and Lindesay, J. (1996). The Southern Tropical Region Experiment (STARE): transport and atmospheric chemistry near the Equator-Atlantic (TRACE A) and southern African fire-atmosphere research initiative (SAFARI): an introduction. *J. Geophys. Res.* **101**, 23,519–23,520.

Arlander, D. W., Cronn, D. R., Farmer, J. C., Menzia, F. A., and Westberg, H. H. (1990). Gaseous oxygenated hydrocarbons in the remote marine troposphere. *J. Geophys. Res.* **95**, 16,391–16,403.

Arnts, R. R., and Meeks, S. A. (1981). Biogenic hydrocarbons contribution to the ambient air in selected areas. *Atmos. Environ.* **15**, 1643–1651.

Atlas, E., Pollock, W., Greenberg, J., Heidt, L., and Thompson, A. M. (1993). Alkyl nitrates, nonmethane hydrocarbons, and halocarbon gases over the Equatorial Pacific Ocean during SAGA 3. *J. Geophys. Res.* **98**, 16933–16947.

Beine, H. J., Jaffe, D. A., Blake, D. R., Atlas, E., and Harris, J. (1996). Measurements of PAN, alkyl nitrates, ozone, and hydrocarbons during spring in interior Alaska. *J. Geophys. Res.* **101**, 12,613–12,619.

Bertrand, I. (1989). Contribution à l'étude des hydrocarbures légers non méthaniques en milieu non pollué. Rapport de stage de Maîtrise–Université de Paris VII.

Blake, D. R. and Rowland, F. S. (1986). Global atmospheric concentrations and source strength of ethane. *Nature.* **321**, 231–233.

Blake, D. R., Hurst, D. F., Smith, T. W., Jr., Whipple, W. J., Chen, T.-Y., Blake, N. J., and Rowland, F. S. (1992). Summertime measurements of selected nonmethane hydrocarbons in the Arctic and Subarctic during the 1988 Arctic Boundary Layer Expedition (ABLE 3A). *J. Geophys. Res.* **97**, 16,559–16,588.

Blake, D. R., Smith, T. W., Jr., Chen, T.-Y., Whipple, W. J., and Rowland, F. S. (1994). Effects of biomass burning on summertime nonmethane hydrocarbon concentrations in the Canadian wetlands. *J. Geophys. Res.* **99**, 1699–1719.

Blake, D. R., Blake, N. J., Smith, T. W., Jr., Wingenter, O. W., and Rowland, F. S. (1996a). Nonmethane hydrocarbon distributions during the Atlantic Stratocumulus Transition Experiment/Marine Aerosol and Gas Exchange (ASTEX-MAGE), June 1992. *J. Geophys. Res.* **101**, 4501–4514.

Blake, D. R., Chen, T.-Y., Smith, T. W., Jr., Wang, C. J.-L., Wingenter, O. W., Blake, N. J., and Rowland, F. S. (1996b). Three-dimensional distribution of nonmethane hydrocarbons and halocarbons over the northwestern Pacific during the 1991 Pacific Exploratory Mission (PEM-West A). *J. Geophys. Res.* **101**, 1763–1778.

Boissard, C. (1992). Distributions tropospériques globales des hydrocarbures nonméthaniques légers: de l'expérimentation à la modélisation, Thèse de $3_{\text{ème}}$ cycle, Université Denis Diderot Paris 7, Paris, 160 pp.

Boissard, C., Bonsang, B., Kanakidou, M., and Lambert, G. (1996). TROPOZ II: global distributions and budgets of methane and light hydrocarbons. *J. Atmos. Chem.* **25**, 115–148.

Bonsang, B. and Lambert, G. (1985). Nonmethane hydrocarbons in an oceanic atmosphere. *J. Atmos. Chem.* **2**, 257–271.

Bonsang, B., Kanakidou, M., Lambert, G., and Monfray, P. (1988). The marine source of C_2–C_6, aliphatic hydrocarbons. *J. Atmos. Chem.* **6**, 3–20.

Bonsang, B., Kanakidou, M., and Lambert, G. (1990). NMHC in the marine atmosphere:preliminary results of monitoring at Amterdam Island. *J. Atmos. Chem.* **11**, 169–178.

Bonsang, B., Martin, D., Lambert, G., Kanakidou, M., Le Roulley, J.-C., and Sennequier, G. (1991). Vertical distribution of nonmethane hydrocarbons in the remote marine boundary layer. *J. Geophys. Res.* **96**, 7313–7324.

Bonsang, B., Kanakidou, M., and Boissard, C. (1994). Contribution of tropical biomass burning to the global budget of hydrocarbons, carbon monoxide and tropospheric ozone. *In* Non-CO_2

Greenhouse Gases. (J. Van Ham *et al.*, eds.) Kluwer Academic Publishers: Dordrecht, pp. 261–270.

Bonsang, B., Boissard, C., and Le Cloarec, M.-F. (1995). Methane, carbon monoxide and light non-methane hydrocarbon emissions from African savanna burnings during the FOS/DECAFE experiment. *J. Atmos. Chem.* **22**, 149–162.

Bottenheim, J. W., and Shepherd, M. F. (1995). C2-C6 hydrocarbon measurements at four rural locations across Canada, Atmos. Env., **29**, 647–664.

Boundries, H., Toupance, G., and Dutot, A. (1994). Seaasonal variation of atmospheric nonmethane hydrocarbons on the western coast of Brittany, France. *Atmos. Environ.* **28**, 1095–1112.

Brocco, D., Fratarcangeli, R., Lepore, L., Petricca, M., and Ventore, I. (1997). Determination of aromatic hydrocarbons in urban air of Rome. *Atmos. Environ.* **31**, 557–566.

Carsey, T. P., Churchill, D. D., Farmer, M. L., Fischer, C. J., Pszenny, A. A., Ross, V. B., Saltzman, E. S., Springer-Young, M., and Bonsang, B. (1997). Nitrogen oxides and ozone production in the north Atlantic marine boundary layer. *J. Geophys. Res.* **102**, 10,653–10,665.

Charméders, W. L., Fehsenfeld, F., Rodgers, M. O., Cardelino, C., Martinez, J., Parrish, D., Lonneman, W., Lawson, D. R., Rasmussen, R. A., Zimmerman, P., Greenberg, J., Middleton, P., and Wang, T. (1992). Ozone precursor relationships in the ambient atmosphere. *J. Geophys. Res.* **97**, 6037–3055.

Cheng, L., Fu, L., Angle, R. P., and Sandhu, H. S. (1997). Seasonal variations of volatile organic compounds in Edmonton, Alberta. *Atmos. Environ.* **31**, 239–246.

Clarkson, T. S., Martin, R. J., Rudolph, J., and Graham, B. W. L. (1996). Benzene and toluene in New Zealand air. *Atmos. Env.* **30**, 569–577.

Clarkson, T. S., Martin, R. J., and Rudolph J. (1997). Ethane and propane in the southern marine troposphere. *Atmos. Env.* **31**, 3763–3771.

Cronn, D., and Robinson, E. (1979). Tropospheric and lower stratospheric vertical profiles of ethane and acetylene. *J. Geophys. Res.* **6**, 641–644.

Cronn, D. R., and Harsch, D. E. (1980). Smoky mountain ambient halocarbon and hydrocarbon monitoring. Washington State University, Pullman, WA, September 1978. U.S. EPA contract No. RO804033-03-2.

Davis, K. J., Lenschow, D. H., and Zimmerman, P. R. (1994). Biogenic nonmethane hydrocarbon emissions estimated from tethered balloon observations. *J. Geophys. Res.* **99**, 25,587–25,598.

Dollard, G. J., Davies, T. J., Jones, B. M. R., Nason, P. D., Chandler, J., Dumitrean, P., Delaney, M., Watkins, D., and Field, R. A. (1995). The UK hydrocarbon monitoring network. *In* "Volatile Organic Compounds in the Atmosphere." (R. E. Hester and R. M. Harrison, eds.), Royal Society of Chemistry, London.

Donahue, N. M., and Prinn, R. G. (1993). *In situ* nonmethane hydrocarbon measurements on SAGA 3. *J. Geophys. Res.* **98**, 16,915–16,932.

Donoso, L., Romero, R., Rondon, A., Fernandez, E., Oyola, P., and Sanhueza, E. (1996). Natural and anthropogenic C_2 to C_6 hydrocarbons in the Central-Eastern Venezuelan atmosphere during the rainy season. *J. Atmos. Chem.* **25**, 201–214.

Doskey, P. V., and Gaffney, J. S. (1992). Non-methane hydrocarbons in the Arctic atmosphere at Barrow, Alaska. *J. Geophys. Res.* **19**, 381–384.

Ehhalt, D. H., Rudolph, J., Meixner, F., and Schmidt, U. (1985). Measurements of selected C_2–O_5 hydrocarbons in the background troposphere: vertical and latitudinal variations. *J. Atmos. Chem.* **3**, 29–52.

Fehsenfeld, F., Calvert, J., Fall, R., Goldan, P., Guenther, A. B., Hewitt, C. N., Lamb, B., Shaw, L., Trainer, M., Westberg, H., and Zimmerman, P. R. (1992). Emissions of volatile organic compounds from vegetation and the implications for atmospheric chemistry. *Global Biogeoch. Cycles.* **6**, 389–430.

Ferman, M. A. (1981). Rural nonmethane hydrocarbon concentration and composition. *In* "Atmospheric Biogenic Hydrocarbons," Vol. 2, Ambient Concentrations and Atmosphere Science. (J. J. Bufalini, and R. R., Arnts, eds.), Ann Arbor, MI pp. 333–367.

Fuentes, J. D., Wang, D., Neumann, H. H., Gillespie, T. J., Hartog, G. D., and Dann, T. F. (1996). Ambient biogenic hydrocarbons and isoprene emissions from a mixed deciduous forest. *J. Atmos. Chem.* 25, 67–95.

Goldan, P. D., Kuster, W. C., Fehsenfeld, F. C., and Montzka, S. A. (1995). Hydrocarbon measurements in the southeastern United States: the Rural Oxidants in the Southern Environment (ROSE) Program 1990. *J. Geophys. Res.* 100, 25,945–25,963.

Goldan, P. D., Kuster, W. C., and Fehsenfeld, F. C. (1997). Nonmethane hydrocarbon measurements during the Tropospheric OH Photochemistry Experiment. *J. Geophys. Res.* 102, 6315–6324.

Greenberg, J. P., and Zimmerman, P. R. (1984). Nonmethane hydrocarbons in remote tropical, continental, and marine atmospheres. *J. Geophys. Res.* 89, 4767–4778.

Greenberg, J. P., Zimmerman, P. R., and Haagen, P. (1990). Tropospheric hydrocarbon and CO profiles over the U.S. West Coast and Alaska. *J. Geophys. Res.* 95, 14,015–14,026.

Greenberg, J. P., Zimmerman, P. R., Pollock, W. F., Lueh, R. A., and Heidt, L. E. (1992). Diurnal variability of atmospheric methane, nomethane hydrocarbons, and carbon monoxide at Mauna Loa. *J. Geophys. Res.* 97, 10,395–10,413.

Greenberg, J. P., Helming, D., and Zimmerman, P. R. (1996). Seasonal measurements of non methane hydrocarbons and carbon monoxide at the Mauna Loa observatory during the Mauna Loa Observatory Photochemistry Experiment 2. *J. Geophys. Res.* 101, 14581–14598.

Gros, V., Martin, D., Poisson, N., Kanakidou, M., Bonsang, B., Le Guern, F., and Demont, E. (1998). Ozone, C_2–C_5 hydrocarbons and radon-222 measurements in the marine boundary layer between 45° S and 77° S: EREBUS94. *Tellus.* (submitted).

Guenther, A., Hewitt, C. N., Erickson, D., Fall, R., Geron, C., Graedel, T., Harley, P., Klinger, L., Lerdau, M., Makay, W. A., Pierce, T., Scholes, B., Steinbrecher, R., Tallamraju, R., Taylor, J., and Zimmerman, P. R. (1995). A global model of natural volatile organic compound emissions. *J. Geophys. Res.* 100, 8873–8892.

Guenther, A., Baugh, W., Davis, K., Hampton, G., Allwine, E., Dilts, S., Lamb, B., and Westberg, H. (1996). Isoprene fluxes measured by enclosure, relaxed eddy accumulation, surface layer gradient, and mixed layer mass balance techniques. *J. Geophys. Res.* 101, 18,555–18,567.

Holdren, M. W., Westberg, H. H., and Zimmerman, P. R. (1979). Analysis of monoterpene hydrocarbons in rural atmospheres. *J. Geophys. Res.* 84, 5083–5088.

Hov, O., Schmidbauer, N., and Oehme, M. (1989). Light hydrocarbons in the Norwegian Arctic. *Atmos. Environ.* 11, 2471–2482.

Hov, O., Chmidbauer, N., and Oehme, M. (1991). C_2–C_5 hydrocarbons in rural south Norway. *Atmos. Environ.* 9, 1981–1999.

Hov, O. (1992). Atmospheric concentrations of nonmethane hydrocarbons at a North European coastal site. *J. Atmos. Chem.* 14, 515–526.

Janson, R. (1992). Monoterpene concentrations in and above a forest of Scots pine. *J. Atmos. Chem.* 14, 385–394.

Jobson, B. T., Niki, H., Yokouchi, Y., Bottenheim, J., Hopper, F., and Leaitch, R. (1994). Measurement of C_2–C_6 hydrocarbons during the Polar sunrise 1992 experiment: evidence for Cl atom and Br atom chemistry. *J. Geophys. Res.* 99, 25,355–25,368.

Kanakidou, M., Bonsang, B., Le Roulley, J.-C., Lambert, G., Martin, D., and Sennequier, G. (1988). Marine source of acetylene. *Nature.* 333, 51–52.

Kanakidou, M., Bonsang, B., and Lambert, G. (1989). Light hydrocarbons, vertical profiles and fluxes in a French rural area. *Atmos. Environ.* 23, 921–927.

Kanakidou, M., Singh, H. B., Valentin, K. M., and Crutzen, P. (1991). A two-dimensional study of ethane and propane oxidation in the troposphere. *J. Geophys. Res.* 96, 15,395–15,413.

Koppman, R., Bauer, R., Johnen, F. J., Plass, C., and Rudolph, J. (1992). The distribution of light nonmethane hydrocarbons over the mid-Atlantic: results of the Polarstern cruise ANT VII/1. *J. Atmos. Chem.* **15**, 215–234.

Lamb, B., Guenther, A., Gay, D., and Westberg, H. (1987). A national inventory of biogenic hydrocarbon emissions. *Atmos. Environ.* **21**, 1695–1705.

Lawrimore, J. H., Das, M., and Aneja, V. P. (1995). Vertical sampling and analysis of nonmethane hydrocarbons for ozone control in urban North Carolina. *J. Geophys. Res.* **100**, 22,785–22,793.

Lightman, P., Kallend, A. S., and Marsh, A. R. W. (1990). Seasonal variation of hydrocarbons in the free troposphere at mid-latitudes. *Tellus.* **4**, 408–422.

Lindesay, J. A., Andreae, M. O., Goldammer, J. G., Harris, G., Annegarn, H. J., Garstang, M., Scholes, R. J., and van Wilgen, B. W. (1996). International geosphere-biosphere programme/international global atmospheric chemistry SAFARI-92 field experiment: background and overview. *J. Geophys. Res.* **101**, 23,521–23,530.

Lindskog, A. (1996). Nonmethane hydrocarbons and nitrogen oxides in the background air, Department of Analytical Chemistry, Stockholm University, Sweden.

Logan, J. A., Prather, M. J., Wolfy, S. C., and McElroy, M. B. (1981). Tropospheric chemistry: a global perspective. *J. Geophys. Res.* **86**, 7210–7254.

Lonneman, W. A., Seila, R. L., and Meeks, S. A. (1977). Proceedings of Symposium on 1975 Northeast Oxidant Transport, Available through National Information Service, Springfield, VA 22161.

Lonneman, W. A., Seila, R. L., and Bufalini, J. J. (1978). Ambient air hydrocarbon concentrations in Florida. *Environ. Sci. Technol.* **12**, 459–463.

Lübkert B, and Shoepp, W. (1989). A model to calculate natural VOC emission from forests in Europe Laxenburg, Austria, IIASA working paper WP-89-082.

Martinez J. R., and Singh, H. B. (1979). Survey of the role of NO_x in nonurban ozone formation. SRI International Report to U.S. Environmental Protection Agency. Monitoring and Data Analysis Division, Office of Air Quality Planning and Standards, Research Triangle Park, Raleigh, NC.

Müller, J.-F. (1992). Geographical distribution and seasonal variation of surface emissions and deposition velocities of atmospheric trace gases. *J. Geophys. Res.* **97**, 3787–3804.

Mount, G. H., and Williams, E. J. (1997). An overview of the tropospheric OH photochemistry experiment, Fritz/Idaho Hill, Colorado, fall 1993. *J. Geophys. Res.* **102**, 6171–6186.

Penkett, S. A., Blake, N. J., Lightman, P., Marsh, A. R. W., Anwyl, P., and Butcher, G. (1993). The seasonal variation of nonmethane hydrocarbons in the free troposphere over the North Atlantic Ocean: possible evidence for extensive reaction of hydrocarbons with the nitrate radical. *J. Geophys. Res.* **98**, 2865–2885.

Plass-Dülmer, C., Koppmann, R., Ratte, M., and Rudolph, J. (1995). Light nonmethane hydrocarbons in seawater. *Global Geochemical Cycles*, **9**, 79–100.

Quarles, T., Lamb, B., and Robinson, E. R. (1980). Measurement of isoprene fluxes from a north western deciduous forest. Presented at 73rd Annual Meeting of the Air Pollution Control Association, Montreal, Canada.

Rasmussen, R. A., Chatfield, R. B., Holdren, M. H., and Robinson, E. R. (1976). Hydrocarbon levels in a midwest open-forested area. Technical, Report to Coordinating Research Council Inc., 219 Perimeter Center Parkway, Atlanta, GA 30346.

Rasmussen, R. A., Chatfield, R. B., and Holdren, M. (1977). Hydrocarbon and oxidant chemistry observed at a site near St. Louis. U.S. Environmental Protection Agency, Environmental Sciences Research Laboratory, Research Triangle Park, Raleigh, NC, EPA-600/7-77-056.

Rasmussen, R. A., and Khalil, M. A. K. (1983). Altitudinal and temporal variation of hydrocarbons and other gaseous tracers of Arctic haze. *Geophys. Res. Lett.* **2**, 144–147.

Rasmussen, R. A., and Khalil, M. A. K. (1988). Isoprene over the Amazon basin. *J. Geophys. Res.* **93**, 1417–1421.

Rinsland, C. P., Zander, R., Farmer, C. B., Norton, R. H., and Russell, J. M. (1987). Concentrations of ethane (C_2H_6) in the lower stratosphere and upper troposphere and acetylene (C_2H_2) in the upper troposphere deduced from atmospheric trace molecule spectroscopy/Spacelab 3 Spectra. *J. Geophys. Res.* **92**, 11,951–11,964.

Roberts, J. M., Fehsenfeld, F. C., Albritton, D. L., and Sievers, R. E. (1983). Measurement of monoterpene hydrocarbons at Niwot Ridge, Colorado. *J. Geophys. Res.* **88**, 667–10,678.

Rudolph, J. (1988). Two-dimensional distribution of light hydrocarbons: results from the STRATOZ III experiment. *J. Geophys. Res.* **93**, 8367–8377.

Rudolph, J. (1995). The tropospheric distribution and budget of ethane. *J. Geophys. Res.* **100**, 11,639–11,381.

Rudolph, J., Ehhalt, D. H., and Gravenhorst, G. (1980). Recent measurements of light hydrocarbons in remote areas. Proceedings of the First European Symposium on the Physico-Chemical Behaviour of Atmospheric Pollutants. (B. Versino and H. Ott, eds.). pp. 41–51. Commission of the European Communities: Ispra, Italy.

Rudolph, J., Ehhalt, D. H., and Toñisseen, A. (1981a). Vertical profiles of ethane and propane in the stratosphere. *J. Geophys. Res.* **86**, 7267–7272.

Rudolph, J., and Ehhalt, D. H. (1981b). Measurements of C_2–C_5 hydrocarbons over the North Atlantic. *J. Geophys. Res.* **86**, 11,959–11,964.

Rudolph, J., Ehhalt, D. H., and Khedim, A. (1984). Vertical profiles of acetylene in the troposphere and stratosphere. *J. Atmos. Chem.* **2**, 117–124.

Rudolph, J., Khedim, A., and Wagenbach, D. (1989). The seasonal variation of light nonmethane hydrocarbons in the Antarctic troposphere. *J. Geophys. Res.* **94**, 13,039–13,044.

Rudolph, J., and Johnen, F. J. (1990). Measurements of light atmospheric hydrocarbons over the Atlantic in regions of low biological activity. *J. Geophys. Res.* **95**, 20,583–20,591.

Rudolph, J., Khedim, A., and Bonsang, B. (1992a). Light hydrocarbons in the troposphere boundary layer over tropical Africa. *J. Geophys. Res.* **97**, 6181–6186.

Rudolph, J., Khedim, A., Clarkson, T., and Wagenbach, D. (1992b). Long-term measurements of light alkanes and acetylene in the Antarctic troposphere. *Tellus.* **44B**, 252–261.

Schjoldager, J., and Wathne, B. M. (1978). Preliminary study of hydrocarbons in forest. *Norsk Institute of Air Quality Studies.* 1–26.

Seila, R. L. (1984). Atmospheric volatile hydrocarbon composition at five remote sites in Northwestern North Carolina. *In* "Impact of Natural Emissions." V. Aneja, ed. Air Pollution Control Assoc: Pittsburgh, PA, pp. 125–140.

Simon, V., Clement, B., Riba, M.-L., and Torres, L. (1994). The Landes experiment: monoterpenes emitted from the maritime pine. *J. Geophys. Res.* **99**, 16,501–16,510.

Singh, H. B., and Salas, L. J. (1982). Measurement of selected light hydrocarbons over the Pacific Ocean: latitudinal and seasonal variations. *Geophys. Res. Lett.* **8**, 842–845.

Singh, H. B., Viezee, W., and Salas, L. J. (1988). Measurements of selected C_2–C_5 hydrocarbons in the troposphere: latitudinal, vertical, and temporal variations. *J. Geophys. Res.* **93**, 15,861–15,878.

Singh, H. B., and Zimmerman, P. B. (1992). Atmospheric distribution and sources of nonmethane hydrocarbons. *In* "Gaseous Pollutants: Characterization and Cycling." Nriagu, J. O. (ed.), John Wiley & Sons: New York.

Singh, H. B., Kanakidou, M., Crutzen, P. J., and Jacob, D. J. (1995). High concentrations and photochemical fate of oxygenated hydrocarbons in the global troposphere. *Nature.* **378**, 50–54.

Solberg, S., Dye, C., Schmidbauer, N., Herzog, A., and Gehrig, R. (1996). Carbonyls and nonmethane hydrocarbons at rural European sites from the Mediterranean to the arctic. *J. Atmos. Chem.* **25**, 33–66.

Stevens, R. K., Dzubay, T. G., and Shaw, R. W. (1981). Characterization of the haze in the Great Smoky and Abastumani mountains. In "Atmospheric Biogenic Hydrocarbons, Vol. 2, Ambient Concentrations and Atmosphere Science." (J. J. Bufalini, and R. R. Arnts, eds.), Ann Arbor, MI.

Tille, K. J. W., Savelsberg, M., and Bächmann, K. (1985). Airborne measurements of nonmethane hydrocarbons over western Europe: vertical distributions, seasonal cycles of mixing ratios and source strengths. Atmos. Env. 19, 1751–1760.

Van Valin, C. C., and Luria, M. (1988). O_3, CO, hydrocarbons and dimethyl sulfide over the western Atlantic Ocean. Atmos. Environ. 22, 2401–2409.

Yokouchi, Y., and Ambe, Y. (1988). Diurnal variations of atmospheric isoprene and monoterpene hydrocarbons in an agricultural area in summertime. J. Geophys. Res. 93, 3751–3759.

Young, V. L., Kieser, B. N., Chen, S. P., and Niki, H. (1997). Seasonal trends and local influences on nonmethane hydrocarbon concentrations in the Canadian boreal forest. J. Geophys. Res. 102, 5916–5918.

Zimmerman, P. R. (1979). Determination of emission rates of hydrocarbons from indigenuous species of vegetation in the Tampa/St. Petersburg, Florida area. Environmental Protection Agency, Washington, DC, EPA-904/9-77-028.

Zimmerman, P. R., Greenberg, J. P., and Westberg, C. E. (1988). Measurements of atmospheric hydrocarbons and biogenic emission fluxes in the Amazon boundary layer. J. Geophys. Res. 93, 1407–1416.

CHAPTER 7

Reactive Hydrocarbons and Photochemical Air Pollution

R. G. DERWENT

Atmospheric Processes Research, Meteorological Office, Bracknell, Berkshire, United Kingdom

---

   I. Introduction
  II. Reactive Hydrocarbons and Photochemical Ozone
      Formation
      A. Chemical Processes in the Fast Photochemical
         Balance
      B. Photochemical Generation of Hydroxyl and Per-
         oxy Radicals
      C. Reactions That Convert Hydroxyl Radicals into
         Hydroperoxy and Organic Peroxy Radicals
      D. Reactions That Convert Peroxy Radicals into Hy-
         droxyl Radicals
      E. Reactions That Remove Free Radicals
      F. Construction of the Fast Photochemical Balance
      G. Photostationary State and Ozone Production
      H. Oxidation Mechanisms for Reactive Hydrocar-
         bons
      I. Hydroxyl to Peroxy Radical Interconversion and
         Photochemical Ozone Formation
      J. Reactive Hydrocarbons and Photochemical
         Ozone Formation
 III. Reactivity Scales
      A. Introduction to the Different Reactivity Concepts
      B. Smog Chamber Reactivity Scales
      C. Hydroxyl Reactivity Scales

Reactive hydrocarbons play a fundamental role in photochemical air pollution that has been recognized for the last 40 or so
years. With the growing understanding of the fast photochemical reactions occurring in the sunlit atmospheric boundary
layer, it is now possible to relate ozone formation quantitatively
to the atmospheric oxidation of reactive hydrocarbons. This
quantitative relationship necessarily implies that individual reactive hydrocarbons make a different contribution to ozone formation and hence it is possible to define a reactivity scale that
orders reactive hydrocarbons according to their relative contributions. With this detailed understanding of reactivity, policymakers should be able to focus action on those reactive hydrocarbons that contribute most to photochemical ozone formation.

I. INTRODUCTION

Photochemical air pollution or photochemical smog was first observed in Los
Angeles in the late 1940s and early 1950s (Haagen-Smit *et al.* 1953) and over
the succeeding decades has been subsequently observed in all the major
industrial and population centers, at increasing latitudes. The first observations of photochemical air pollution in Europe were made in the Netherlands
in 1965 (Ten Houten, 1966) and since then elevated ozone concentrations
have been found in every European country where continuous measurements
have been made (Sluyter and van Zantvoort, 1996).

Early studies showed that the main photochemical oxidant present in
photochemical air pollution was ozone and that the formation process involved sunlight, and oxides of nitrogen and hydrocarbons (Haagen-Smit *et
al.*, 1953). It has been recognized for the last 40 years that each hydrocarbon
makes a different quantitative impact on photochemical ozone production.
Early smog chamber studies were used to develop the concept of reactivity

and various reactivity scales have been compiled (Huess and Glasson, 1968; Dimitriades and Joshi, 1977). There has always been a policy focus within North America to identify those reactive hydrocarbons that contribute most to ozone formation on the urban scale (Dodge, 1984). A distinction has been made between hydrocarbons that were thought to be of "neglible" reactivity and those of "low" reactivity. Increasingly, reactivity is being used in the policy process and hydrocarbons of negligible reactivity are exempt from regulations (Dimitriades, 1996).

Europeans were faced with the problems of photochemical air pollution much later than were North Americans. Long-range transport and multiday photochemistry on the regional scale have been perceived to be the more relevant policy issues (Nordic Council of Ministers, 1991), rather than urban-scale photochemical ozone formation. Other than this divergence in policy approaches, no major differences in mechanism or phenomenology have emerged from the North American and European studies of photochemical air pollution. Such is the concern among European member states that they have agreed to combat the regional ozone problem through international action within the scope of the United Nations Economic Commission for Europe (UN ECE) and its international convention on Long-Range Trans-boundary Air Pollution (UN ECE, 1991). European countries have agreed to cut emissions of man-made hydrocarbons by up to 30% and to freeze their NO_x emissions as a first step toward reducing episodic peak ozone concentrations. This action focuses attention on controlling those emissions that contribute most to regional-scale ozone formation. The most reactive hydrocarbons therefore must be clearly and unambiguously identified and control actions must be focused on reducing their emissions.

II. REACTIVE HYDROCARBONS AND PHOTOCHEMICAL OZONE FORMATION

A. CHEMICAL PROCESSES IN THE FAST PHOTOCHEMICAL BALANCE

Reactive hydrocarbons and photochemical ozone formation are inextricably linked through the set of fast free-radical reactions that occur in the sunlit atmospheric boundary layer. These reactions establish steady-state concentrations of a number of highly reactive free-radical species that control the formation and destruction of a wide range of major tropospheric trace gases. The free-radical species include hydroxyl (OH), hydroperoxy (HO_2), methylperoxy (CH_3O_2), and a whole range of organic peroxy radicals (RO_2). The system of sunlight-driven chemical reactions is termed the fast photochemical balance and its importance was first recognized by Levy (1971).

The fast photochemical balance in the sunlit atmospheric boundary layer is dominated by four types of processes:

- Photochemical reactions that generate the major free radicals
- Chemical reactions that convert OH radicals to HO_2 and RO_2 radicals
- Chemical reactions that convert HO_2 and RO_2 radicals to OH radicals
- Chemical reactions that remove or recombine free radicals.

The individual elementary chemical reactions that contribute to each of these processes are characterized in the sections that follow.

B. PHOTOCHEMICAL GENERATION OF HYDROXYL AND PEROXY RADICALS

The chemistry of the troposphere is driven by the sunlight photochemical destruction of a whole range of labile molecules. The major trace constituents that absorb ultraviolet (UV) radiation to produce OH, HO_2, and organic peroxy free radicals are ozone, formaldehyde, higher aldehydes, ketones, gaseous nitric acid, hydrogen peroxide, and higher peroxides. In each case, the free-radical production is obtained by multiplying together a photolysis coefficient (J-value, J) and the concentration of the photochemically labile trace gas. J values are calculated by folding together the wavelength-dependent solar actinic irradiance, the absorption cross section of the photochemically labile species, and the quantum yield for the particular process over a succession of discrete wavelength intervals. The solar actinic irradiance values (Peterson, 1976) at each wavelength can be taken from observed data or from a model description of the transmission of UV and visible solar radiation through the atmosphere (Demerjian et al., 1980; Finlayson-Pitts and Pitts, 1986).

Table 1 shows the J values for a number of important photochemical processes in the boundary layer for the solar actinic irradiance conditions of July 1, midday, 53° N, land albedo, 350 Dobson units overhead ozone and no clouds. Concentrations of the photochemically labile species are tabulated for typical polluted conditions over Europe. From the product of the J values and the species concentrations it is straightforward to estimate the rate of free-radical production, whether OH or HO_2 or whatever. This can be illustrated

TABLE 1   Photochemical Generation of OH, HO$_2$, and Organic Peroxy Radicals in the Sunlit Polluted Boundary Layer

Photochemical process	J value, (per sec[a])	Concentration of photochemically labile species (ppb)	Reaction flux, (molecules/cm^3/sec)
O$_3$ = O^{1}D + O$_2$[b]	2.08 × 10^{-5}	65[b,c,d]	6.1 × 10^6
HNO$_3$ = OH + NO$_2$	5.4 × 10^{-7}	10[e]	1.4 × 10^5
HCHO = H + HCO	2.7 × 10^{-5}	2.3[e]	3.1 × 10^6
H$_2$O$_2$ = OH + OH	6.6 × 10^{-6}	4.6[e]	1.5 × 10^6
CH$_3$OOH =	3.6 × 10^{-6}	1.3[e]	2.3 × 10^5
CH$_3$CHO =	3.8 × 10^{-6}	0.8[e]	1.4 × 10^5
C$_2$H$_5$CHO =	1.6 × 10^{-5}	0.1[e]	1.0 × 10^5
CH$_3$COCH$_3$ =	4.6 × 10^{-7}	1.0[e]	2.4 × 10^4
Total radical production rate, (molecule/cm^3/sec)			1.13 × 10^7

[a]Land albedo, solar zenith angle 30°, 53°N, 288 K, ozone amount 350 Dobson Units, no clouds, July 1, (Hough, 1988).
[b]Based on [H$_2$O] = 3.0 × 10^{17} molecules/cm^3 [N$_2$] = 1.9 × 10^{19} molecules/cm^3 and [O$_2$] = 4.8 × 10^{18} molecules/cm^3, with rate coefficients for O^{1}D reactions taken from Atkinson et al. (1992).
[c]UK PORG (1993).
[d]Simmonds et al. (1993).
[e]Photochemical model calculation from Derwent and Jenkin (1991).

with the case of the photolysis of nitric acid, see Table 1, according to the Eq. 1:

$$HNO_3 + radiation \text{ (wavelengths 200 to 400 nm)} = OH + NO_2. \quad (1)$$

By applying the Law of Mass Action:

$$-d/dt\,[HNO_3] = d/dt[OH] = J_1[HNO_3]. \quad (2)$$

At this stage, it matters little which particular free radical is being generated since the aim is to calculate the rate at which new free radicals enter the free-radical pool.

The most important free-radical source by a significant margin is ozone photolysis (see Table 1). The total free-radical production rate at midday, July 1, 53° N in the polluted boundary layer is estimated to be about 1.1 × 10^7 molecules per cubic centimeter per second (cm^3/sec). Photolysis of reactive hydrocarbons, such as formaldehyde, higher aldehydes, and ketones, are second in importance to ozone photolysis as free-radical sources.

C. Reactions That Convert Hydroxyl Radicals into Hydroperoxy and Organic Peroxy Radicals

The hydroxyl radical is a highly reactive species, reacting rapidly with most atmospheric trace constituents. These reactions at first sight look like OH radical loss processes and if this is all that they are, then the hydroxyl radical would be of little overall importance in tropospheric chemistry. It is important to look at the products of the hydroxyl radical reactions to see if any are reactive free radicals that can keep the reaction system going. In fact, many of the hydroxyl radical reactions go through highly reactive intermediates whose fates are to form hydrogen atoms, H, hydroperoxy radicals, HO_2, or alkylperoxy radicals, RO_2. As a result, many of the hydroxyl radical reactions have the ability to form ultimately hydroperoxy radicals. Hence, such reactions are not hydroxyl radical loss processes but reactions that convert hydroxyl radicals into hydroperoxy and organic peroxy radicals.

The simplest reactions that convert hydroxyl radicals into hydroperoxy radicals involve the inorganic atmospheric trace gas constituents, carbon monoxide, hydrogen, ozone, and sulfur dioxide.

$$OH + CO = H + CO_2, \tag{3}$$

$$H + O_2 + M = HO_2 + M, \tag{4}$$

$$OH + H_2 = H_2O + H, \tag{5}$$

$$OH + O_3 = HO_2 + O_2, \tag{6}$$

$$OH + SO_2 + M = HOSO_2 + M, \tag{7}$$

and

$$HOSO_2 + O_2 = HO_2 + SO_3, \tag{8}$$

where M is an unreactive molecule, normally nitrogen N_2. The reactions of OH radicals with the many reactive hydrocarbons present in the polluted boundary layer have the same result: they convert OH into HO_2 and organic peroxy radicals through the mechanisms described in some detail in a later section, involving a cascade of organic free radicals and carbonyl species.

In Table 2, the rate coefficient and concentration data are collected for this important class of reactions for conditions appropriate to the polluted boundary layer over Europe. The interconversion of hydroxyl radicals to hydroperoxy and organic peroxy radicals is thus seen to be dominated by its reaction with carbon monoxide. The total loss coefficient for hydroxyl radicals in Table 2 is estimated to be 4.2 per sec, which implies a time constant or lifetime of

TABLE 2   The Reactions That Convert OH into HO_2 and Organic Peroxy Radicals in the Sunlit Polluted Boundary Layer

OH + X	Rate coefficient for OH + X (cm^3/ molecule/sec)	[X](ppb)	Reaction loss coefficient, (per sec)
CO	2.4×10^{-13}	200^a	1.2
H_2	5.2×10^{-15}	550^b	0.072
SO_2	1.5×10^{-12}	2.6^c	0.098
O_3	5.9×10^{-14}	65^d	0.096
H_2O_2	1.7×10^{-12}	4.6^e	0.196
CH_4	7.0×10^{-15}	1800^a	0.32
Reactive hydrocarbons	See Table 5^f	see Table 5^f	2.3
Total loss coefficient, (per sec)		4.2	

[a]Mace Head, Ireland data from Derwent, et al. (1994).
[b]Schmidt (1974).
[c]Mean value for 25 sites in rural United Kingdom, June to August 1990.
[d]UK PORG (1993).
[e]Photochemical model calculation (Derwent and Jenkin, 1991).
[f]OH + RH is set at one-half the rate of NO to NO_2 conversion from Table 5.
[g]Rate coefficient data from Atkinson et al. (1992).

about two-tenths of a second for the hydroxyl radical. Reactive hydrocarbons appear to account for 60% of the OH to peroxy radical interconversions. In a later section we will show how this large contribution is accounted for in terms of the contributions from individual reactive hydrocarbons.

D. REACTIONS THAT CONVERT PEROXY RADICALS INTO HYDROXYL RADICALS

In contrast with the hydroxyl radical, the hydroperoxy radical, HO_2 is much less reactive and then only with a limited range of trace gases. The two main reactions are with nitric oxide, NO, and ozone, O_3.

$$NO + HO_2 = NO_2 + OH, \tag{9}$$

and

$$O_3 + HO_2 = O_2 + O_2 + OH. \tag{10}$$

In the polluted boundary layer, the nitric oxide reaction dominates over the ozone reaction (Table 3). The overall loss coefficient for hydroperoxy radicals is seen to be 0.22 per sec, which implies a lifetime of 5 sec. This is considerably longer than the corresponding lifetime for the hydroxyl radical.

TABLE 3   The Reactions That Convert HO_2 into OH Radicals in the Sunlit Polluted Boundary Layer

HO_2 + X	Rate coefficient for HO_2 + X (cm^3/ molecule/sec)	[X](ppb)	Reaction loss coefficient, (per sec)
NO	8.5×10^{-12}	1.0^a	0.213
O_3	1.7×10^{-15}	65^b	0.003
Total loss coefficient, (per sec)		0.216	

[a]Based on median summertime concentrations of 1.1 and 0.9 ppb, respectively, at Ladybower and Lullington Heath, United Kingdom (Broughton et al., 1993).
[b]UK PORG (1993).

The emerging picture, therefore, is one of a few limited reactions that generate a steady state of reactive free-radical species in the sunlit polluted boundary layer. Every 5 sec the free-radical species changes to the more reactive hydroxyl radical for two-tenths of a second before converting back to the less reactive hydroperoxy radical for 5 sec and so on.

E. REACTIONS THAT REMOVE FREE RADICALS

A final set of reactions control the fast photochemical balance in the troposphere because they remove free radicals from the reaction system without recycling them. The two main processes involve the formation of nitric acid and hydrogen peroxide:

$$OH + NO_2 + M = HNO_3 + M, \qquad (11)$$

and

$$HO_2 + HO_2 + M = H_2O_2 + O_2 + M. \qquad (12)$$

Table 4 consolidates the information on the rate coefficients and concentration of the species involved in the free-radical loss processes for conditions appropriate to the polluted boundary layer

F. CONSTRUCTION OF THE FAST PHOTOCHEMICAL BALANCE

The first step in the construction of the fast photochemical balance of the troposphere is to equate the rate of interconversion of OH into HO_2 and organic peroxy radicals with the rate of interconversion of HO_2 and organic

TABLE 4   The Reactions That Remove Free Radicals in the Sunlit Polluted Boundary Layer

Radical loss reaction	Rate coefficient (cm³/molecule/sec)	Reaction flux (molecules/cm⁻³/sec)
$HO_2 + HO_2 + M = H_2O_2$ $+ O_2 \,-\!+ M$	$(1.6 \times 10^{-12}$ $+ 5.2 \times 10^{-32} [N_2]$ $+ 4.5 \times 10^{-32} [O_2])$ $\times (1 + 1.4 \times 10^{-21}$ $\exp (2200/T)[H_2O])$	$9.4 \times 10^{-12} [HO_2]^2$
$OH + NO_2 + M = HNO_3 + M$	1.4×10^{-11}	$1.89 [OH]$
$OH + HO_2 = H_2O + O_2$	1.1×10^{-10}	$1.1 \times 10^{-10} [OH][HO_2]$
Total radical loss rate (molecule/cm⁻³/sec)	$9.4 \times 10^{-12}[HO_2]^2 + 1.89 [OH] + 1.1 \times 10^{-10}$ $[OH][HO_2]$	

[a][H₂O]: 3.0×10^{17} molecules/cm³.
[b][NO₂]: 5.4 ppb based on median summertime concentrations at Ladybower and Lullington Heath of 6.2 and 4.6 ppb, respectively (Broughton et al., 1993).

peroxy radicals into OH. This generates a relationship between the concentrations of the OH and HO_2 radicals:

$$4.2 \times [OH] = 0.22 \times [HO_2]. \tag{13}$$

The second step is to equate the rates of free-radical production and loss:

$$1.1 \times 10^7 = 1.9 \times [OH] + 1.1 \times 10^{-10}$$
$$\times [OH][HO_2] + 9.4 \times 10^{-12} \times [HO_2]^2. \tag{14}$$

These two relationships can be solved to yield the instantaneous free-radical concentrations when the fast photochemical balance is established:

$$[OH] = 5.0 \times 10^6 \text{ molecules/cm}^3,$$
$$[HO_2] = 1.2 \times 10^8 \text{ molecules/cm}^3,$$

and

$$[CH_3O_2] = 3.0 \times 10^7 \text{ molecules/cm}^3,$$

for noontime conditions in the polluted boundary layer over Europe.

G. Photostationary State and Ozone Production

Nitrogen oxides ($NO_x = NO + NO_2$) play an essential role in photochemical ozone formation through the reactions of hydroperoxy and organic peroxy radicals with NO in Reactions 4 and 5:

$$HO_2 + NO = NO_2 + OH, \qquad (9)$$

and

$$RO_2 + NO = NO_2 + RO. \qquad (15)$$

All the peroxy radical reactions exemplified by Reactions 9 and 15 result in the conversion of NO to NO_2. This conversion process is an essential element of the photochemical generation of ozone in the polluted urban boundary layer. During daylight, the main fate of the NO_2 formed is to absorb solar UV radiation and to undergo photolysis, reforming the NO and O_3 from which it was made, through Reactions 16 and 17:

$$NO_2 + \text{radiation (wavelengths 200 to 420 nm)} = NO + O, \qquad (16)$$

and

$$O + O_2 + M = O_3 + M, \qquad (17)$$

where M represents a nitrogen N_2 or an oxygen O_2 molecule that acts as a so-called "third body." The rate of photolysis of NO_2 depends on the solar actinic irradiance; this in turn depends on the height of the sun in the sky and hence time of day, latitude, and season, as well as on the amount and height of any cloud and haze that may obscure the sun. For much of the daylight portion of the year, the lifetime of NO_2 is only a matter of minutes before it is photolyzed back to NO.

During the warm, sunny summertime anticyclonic conditions associated with photochemical ozone formation, peroxy radical concentrations become elevated by photochemical activity and the local rate of ozone production increases dramatically through the NO to NO_2 interconversion reactions of the form:

$$HO_2 + NO = OH + NO_2 \qquad (9)$$

$$NO_2 + \text{radiation} = NO + O \qquad (16)$$

$$\underline{O + O_2 + M = O_3 + M} \qquad (17)$$

$$\text{Net: } RO_2 + O_2 = RO + O_3. \qquad (18)$$

$$RO_2 + NO = RO + NO_2 \qquad (15)$$

$$NO_2 + \text{radiation} = NO + O \qquad (16)$$

$$\underline{O + O_2 + M = O_3 + M} \qquad (17)$$

$$\text{Net: } RO_2 + O_2 = RO + O_3. \qquad (19)$$

In these reactions, NO and NO_2 appear to be left unchanged and act as catalysts. By shifting the balance in the $NO–NO_2–O_3$ photostationary state, the peroxy radicals become an important photochemical source of ozone. The peroxy radicals themselves are irreversibly degraded into alkoxy radicals (RO). Generally speaking, alkoxy radicals are highly reactive with oxygen, forming an HO_2 radical and a carbonyl compound:

$$RO + O_2 = HO_2 + R'COR''. \tag{20}$$

The HO_2 radical may go on to react with NO, producing another NO to NO_2 conversion step that results in the production of another ozone molecule:

$$HO_2 + NO = NO_2 + OH. \tag{9}$$

The hydroperoxy radical is thus rapidly converted into a hydroxyl radical (OH).

The rate of ozone production then can be estimated by applying the Law of Mass Action to the Eqs 9 + 15 + 16 + 17:

$$d[O_3]/dt = k_{NO+HO_2}[NO][HO_2] + k_{NO+CH_3O_2}[NO][CH_3O_2]$$
$$+ \text{ sum of terms of the type } k_{NO+RO_2}[NO][RO_2], \tag{21}$$

and by substituting typical values of the terms for summertime at noon:

$$[OH] \quad 5.0 \times 10^6 \text{ molecules/cm}^3,$$
$$[HO_2] \quad 1.2 \times 10^8 \text{ molecules/cm}^3,$$
$$[CH_3O_2] \quad 3.0 \times 10^7 \text{ molecules/cm}^3,$$
$$[RO_2] \quad 1.8 \times 10^8 \text{ molecules/cm}^3,$$
$$d[O_3]/dt = 1 + 0.2 + 1.2 \text{ ppt/sec} = 8 \text{ ppb/hr}. \tag{22}$$

This analysis implies that to reach the ozone concentrations typically found in regional-scale ozone episodes of about 100 ppb, an elevation of about 50 to 70 ppb above the Northern Hemisphere background level, 6 to 18 h of intense, sustained photochemical activity is required. This might imply 2 to 4 days total reaction time and of the order of 500 to 1000 km of travel. Long-range transport is therefore anticipated to be an important dimension to regional-scale photochemical episodes.

Photochemical ozone formation can therefore be described as the process by which organic peroxy radicals shift the balance in the $NO-NO_2-O_3$ photostationary state in favor of ozone production. The peroxy radicals themselves are converted into carbonyl compounds such as aldehydes and ketones. To complete the picture, an explanation is required of the origins of the organic peroxy radicals that are essential in regional-scale ozone production.

H. OXIDATION MECHANISMS FOR REACTIVE HYDROCARBONS

Organic peroxy radicals are almost exclusively formed by the attack of hydroxyl radicals on the reactive hydrocarbons and other organic compounds ubiquitously present in the polluted atmospheric boundary layer. These reactions may be represented as:

$$OH + RH = R + H_2O, \tag{23}$$

and

$$R + O_2 + M = RO_2 + M. \tag{24}$$

The detailed mechanisms of the reaction that convert organic compounds into their corresponding peroxy radicals depend on the structure of the individual organic compounds involved (Atkinson, 1990, 1994). Most reactive hydrocarbons react with hydroxyl radicals either by H-abstraction or by addition, if they contain carbon–carbon multiple bonds. In the case of alkanes, cycloalkanes, carbonyls, and oxygenated hydrocarbons, the hydroxyl radical removes a hydrogen atom originally connected to the carbon skeleton of the parent compound, forming a carbon radical and water vapor. These carbon radicals quickly react with oxygen to form the corresponding peroxy radical. For example, if the parent organic compound were methane, the organic radical formed would be the methyl radical. This radical rapidly combines with oxygen forming a methyl peroxy radical:

$$OH + CH_4 = CH_3 + H_2O, \tag{25}$$

and

$$CH_3 + O_2 + M = CH_3O_2 + M. \tag{26}$$

In the case of alkenes, alkynes, and aromatics, the hydroxyl radical adds to the multiple bond, producing a carbon radical, which again almost invariably reacts with oxygen in the analogous process to form a peroxy radical. For example, with ethylene (ethene) the reaction sequence would look like:

$$OH + C_2H_4 + M = HOC_2H_4 + M, \tag{27}$$

and

$$HOC_2H_4 + O_2 = HOC_2H_4O_2. \tag{28}$$

By coupling the OH attack on the reactive hydrocarbon with the conversion of the peroxy radicals to alkoxy radicals and of the alkoxy radicals to carbonyls, the ozone production begins to take shape for methane as follows:

$$OH + CH_4 = CH_3 + H_2O \tag{25}$$

$$CH_3 + O_2 + M = CH_3O_2 + M \tag{26}$$

$$CH_3O_2 + NO = CH_3O + NO_2 \tag{29}$$

$$NO_2 + radiation = NO + O \tag{16}$$

$$O + O_2 + M = O_3 + M \tag{17}$$

$$CH_3O + O_2 = HO_2 + HCHO \tag{30}$$

$$HO_2 + NO = OH + NO_2 \tag{9}$$

$$NO_2 + radiation = NO + O \tag{16}$$

$$\underline{O + O_2 + M = O_3 + M} \tag{17}$$

$$\text{Net: } CH_4 + 4O_2 = HCHO + O_3 + O_3 + H_2O. \tag{31}$$

For ethylene (ethene):

$$OH + C_2H_4 + M = HOC_2H_4 + M \tag{27}$$

$$HOC_2H_4 + O_2 = HOC_2H_4O_2 \tag{28}$$

$$HOC_2H_4O_2 + NO = NO_2 + HOC_2H_4O \tag{32}$$

$$NO_2 + radiation = NO + O \tag{16}$$

$$O + O_2 + M = O_3 + M \tag{17}$$

$$HOC_2H_4O + O_2 = HO_2 + HCHO + HCHO \tag{33}$$

$$HO_2 + NO = OH + NO_2 \tag{9}$$

$$NO_2 + radiation = NO + O \tag{16}$$

$$\underline{O + O_2 + M = O_3 + M} \tag{17}$$

$$\text{Net: } C_2H_4 + 4O_2 = HCHO + HCHO + O_3 + O_3 + H_2O. \tag{34}$$

In this series of rapid consecutive reactions the OH radical is recycled, the nitric oxide and nitrogen dioxide are recycled, two molecules of ozone are produced, and the reactive hydrocarbons are converted into carbonyl compounds. In this way, small steady-state concentrations of the highly reactive hydroxyl radicals can degrade substantial concentrations of organic compounds, producing ozone as an important reaction product. To estimate the rate of ozone production, we need to understand the rate of reaction of the individual reactive hydrocarbons with OH radicals.

I. Hydroxyl Peroxy Radical Interconversion and Photochemical Ozone Formation

The fast photochemical balance leads to the generation of steady-state concentrations of highly reactive radicals such as hydroxyl and peroxy radicals. The concentrations of organic peroxy radicals are sufficient to sustain the ozone production rates required to explain the elevated ozone concentrations observed in the polluted atmospheric boundary layer. To understand the role of reactive hydrocarbons in photochemical ozone formation, the relationship between peroxy radicals and reactive hydrocarbons must be explored further.

Each reactive hydrocarbon will react with OH at a particular rate, generating peroxy radicals that will drive ozone formation and also producing carbonyl molecules. These carbonyls are themselves highly reactive hydrocarbons and will be attacked by OH producing more peroxy radicals. Some carbonyls are photochemically labile and may be photolyzed to give more peroxy radicals. The simpler peroxy radicals, such as methylperoxy, may be produced by the degradation of many hydrocarbons and so it is difficult to unravel the contribution from particular reactive hydrocarbons.

The rate-determining stage for the generation of ozone by a particular reactive hydrocarbon is its reaction rate with OH, since the reactions of the subsequent carbon radicals, peroxy radicals, and alkoxy radicals are relatively rapid. If all the peroxy radicals react exclusively with NO, then by applying the steady-state hypothesis to Reactions 25 + 26 + 16 + 17 + 30:

$$d[OH]/dt = \cdot k_{OH+HO}[OH][HC], \tag{35}$$

$$d[RO_2]/dt = k_{OH+HC}[OH][HC] \cdot k_{RO_2+NO}[RO_2][NO], \tag{36}$$

$$d[O_3]/dt = k_{RO_2+NO}[RO_2][NO] + k_{RO_2+NO}[HO_2][NO], \tag{37}$$

$$= 2 \times k_{OH+HC}[OH][HC]. \tag{38}$$

To a first approximation then, the relative rates of photochemical ozone production from a range of reactive hydrocarbons can be represented by the rates at which they are attacked by OH, multiplied by a stoichiometric factor that is related to the number of NO to NO_2 interconversions in the degradation pathway. Rates of OH attack on each reactive hydrocarbon can be estimated from the elementary rate coefficients for the attack by OH on the reactive hydrocarbon, the reactive hydrocarbon concentration, and the hydroxyl radical concentration. These rates of OH attack should give a guide to the likely contribution to photochemical ozone formation from each reactive hydrocarbon.

J. REACTIVE HYDROCARBONS AND PHOTOCHEMICAL OZONE FORMATION

Table 5 consolidates the information required to estimate the rates of photochemical ozone formation from 85 reactive hydrocarbons under conditions

TABLE 5   Mean Summertime Concentrations for 85 Reactive Hydrocarbons, Rate Coefficient for Their OH Reactions,[a] and Contributions to Ozone Production[b] for Conditions Appropriate to the Polluted Boundary Layer over the United Kingdom

Reactive hydrocarbon	Mean conc (ppb)	Rate coefficient ($k_{OH} \times 10^{12}$)	Ozone production (ppb/h)
Isobutene[c]	0.21	51.4	0.387
Propylene[d]	0.27	26.3	0.256
Ethylene[d]	0.67	8.52	0.206
Isoprene[d]	0.05	101	0.182
1,2,4-Trimethylbenzene[e]	0.15	32.5	0.176
(m + p)-Xylene[d]	0.21	19	0.144
1,3,5-Trimethylbenzene[e]	0.06	57.5	0.128
trans-But-2-ene[d]	0.05	64	0.115
Toluene[d]	0.46	5.96	0.099
trans-Pent-2-ene	0.04	66.9	0.097
1,3-Butadiene[d]	0.04	66.6	0.096
cis-But-2-ene[d]	0.04	56.4	0.081
Isopentane[d]	0.52	3.9	0.073
o-Xylene[d]	0.14	13.7	0.069
But-1-ene[d]	0.05	31.4	0.057
3-Methyl-cis-pent-2-ene[e]	0.02	86.7	0.056
n-Butane[d]	0.59	2.52	0.054
m-Ethyltoluene[e]	0.07	19.2	0.050
3-Methyl-trans-pent-2-ene[e]	0.02	86.7	0.049
2-Methylpent-2-ene[e]	0.02	89	0.048
cis-Pent-2-ene[e]	0.02	65.4	0.047
Hexenes[e]	0.02	47.9	0.041
Acetylene[d]	1.38	0.815	0.041
1,2,3-Trimethylbenzene[e]	0.03	32.7	0.040
3-Methylcyclopentene[e]	0.02	66	0.038
1,2,3,5-Tetramethylbenzene[e]	0.01	77	0.038
Propane[d]	0.74	1.15	0.033
2-Methylpentane[d]	0.15	5.6	0.030
Decane[e]	0.06	11.6	0.026
Ethylbenzene[d]	0.09	7.1	0.023
Styrene[e]	0.01	58	0.023
p-Ethyltoluene[e]	0.05	12.1	0.022
n-Pentane[d]	0.14	3.96	0.020
Isobutane[d]	0.23	2.33	0.020
1,2,4,5-Tetramethylbenzene[e]	0.01	65	0.019
Ethane[d]	1.87	0.257	0.018
o-Ethyltoluene[e]	0.04	12.3	0.017
1,4-Dimethyl-2-ethylbenzene[e]	0.01	32.5	0.017
1,2-Dimethyl-4-ethylbenzene[e]	0.01	32.5	0.016
2,3,4-Trimethylpentane[e]	0.05	8.7	0.015
3-Methylpentane[d]	0.07	5.7	0.015

(continued)

TABLE 5   *(continues)*

Reactive hydrocarbon	Mean conc (ppb)	Rate coefficient ($k_{OH} \times 10^{12}$)	Ozone production (ppb/h)
α-Pinene[e]	0.01	53.7	0.014
2,2,4-Trimethylpentane[e]	0.1	3.6	0.014
1,3-Dimethyl-4-ethylbenzene[e]	0.01	32.5	0.013
Methylcyclopentane[e]	0.04	8.8	0.013
3-Methylhexane[e]	0.05	7.2	0.013
n-Hexane[d]	0.06	5.61	0.012
Undecane[e]	0.02	13.2	0.012
Benzene[d]	0.24	1.32	0.012
2-Methylhexane[e]	0.05	6.8	0.011
n-Nonane[e]	0.03	10.2	0.011
1-Methyl-n-propylbenzene[e]	0.01	21.4	0.010
2,3-Dimethylhexane[e]	0.03	8.6	0.009
Methylcyclohexane[e]	0.02	10.4	0.009
4-Methylnonane[e]	0.02	11.3	0.008
4-Methyloctane[e]	0.02	10	0.008
n-Heptane[d]	0.03	7.2	0.008
2-Methylnonane[e]	0.02	11	0.008
2-Methyloctane[e]	0.02	9.6	0.008
n-Butylbenzene[e]	0.03	6.4	0.008
n-Octane[e]	0.02	8.7	0.007
2,4-Dimethylhexane[e]	0.02	8.6	0.006
2-Methylheptane[e]	0.02	8.2	0.006
Indane[e]	0.02	9.2	0.006
2,3-Dimethylpentane[e]	0.02	7.2	0.006
Dodecane[e]	0.01	14.2	0.006
3-Methylheptane[e]	0.02	8.6	0.006
Propylbenzene[e]	0.02	6	0.006
1,3-Diethylbenzene[e]	0.01	21.4	0.006
Dimethylheptanes[e]	0.01	10	0.005
trans-Dimethylcyclopentane[e]	0.01	10.6	0.005
Propylcyclohexane[e]	0.01	12.7	0.004
2,5-Dimethylhexane[e]	0.01	8.3	0.004
n-Nonane[e]	0.01	10.2	0.004
2,4-Dimethylpentane[e]	0.02	6.9	0.004
Dimethyloctanes[e]	0.01	11.4	0.004
cis-1,3-Dimethylcyclopentane[e]	0.01	10.6	0.004
3-Methyloctane[e]	0.01	10	0.004
2,3,3-Trimethylpentane[e]	0.02	4.4	0.004
cis-1,3-Dimethylcyclohexane[e]	0.01	12	0.003
Cyclohexane[e]	0.01	7.5	0.003
4-Methylheptane[e]	0.01	8.6	0.003
1,4-Dimethylcyclohexane[e]	0.01	12	0.003

(continued)

TABLE 5 *(continues)*

Reactive hydrocarbon	Mean conc (ppb)	Rate coefficient ($k_{OH} \times 10^{12}$)	Ozone production (ppb/h)
Dimethylcyclohexanes[e]	0.01	12	0.003
1,1-Dimethycyclohexane[e]	0.01	12	0.003
Isopropylbenzene[e]	0.01	6.5	0.002
2,2,5-Trimethylbenzene[e]	0.01	6.1	0.002
Total			3.302

[a]Reaction rate coefficients for OH attack on each hydrocarbon taken for 1013 mb and 298 K (Atkinson, 1994) in cm³/molecule/s.

[b]Ozone production rate calculated using the hydrocarbon concentration, the OH rate coefficient and an [OH] of 5×10^6 molecules/cm³, assuming that there are two NO to NO_2 steps for each reactive hydrocarbon molecule degraded (see text).

[c]Mean hydrocarbon concentrations for West Beckham and Great Dun Fell sites (UK PORG, 1993).

[d]April 1 to August 31, 1996 mean hydrocarbon concentrations for Harwell (Dollard *et al.*, 1997).

[e]Mean hydrocarbon concentrations expressed relative to toluene from hydrocarbon monitoring at eight sites in Leeds (Bartle *et al.*, 1995).

typical of the polluted summertime boundary layer in rural southern England. The estimates of individual contributions are based on measured rural concentrations (UK PORG, 1993; Dollard *et al.*, 1997) or on urban measurements (Bartle *et al.*, 1995), scaled to rural values using concentration ratios relative to toluene. Together, the total of individual reactive hydrocarbon contributions amounts to an ozone production of just over 3 ppb/h. This is an entirely reasonable estimate for photochemical ozone production driven by the oxidation of reactive hydrocarbons under the conditions studied.

The main interest in Table 5 lies in the understanding it provides of the roles played by individual reactive hydrocarbons, based entirely on observations. Alkenes account for the top 4 positions and dominate the top 12 positions as a class. Aromatic hydrocarbons also make a relatively strong showing at the top of the table, through the trimethylbenzenes in positions five and seven. The highest ranked alkane is isopentane, which is well down the rankings at position 13. Isoprene appears fourth in overall order of importance and is the highest ranked reactive hydrocarbon with appreciable natural biogenic sources.

III. REACTIVITY SCALES

A. INTRODUCTION TO THE DIFFERENT REACTIVITY CONCEPTS

It is apparent from Table 5 that each of approaching 100 individual reactive hydrocarbons makes some contribution to photochemical ozone formation based on available reactive hydrocarbon measurements in the United Kingdom. Studies in the Los Angeles basin suggest that there may be up to 300 individual reactive hydrocarbons taking part in ozone formation (Bowman and Seinfeld, 1994). Each reactive hydrocarbon makes a different contribution because of differences in emissions, in the rates of their photochemical reactions, and in their propensities to produce ozone during these reactions. Some reactive hydrocarbons are more reactive than others and strategies aimed at reducing ozone exposure levels should address the most reactive organic compounds. On this basis, a reactivity scale is seen as an ordered list of reactive hydrocarbons in which the ranking is based on the amount of ozone formed from each reactive hydrocarbon under particular atmospheric conditions.

Reactivity can be quantified in many ways but here we take it to refer to the amount of ozone formed during a predefined period, following the release of a given mass of a reactive hydrocarbon, i, into the atmosphere. Because of concerns that the photochemical formation of ozone on the regional scale might be highly nonlinear, it is envisaged that the release to the atmosphere of the reactive hydrocarbon not only is small enough so as not to disturb significantly the chemistry occurring but also is such that the ozone response is large enough to be detectable. Inevitably, therefore, attention is directed to some measure of incremental reactivity and not to some absolute measure of reactivity. Incremental reactivity, IR_i can be defined as:

$$IR_i = \frac{\text{ozone response in ppb}}{\text{incremental release of reactive hydrocarbon, } i, \text{ in mass units}} \quad (38)$$

To remove some of the inherent variability in incremental reactivities sometimes they are expressed relative to that of a chosen compound, ethylene, and so producing a photochemical ozone creation potential, $(POCP_i)$, as follows:

$$POCP_i = \frac{IR_a}{IR_{\text{ethylene}}} \times 100. \quad (39)$$

In this way, the potential for photochemical ozone formation for a particular emission source can be expressed in terms of ethylene equivalents and directly compared with other emission sources. The ranked list of reactive hydrocarbons and their incremental reactivities or POCP values comprises a reactivity scale.

There appear to be four candidate methods of producing reactivity scales for the reactive hydrocarbons and these involve:

1. Smog chamber studies
2. Rate coefficient data for OH attack on the reactive hydrocarbons
3. Air quality simulation models
4. Explicit chemical mechanisms

B. SMOG CHAMBER REACTIVITY SCALES

Smog chamber studies have been fundamental to the development of the understanding of reactivity and of the development of reactivity scales from the early work of Haagen-Smit *et al.* (1953) and onward (Dimitriades, 1972). These early studies led to the promulgation of Rule 66 (Dodge, 1984) and its emphasis on limiting emissions of highly reactive organic compounds in the Los Angeles area. Carter and Atkinson (1987) measured the incremental reactivities of a set of organic compounds by observing the effects of small additions of each organic compound to an irradiated mixture of organic compounds and NO_x in smog chamber system. This method has been considerably extended using computer modeling (Carter and Atkinson, 1989) and is now the basis of the widely accepted maximum incremental reactivity (MIR) scale (Carter, 1994, 1995; Carter *et al.*, 1995).

C. HYDROXYL REACTIVITY SCALES

It is straightforward to assemble a reactivity scale for organic compounds based on the rate coefficients of the respective reactions with the hydroxyl radical in reactions such as:

$$OH + \text{organic compound} + O_2 \longrightarrow \text{peroxy radicals.}$$

These OH reaction rate coefficients are intrinsic properties of the organic compounds and as geophysical constants, represent an ideal means of classifying quantitatively the organic compounds present in emissions or in the atmosphere (U.S. EPA, 1977). There is, however, no unique relationship between the competitive reaction rates of a set of organic compounds with hydroxyl radicals and their ability to produce ozone because the latter depends on the subsequent reaction mechanisms of the products of the OH radical attack (Carter and Atkinson, 1987). Reactivity scales based on OH rate coefficient data have, therefore, been dismissed as being too simplistic (Dimitriades, 1996).

D. REACTIVITY SCALES FROM AIR QUALITY SIMULATION MODELS

In the same way as chemical reaction mechanisms can be used to simulate increments in ozone formation in a smog chamber from small additions of organic compounds, they can also be used to calculate analogous increments in an air quality simulation model describing ozone formation in urban area. Russell *et al.* (1995) describe how a three-dimensional airshed model of the Los Angeles air basin can be used with a detailed chemical mechanism to assess the impacts on ambient ozone of the modifications to the emissions of the organic compounds brought about by reactivity-based controls. The correspondence between the incremental reactivities of a wide range of organic compounds derived from an air quality simulation model (Russell *et al.*, 1995) and from smog chambers (Carter, 1994), is generally good.

E. PHOTOCHEMICAL OZONE CREATION POTENTIALS

POCPs have been determined for 120 reactive hydrocarbons using a highly detailed and explicit chemical mechanism, the master chemical mechanism (MCM), and a photochemical trajectory model, employing a realistic air mass trajectory for regional-scale ozone formation across northwest Europe (Derwent *et al.*, 1998). In this way, it has been possible to estimate the incremental change in ozone produced by an additional incremental mass emission of each organic compound. POCP values have been estimated from the respective ozone responses to the increased emissions of each organic compound and their sensitivity to NO_x emissions have been studied.

Incremental reactivities for the U.S. conditions, POCP values estimated with explicit mechanisms, and analogous values obtained for Swedish conditions (Andersson-*Skold et al.*, 1992) are provided as annex to the VOC protocol of the international convention on long-range transboundary air pollution (UN ECE, 1991) to provide a basis for targeting emission controls onto the more important organic compounds.

F. COMPARISON OF THE MAXIMUM INCREMENTAL REACTIVITY AND PHOTOCHEMICAL OZONE CREATION POTENTIAL REACTIVITY SCALES

Figure 1 compares the POCPs derived with the MCM for some of the more important reactive hydrocarbons, on a mass emissions basis, across northwest Europe with the corresponding MIR values determined from smog chamber mechanism studies for U.S. conditions. The two reactivity scales show a striking degree of agreement over a wide range of reactivity. Both scales indicate that the alkanes are the least reactive and that the alkenes and aromatics are the most reactive. Indeed, both scales point to methane as the least reactive and 1,3,5-trimethylbenzene as the most reactive hydrocarbon. For the large majority of reactive hydrocarbons, the general levels of reactivity indicated by both scales are remarkably consistent. This is quite a surprising conclusion since the two methods are completely independent and have been developed for entirely different facets of the ozone problem. There are, however, a number of noticeable discrepancies that are marked on Fig. 1. Each of the discrepancies appears to have been caused by significantly lower reactivities on the POCP scale compared with the MIR scale. They involve ethyl t-butyl ether, formaldehyde, 1,3-butadiene, and butylene and must reflect some aspect of the different background situations that have been assumed in the different estimation procedures.

IV. CONCLUSIONS

The fundamental role played by reactive hydrocarbons in photochemical ozone formation has been recognized for over 40 years. Over the intervening period, understanding has grown concerning the fast photochemical reactions occurring in the sunlit atmospheric boundary layer. It is now understood how these reactions generate small steady-state concentrations of the highly reactive hydroxyl, hydroperoxy, and organic peroxy radicals. Together, these reactions form the fast photochemical balance of the sunlit troposphere.

From a quantitative understanding of the fast photochemical balance, it is possible to relate the reactions that convert hydroxyl radicals into organic peroxy radicals to those that drive the photochemical ozone formation. We have shown here how to relate typical observed rural concentrations of reactive hydrocarbons to the regional-scale ozone production rate. This understanding necessarily implies that each reactive hydrocarbon makes a different

288

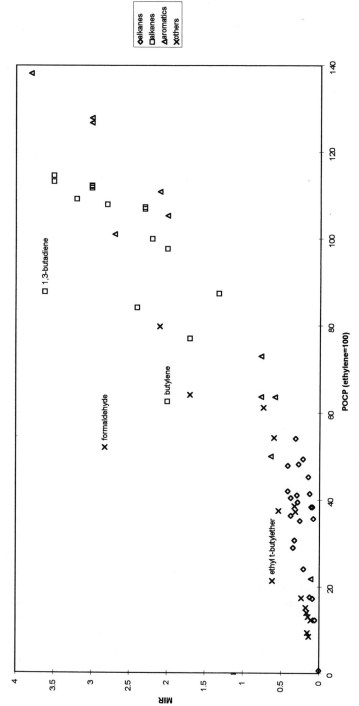

FIGURE 1   The relationship between the MIR and POCP reactivity scales.

contribution to photochemical ozone formation. A list of the reactive hydrocarbons, ordered by their respective contributions to ozone formation, is called a reactivity scale. Two such reactivity scales have been presented in this chapter and their similarities and differences have been highlighted. By using these reactivity scales, policymakers should be able to focus control actions onto those reactive hydrocarbons that contribute most to photochemical ozone formation.

ACKNOWLEDGMENTS

This review was prepared with support from the Air Quality Research Program of the Department of the Environment under contract EPG 1/3/70. The assistance of Geoff Dollard and Michael Jenkin of NETCEN, AEA Technology, and Alastair Lewis of the University of Leeds, with the preparation of Table 5 is deeply appreciated.

REFERENCES

Andersson-Skold, Y., Grennfelt, P., and Pleijel K. (1992). Photochemical ozone creation potentials: a study of different concepts. *J. Air Waste Manage. Assoc.* **42**, 1152–1158.

Atkinson, R. (1990). Gas-phase tropospheric chemistry of organic compounds: a review. *Atmos. Environ.* **24A**, 1–42.

Atkinson, R. (1994). Gas-phase tropospheric chemistry of organic compounds. *J. Phys. Chem. Ref. Data.* Monograph No. **2**, 1–216.

Atkinson, R., Baulch, D. L., Cox, R. A., Hampson, R. F., Kerr, J. A., and Troe, J. (1992). *J. Phys. Chem. Ref. Data.* **21**, 1125–1568.

Bartle, K. D., Denha, A., Hassoun, S., Kupiszewska, D., Lewis, A. C., and Pilling, M. J. (1995). A catalogue of urban hydrocarbons, January to December 1994. Report on contract PECD 7/12/111. School of Chemistry, University of Leeds, UK.

Bowman, F. M. and Seinfeld, J. H. (1994). Ozone productivity of atmospheric organics. *J. Geophys. Res.* **99**, 5309–5324.

Broughton, G. F. J., Bower, J. S., Stevenson, K. J., Lampert, J. E., Sweeney, B. P., Wilken, J., Eaton, S. W., Clark, A. G., Willis, P. G., Stacey, B. R. W., Driver, G. S., Laight, S. E., Berwick, R., and Jackson, M. S. (1993). Air quality in the UK: a summary of results from instrumented air monitoring networks in 1991/92. Warren Spring Laboratory, Stevenage, Herts, U.K.

Carter, W. P. L. (1994). Development of ozone reactivity scales for volatile organic compounds. *J. Air Waste Manage. Assoc.* **44**, 881–899.

Carter, W. P. L. (1995). Computer modelling of environmental chamber measurements of maximum incremental reactivities of volatile organic compounds. *Atmos. Environ.* **29**, 2513–2527.

Carter, W. P. L., and Atkinson, R. (1987). An experimental study of incremental hydrocarbon reactivity. *Environ. Sci. Technol.* **21**, 670–679.

Carter, W. P. L., and Atkinson, R. (1989). Computer modelling study of incremental hydrocarbon reactivity. *Environ. Sci. Technol.* **23**, 864–880.

Carter, W. P. L., Pierce, J. A., Luo, D., and Malkina, I. L. (1995). Environmental chamber study of maximum incremental reactivities of volatile organic compounds. *Atmos. Environ.* **29**, 2499–2511.

Demerjian, K. L., Schere, K. L., and Peterson J. T. (1980). Theoretical estimates of actinic (spherically integrated) flux and photolytic rate constants of atmospheric species in the lower atmosphere. *Adv. Environ. Sci. Technol.* **10**, 369–459.

Derwent, R. G., and Jenkin, M. E. (1991). Hydrocarbons and the long range transport of ozone and PAN across Europe. *Atmos. Environ.* **25A**, 1661–1678.

Derwent, R. G., Jenkin, M. E., and Saunders, S. M. (1996). Photochemical ozone creation potentials for a large number of reactive hydrocarbons under European conditions. *Atmos. Environ.* **30**, 181–199.

Derwent, R. G., Jenkin, M. E., Saunders, S. M., and Pilling, M. J. (1998). Photochemical ozone creation potentials for organic compounds in northwest Europe calculated with a master chemical mechanism. *Atmos. Enviroin.* **32**, in press.

Derwent, R. G., Simmonds, P. G., and Collins, W. J. (1994). Ozone and carbon monoxide measurements at a remote maritime location. Mace Head, Ireland from 1990–1992. *Atmos. Environ.* **28**, 2623–2637.

Dimitriades, B. (1972). Effects of hydrocarbons and nitrogen oxides on photochemical smog formation. *Environ. Sci. Technol.* **6**, 253–260.

Dimitriades, B. (1996). Scientific basis for the VOC reactivity issues raised by Section 183(e) of the Clean Air Act Amendments of 1990. *J. Air Waste Manage. Assoc.* **46**, 963–970.

Dimitriades, B., and Joshi, S. B. (1977). Application of reactivity criteria in oxidant related control in the U.S.A. Proceedings of International Conference on Photochemical Oxidant Pollution and its Control, Vol 2. Research Triangle Park, Raleigh, NC, U.S. EPA Report EPA-600/3-77-001b.

Dodge, M. C. (1984). Combined effects of organic reactivity and NMHC/NO$_x$ ratio on photochemical oxidant formation—a modelling study. *Atmos. Environ.* **18**, 1657–1665.

Dollard, G. J., Davies, T. J., Jones, B. M. R., Nason, P. D., Chandler, J., Dumitrean, P., Delaney, M., Watkins, D., and Field, R. A. (1997). Automatic hydrocarbon network 1994–1996. NETCEN, AEA Technology, Culham Laboratory, Oxfordshire, UK.

Finlayson-Pitts, B. J., and Pitts, J. N. (1986). "Atmospheric Chemistry: Fundamentals and Experimental Techniques." Wiley-Interscience: New York.

Haagen-Smit, A. J., Bradley, C. E., and Fox, M. M. (1953). Ozone formation in photochemical oxidation of organic substances. *Ind. Eng. Chem.* **45**, 2086–2089.

Hough, A. M. (1988). The calculation of photolysis rates for use in global tropospheric modelling studies. Her Majesty's Stationery Office, London, AERE Report R-13259.

Huess, J. M., and Glasson, W. A. (1968). Hydrocarbon reactivity and eye irritation. *Environ. Sci. Technol.* **2**, 1109–1116.

Levy, H. (1971). Normal atmosphere: large radical and formaldehyde concentrations predicted. *Science.* **173**, 141–143.

Nordic Council of Ministers. (1991). Photochemical oxidants in the atmosphere. Nord 1991:7, Copenhagen, Denmark.

Peterson, J. T. (1976). Calculated actinic fluxes (290–700 nm) for air pollution photochemistry applications. U.S. Environmental Protection Agency, EPA-600/4-76-025.

Russell, A., Milford, J., Bergin, M. S., McBride, S., McNair, L., Yang, Y., Stockwell, W. R. T., and Croes, B. (1995) Urban ozone control and atmospheric reactivity of organic gases. *Science.* **269**, 491–495.

Schmidt, U. (1974). Molecular hydrogen in the atmosphere. *Tellus.* **26**, 78–90.

Simmonds, P. G., Cunnold, D. M., Dollard, G. J., Davies, T. J., McCulloch, A., and Derwent, R. G. (1993). Evidence for the phase-out of CFC use in Europe over the period 1987–1990. *Atmos. Environ.* **27**, 1397–1407.

Sluyter, R., and van Zantvoort, E. (1996). Information concerning air pollution by ozone. European Environment Agency, Copenhagen, Denmark.

Ten Houten, J. G. (1966). *Landbouwk T.* **78**, 2–13.

UN ECE (1991). Protocol to the 1979 Convention on Long-Range Transboundary Air Pollution Concerning the Control of Emissions of Volatile Organic Compounds or Their Transboundary Fluxes. United Nations Economic Commission for Europe, Geneva, Switzerland, ECE/EB.AIR/30.

U.S. EPA (1977). Recommended policy on control of volatile organic compounds. United States Environmental Protection Agency. *Fed. Regist.* **1977**, 42,35314–35316.

UK PORG (1993). Ozone in the United Kingdom 1993. Third report of the United Kingdom Photochemical Oxidants Review Group, Department of the Environment, London.

CHAPTER 8

Global Atmospheric Chemistry of Reactive Hydrocarbons

JOHN H. SEINFELD

California Institute of Technology, Pasadena, California

The chemistry of the global troposphere is controlled by its most abundant hydrocarbon, methane. Ozone can be considered to be the principal product of tropospheric chemistry, and understanding the relative importance of the two main sources of ozone in the troposphere—transport from the stratosphere versus local photochemistry—is one of the major problems in atmospheric chemistry. A critical nitrogen oxide NO_x (NO_x = $NO + NO_2$) concentration exists below which ozone is consumed and above which ozone is produced. Although the chemistry of the background troposphere is controlled by CH_4 and its oxidation product CO, a number of other organic species play important roles in global-scale chemistry. These include alkane molecules with lifetimes longer than a few days, acetone, and peroxyacetyl nitrate (PAN).

The predominant hydrocarbon molecule in the troposphere is methane, and the chemistry of the background troposphere is fueled by methane. In the urban and continental troposphere a large number of other anthropogenic and biogenic hydrocarbons and organic species are present, as has already been discussed in some detail in earlier chapters. Once thought to be relatively isolated from each other, urban regions and the "undisturbed" global troposphere are now recognized to be just two extremes of a continuum that links all of tropospheric chemistry. Although methane is the dominant global hydrocarbon, the chemistry of the global troposphere depends importantly on a variety of other organic species. The troposphere is an oxidative medium; the tendency is for species to be eventually converted to a more oxidized state. Hydrocarbons are reacted to aldehydes, then to acids, and finally to CO_2. Though often present at mixing ratios of only a part per billion (ppb) or less, oxides of nitrogen play a central role in the chemistry of the troposphere. The goal of this chapter is to present a self-contained treatment of the global chemistry of the troposphere, focusing especially on the role of reactive hydrocarbons and other organic species.

I. HYDROXYL RADICAL

Ozone can be considered as the principal product of tropospheric chemistry. Ozone photolyzes at wavelengths less than 319 nm to produce both ground-state (O) and excited singlet ($O(^1D)$) oxygen atoms,

$$O_3 + h\nu \longrightarrow O + O_2 \qquad (1)$$

$$\longrightarrow O(^1D) + O_2. \qquad (2)$$

The ground-state O atom recombines rapidly with O_2 to form O_3,

$$O + O_2 + M \longrightarrow O_3 + M, \qquad (3)$$

where M represents N_2 or O_2 or another third molecule that absorbs the excess vibrational energy and thereby stabilizes the O_3 molecule formed.

Reaction 1 followed by Reaction 3 has no net chemical effect. However, when $O(^1D)$ is formed, the $O(^1D) \rightarrow O$ transition is forbidden. Thus $O(^1D)$ must react with another atmospheric species. Most often $O(^1D)$ encounters N_2 or O_2 removing the excess energy and quenching $O(^1D)$ to its ground state,

$$O(^1D) + M \longrightarrow O + M. \tag{4}$$

If this occurs, the set of Reactions, 2, 3, and 4, is just a cycle with no net chemical effect. Occasionally, however, $O(^1D)$ collides with a water molecule and produces two hydroxyl radicals,

$$O(^1D) + H_2O \longrightarrow 2OH \tag{5}$$

Reaction 5 is important from two aspects: it represents a primary source of OH radicals and it serves as a sink for O_3.

A question of great importance is how frequently does Reaction 5 occur relative to the quenching Reaction 4? The ratio of the rates of the two reactions is

$$\frac{R_5}{R_4} = \frac{k_5[O(^1D)][H_2O]}{k_4[O(^1D)][M]}$$
$$= \frac{k_5[H_2O]}{k_4[M]}$$

At 298 K, in air, $k_4 = 2.9 \times 10^{-11}$ cm^3/molecule/sec and $k_5 = 2.2 \times 10^{-10}$ cm^3/molecule/sec (see Table 1), so $k_5/k_4 = 7.58$. Thus, Reaction 5 has a rate constant a factor of 7.58 larger than the quenching Reaction 4 in air. Water vapor mixing ratios in the lower troposphere have values as large as 10^4 ppm (1% by volume), so $[H_2O]/[M] = 0.01$ under these conditions. Then $R_5/R_4 = 0.0758$, that is, as much as 7.6% of the $O(^1D)$ produced reacts with H_2O to generate OH. Because two OH radicals are formed in Reaction 5, this leads to an OH yield of approximately 0.15 molecules of OH per O_3 molecule photolyzed.

The key to understanding tropospheric chemistry lies in the reactions of the hydroxyl radical. This radical, unlike most molecular fragments formed from carbon-containing molecules, is unreactive toward O_2, and, as a result, it survives to react with most atmospheric trace species. Despite tropospheric concentrations that are only about 4 molecules of OH for every 10^{14} molecules of air, high reactivity of the hydroxyl radical makes it the atmosphere's detergent. As a result, the reactions that can be considered to trigger background tropospheric chemistry are Reactions 2 and 5. Because the H_2O vapor mixing ratio decreases with increasing altitude, and the O_3 mixing ratio generally increases with increasing altitude, the OH radical production rate is reasonably independent of altitude. There is less total overhead ozone in the tropics than anywhere else on the earth, so more ultraviolet (UV) radiation penetrates to the troposphere, promoting Reactions 1 and 2. Since there is also more water vapor in the tropical atmosphere, the production of OH radicals is higher there than elsewhere in the troposphere. Although tropospheric ozone comprises only about 10% of all ozone in the atmosphere, through the formation of OH it determines the oxidizing efficiency of the troposphere. In so doing, ozone ultimately controls the chemical composition of the troposphere.

II. METHANE OXIDATION

A. METHANE OXIDATION CHAIN

Hydroxyl radicals react with CH_4,

$$CH_4 + OH \cdot \longrightarrow CH_3 \cdot + H_2O. \tag{6}$$

The methyl radical, CH_3, reacts virtually instantaneously with O_2 to yield the methyl peroxy radical, CH_3O_2,

$$CH_3 \cdot + O_2 + M \longrightarrow CH_3O_2 \cdot + H_2O, \tag{7}$$

so that the CH_4–OH reaction may be written concisely as

$$CH_4 + OH \cdot \xrightarrow{O_2} CH_3O_2 \cdot + H_2O. \tag{8}$$

Under tropospheric conditions, the methyl peroxy radical can react with NO, NO_2, and HO_2 radicals. The reaction with NO leads to formation of the methoxy radical, CH_3O,

$$CH_3O_2 \cdot + NO \longrightarrow CH_3O \cdot + NO_2. \tag{9}$$

The only important reaction for the methoxy radical under tropospheric conditions is with O_2 to form formaldehyde, HCHO, and the HO_2 radical,

$$CH_3O \cdot + O_2 \longrightarrow HCHO + HO_2 \cdot. \tag{10}$$

Addition of an H atom to O_2 weakens the O—O bond in O_2, and the resulting HO_2 radical reacts much more freely than does O_2 itself. When NO is present in sufficient quantities, the most important atmospheric reaction that the HO_2 radical undergoes is with NO,

$$HO_2 \cdot + NO \longrightarrow NO_2 + OH \cdot. \tag{11}$$

Formaldehyde, the first major product of CH_4 oxidation with a lifetime longer than a few seconds, is an important molecule in tropospheric chemistry. Formaldehyde undergoes two main reactions in the atmosphere, photolysis;

$$HCHO + h\nu \longrightarrow H\cdot + H\dot{C}O \qquad (12a)$$

$$\longrightarrow H_2 + CO, \qquad (12b)$$

and reaction with OH,

$$HCHO + OH\cdot \longrightarrow H\dot{C}O + H_2O. \qquad (13)$$

The hydrogen atom combines immediately with O_2 to yield HO_2. The formyl radical, HCO, also reacts very rapidly with O_2 to yield an HO_2 radical and CO,

$$H\dot{C}O + O_2 \longrightarrow HO_2\cdot + CO. \qquad (14)$$

Because of the rapidity of both reactions, the formaldehyde reactions may be written concisely as

$$HCHO + h\nu \xrightarrow{2O_2} 2HO_2\cdot + CO \qquad (15a)$$

$$\longrightarrow H_2 + CO. \qquad (15b)$$

and

$$HCHO + OH\cdot \xrightarrow{O_2} HO_2\cdot + CO + H_2O. \qquad (16)$$

Carbon monoxide reacts with the OH radical,

$$CO + OH\cdot \longrightarrow CO_2 + H\cdot, \qquad (17)$$

which, because of the instantaneous combination of the hydrogen atom with O_2, can be written as

$$CO + OH\cdot \xrightarrow{O_2} CO_2 + HO_2\cdot. \qquad (18)$$

Termination of the methane oxidation chain occurs when OH and NO_2 react to form nitric acid,

$$OH\cdot + NO_2 + M \longrightarrow HNO_3 + M. \qquad (19)$$

When NO levels are sufficiently low, peroxy radicals HO_2 and CH_3O_2 will react with each other and themselves. The self-reaction of HO_2 produces hydrogen peroxide,

$$HO_2\cdot + HO_2\cdot \longrightarrow H_2O_2 + O_2. \qquad (20)$$

The HO_2–CH_3O_2 reaction leads to the formation of methyl hydroperoxide, CH_3OOH,

$$CH_3O_2\cdot + HO_2\cdot \longrightarrow CH_3OOH + O_2. \qquad (21)$$

The CH_3O_2 self-reaction can generally be neglected with respect to these two under background tropospheric conditions.

Methyl hydroperoxide can photolyze or react with OH,

$$CH_3OOH + h\nu \longrightarrow CH_3O\cdot + OH\cdot, \qquad (22)$$

$$CH_3OOH + OH\cdot \longrightarrow CH_3O_2\cdot + H_2O \qquad (23a)$$

$$\longrightarrow \cdot CH_2OOH + H_2O, \qquad (23b)$$

$$\cdot CH_2OOH \longrightarrow HCHO + OH\cdot, \qquad (24)$$

where Reaction 24 is fast. Methyl hydroperoxide can be viewed as a temporary reservoir of free radicals, which are released once CH_3OOH photolyzes or reacts with OH.

Another temporary radical reservoir is methyl peroxynitrate, CH_3OONO_2, which is formed when CH_3O_2 reacts with NO_2,

$$CH_3O_2\cdot + NO_2 + M \rightleftharpoons CH_3OONO_2 + M, \qquad (25,26)$$

and thermally dissociates back to the reactants by Reaction 26.

When NO and NO_2 are present in sunlight, ozone formation occurs as a result of the photolysis of NO_2 at wavelengths < 424 nm,

$$NO_2 + h\nu \longrightarrow NO + O, \qquad (27)$$

and

$$O + O_2 + M \longrightarrow O_3 + M. \qquad (28)$$

There are no significant sources of O_3 in the atmosphere other than Reaction 28. Once formed, O_3 reacts with NO to regenerate NO_2,

$$O_3 + NO \longrightarrow NO_2 + O_2. \qquad (29)$$

B. Overall Reaction Sequence

When NO_x levels are sufficiently high that reaction of the peroxy radicals HO_2 and CH_3O_2 with NO predominates over peroxy radical self-reactions, the methane oxidation chain can be written as:

$$CH_4 + OH\cdot \xrightarrow{O_2} CH_3O_2\cdot + H_2O$$

$$CH_3O_2\cdot + NO \longrightarrow CH_3O\cdot + NO_2$$

$$CH_3O\cdot + O_2 \longrightarrow HCHO + HO_2\cdot$$

$$HO_2\cdot + NO \longrightarrow OH\cdot + NO_2$$

$$2(NO_2 + h\nu \longrightarrow NO + O)$$

$$2(O + O_2 + M \longrightarrow O_3 + M)$$

$$\overline{\text{Net: } CH_4 + 4O_2 + 2h\nu \longrightarrow HCHO + 2O_3 + H_2O.}$$

Two molecules of ozone result from each CH_4 molecule. Further oxidation of formaldehyde leads to additional production of ozone:

$$HCHO + h\nu \xrightarrow{O_2} 2HO_2\cdot + CO$$

$$\longrightarrow H_2 + CO,$$

$$HCHO + OH\cdot \xrightarrow{O_2} HO_2\cdot + CO + H_2O,$$

$$HO_2\cdot + NO \longrightarrow OH\cdot + NO_2,$$

$$NO_2 + h\nu \longrightarrow NO + O,$$

$$O + O_2 + M \longrightarrow O_3 + M.$$

Finally, the atmospheric oxidation of CO under conditions where sufficient NO is present is:

$$CO + OH\cdot \xrightarrow{O_2} CO_2 + HO_2\cdot$$

$$HO_2\cdot + NO \longrightarrow OH\cdot + NO_2$$

$$NO_2 + h\nu \longrightarrow NO + O$$

$$O + O_2 + M \longrightarrow O_3 + M$$

$$\overline{\text{Net: } CO + 2O_2 + h\nu \longrightarrow CO_2 + O_3.}$$

Note that neither OH nor HO_2 is consumed in this reaction cycle, which is a catalytic oxidation of CO to CO_2. Net formation of O_3 occurs because the conversion of NO to NO_2 is accomplished by the HO_2 radical rather than by O_3 itself via Reaction 29.

This cycle, as well as the CH_4 and HCHO oxidation chains, can occur repeatedly until one of the active molecules is removed in a termination reaction, such as Reaction 19, 20, 21, or 25. Specifically, production of O_3 in the CH_4 oxidation chain is interrupted if the peroxy radicals HO_2 and CH_3O_2 react with something other than NO, for example, themselves, NO_2, or O_3 itself, or if NO_x (NO + NO_2) is removed from the catalytic cycle by reaction of NO_2 with OH to form HNO_3 (Reaction 19). For the HO_2 radical, in addition to Reactions 20 and 21, another important reaction is:

$$HO_2\cdot\ +\ O_3 \longrightarrow OH\cdot\ +\ 2O_2. \tag{30}$$

Understanding the relative importance of the two main sources of ozone in the troposphere—transport from the stratosphere versus local photochemistry—is one of the major problems in atmospheric chemistry. The principal *in situ* chemical source of ozone in the troposphere is photochemical production through the CH_4 oxidation chain. The level of NO is critical in this chain in dictating the fate of the HO_2 and the CH_3O_2 radicals. To see this, consider the CO oxidation chain. Reaction 11,

$$HO_2\cdot\ +\ NO \longrightarrow NO_2\ +\ OH\cdot, \tag{11}$$

leads to O_3 production; but Reaction 30,

$$HO_2\cdot\ +\ O_3 \longrightarrow OH\cdot\ +\ 2O_2, \tag{30}$$

destroys ozone. The ratio of the rates of Reactions 30 and 11,

$$\frac{R_{30}}{R_{11}} = \frac{k_{30}[O_3]}{k_{11}[NO]},$$

is indicative of which path HO_2 will take. At 298 K, $k_{30}/k_{11} = (2.0 \times 10^{-15})/(8.6 \times 10^{-12}) \approx 2.5 \times 10^{-4}$ (Table 1), so

$$\frac{R_{30}}{R_{11}} \approx 2.5 \times 10^{-4}\,\frac{[O_3]}{[NO]}.$$

The "breakeven" concentration of NO, below which O_3 is destroyed and above which it is produced, can be defined as that when $R_{30}/R_{11} = 1$, or

$$[NO]_{be} \approx 2.5 \times 10^{-4}[O_3].$$

The O_3 mixing ratio near the earth's surface in the remote continental troposphere is about 20 ppb. The breakeven NO mixing ratio corresponding to this O_3 mixing ratio is 5 ppt.

The same concept applies to the overall CH_4 oxidation chain. A competition exists between NO and the HO_2 radical for reaction with the CH_3O_2 radical,

$$CH_3O_2\cdot + NO \longrightarrow CH_3O\cdot + NO_2, \tag{9}$$

$$CH_3O_2\cdot + HO_2\cdot \longrightarrow CH_3OOH + O_2, \tag{21}$$

and the preferred route for CH_3O_2 reaction depends on the concentrations of HO_2 radicals and NO. The ratio of the rates of Reactions 9 and 21 is

$$\frac{R_{21}}{R_9} = \frac{k_{21}[HO_2]}{k_9[NO]}.$$

At 298 K, $k_{21}/k_9 = (5.6 \times 10^{-12})/(7.7 \times 10^{-12}) = 0.73$, so the rate constants are of comparable magnitude. Thus, the breakeven NO concentration based on $R_{21} = R_9$ is achieved when $[NO] \approx [HO_2]$.

Hard et al. (1992) determined HO_2 concentrations at two Oregon sites for continuous periods of 36 to 48 h, one site characterized by clean marine air and the other, an urban site. At both sites, maximum daily $[HO_2]$ was in the range of 1 to 2 $\times$ 10^8 molecules/cm^3 under clear sky conditions. On the basis of a photochemical model, Logan et al. (1981) predicted HO_2 concentrations of 3.3 $\times$ 10^8 molecules/cm^3 (45°N latitude, surface, equinox) at the midday maximum and 1 to 10 $\times$ 10^6 molecules/cm^3 at night. Madrovich and Calvert (1990) predicted noontime HO_2 concentrations of 3.2 $\times$ 10^8 molecules/cm^3. Mihelcic et al. (1990) measured an HO_2 level of 8.3 $\times$ 10^8 molecules/cm^3 at midday at a mountain site in Germany.

Based on calculated HO_2 radical concentrations in the troposphere, Logan et al. (1981) estimated that the CH_3O_2–NO reaction dominates over the CH_3O_2–HO_2 reaction for NO mixing ratios ≥ 30 ppt (equivalent to an NO concentration of 7 $\times$ 10^8 molecules/cm^3); for NO mixing ratios ≤ 30 ppt the CH_3O_2–HO_2 reaction dominates.

Because of ozone loss that occurs by photolysis, Reactions 1 and 2, the NO breakeven concentration at which net O_3 production occurs is somewhat larger than that based just on the R_{21}/R_9 ratio.

The levels of H_2O_2 and CH_3OOH are indicative of the oxidative state of the atmosphere, that is, whether the peroxy radicals HO_2 and CH_3O_2 are reacting with NO or each other, Reactions 9 and 11 versus Reactions 20 and 21. Removal of H_2O_2 and CH_3OOH through precipitation or dry deposition is a sink of free radicals (odd hydrogen, the sum of OH, HO_2, and CH_3O_2). Measurements of H_2O_2 and CH_3OOH in the northern Pacific during the PEM-West A experiment indicated that the $[H_2O_2]/[CH_3OOH]$ ratio increased with altitude and latitude with ratios <1 in the tropical surface layer and >2 at midlatitude high altitude (Heikes et al., 1996). Above 3 km, O_3 concentration increases with decreasing H_2O_2 and CH_3OOH, as expected from the CH_4 oxidation cycle.

TABLE 1 Rate Constants of Atmospheric Chemical Reactions[a]

	Second-order reactions $k(T) = A \exp(-E/RT)$		
Reaction	A (cm³/molecule/sec)	E/R (K)	k (298 K) (cm³/molecule/sec)
$O(^1D) + O_2$	3.2×10^{-11}	-70	4.0×10^{-11}
$O(^1D) + N_2$	1.8×10^{-11}	-110	2.6×10^{-11}
$O(^1D) + H_2O$	2.2×10^{-11}	0	2.2×10^{-10}
$HO_2 + O_3$	1.1×10^{-14}	500	2.0×10^{-15}
$HO_2 + HO_2$	See footnote [b]		
$HO_2 + NO$	3.7×10^{-12}	-250	8.6×10^{-12}
$NO + O_3$	2.0×10^{-12}	1400	1.8×10^{-14}
$OH + CO$	$1.5 \times 10^{-13} (1 + 0.6 p)$	0	$1.5 \times 10^{-13} (1 + 0.6 p)$
$OH + CH_4$	2.65×10^{-12}	1800	6.3×10^{-15}
$HO_2 + CH_3O_2$	3.8×10^{-13}	-800	5.6×10^{-12}
$HCO + O_2$	3.5×10^{-12}	-140	5.5×10^{-12}
$CH_2OH + O_2$			9.1×10^{-12}
$CH_3O + O_2$	3.9×10^{-14}	900	1.9×10^{-15}
$CH_3O_2 + NO$	4.2×10^{-12}	-180	7.7×10^{-12}
$C_2H_5O + O_2$	6.3×10^{-14}	550	1.0×10^{-14}
$C_2H_5O_2 + NO$	8.7×10^{-12}	0	8.7×10^{-12}
$CH_3C(O)O_2 + NO$	2.4×10^{-11}	0	2.4×10^{-11}
$C_2H_6 + OH$			0.257×10^{-12}
$C_3H_8 + OH$			1.15×10^{-12}
$n\text{-Butane} + OH$			2.54×10^{-12}
$n\text{-Pentane} + OH$			3.94×10^{-12}
$n\text{-Hexane} + OH$			5.61×10^{-12}
$C_2H_4 + OH$			8.52×10^{-12}
$C_3H_6 + OH$			26.3×10^{-12}
$1\text{-Butene} + OH$			31.4×10^{-12}
$Toluene + OH$			5.96×10^{-12}
$Isoprene + OH$			101×10^{-12}
$HCHO + OH$			9.37×10^{-12}
$CH_3CHO + OH$			15.8×10^{-12}
$CH_3C(O)CH_3 + OH$			0.219×10^{-12}
$HCOOH + OH$			0.45×10^{-12}
$CH_3COOH + OH$			0.8×10^{-12}
$CH_3OOH + OH$			5.54×10^{-12}
$CH_3C(O)O_2 + HO_2$			1.3×10^{-11d}
$CH_3C(O)O_2 + CH_3O_2$			1.4×10^{-11d}

(continues)

TABLE 1　(*continued*)

| Reaction | Association reactions | | | |
	Low-pressure limit[c] $k_0 = k_0^{300} (T/300)^{-n}$ (cm^6/molecule/sec) k_0^{300}	n	High-pressure limit[c] $k_\infty = k_\infty^{300} (T/300)^{-m}$ (cm^3/molecule/sec) k_∞^{300}	m
$O + O_2 + M$	6.0×10^{-34}	2.3	—	—
$OH + NO_2 + M$	2.6×10^{-30}	3.2	2.4×10^{-11}	1.3
$CH_3 + O_2 + M$	4.5×10^{-31}	3.0	1.8×10^{-12}	1.7
$CH_3O_2 + NO_2 + M$	1.5×10^{-30}	4.0	6.5×10^{-12}	2.0
$CH_3C(O)O_2 + NO_2 + M$	9.7×10^{-29}	5.6	9.3×10^{-12}	1.5

[a]Atkinson (1994); DeMore *et al.* (1997).
[b]The recommended rate constant for the $HO_2 + HO_2$ reaction is (Stockwell, 1995)

$$k = (k_e + k_p)f_w$$
$$k_e = 2 \times 10^{-13} \exp (600/T)$$
$$k_p = 1.7 \times 10^{-33}[M] \exp (100/T)$$
$$f_w = 1 + 1.4 \times 10^{-21} [H_2O] \exp (2200/T),$$

where [M] and [H_2O] are in molecules/cm^3.
[c]Rate constants for third-order reactions of the type

$$A + B \rightleftharpoons [AB]^* \xrightarrow{M} AB$$

are given in the form

$$k_0(T) = k_0^{300} \left(\frac{T}{300}\right)^{-n} cm^6/molecule^a/sec,$$

where k_0^{300} accounts for air as the third body. Where pressure fall-off corrections are necessary, the limiting high-pressure rate constant is given by

$$k_\infty(T) = k_\infty^{300} \left(\frac{T}{300}\right)^{-m} cm^3/molecule/sec.$$

To obtain the effective second-order rate constant at a given temperature and pressure (altitude z) the following formula is used:

$$k(T,z) = \left\{ \frac{k_0(T)[M]}{1 + (k_0(T)[M]/k_\infty(T))} \right\} 0.6^{(1 + [\log_{10}(k_0(T)[M]/k_\infty(T))]^2)^{-1}}.$$

[d]Moortgat *et al.* (1989).

One global methane oxidation chain modeling study led to an estimate of a net annual loss of about 0.22 molecules of OH for each CH_4 molecule destroyed (Tie *et al.*, 1992). The associated average annual yield of CO from CH_4 oxidation is about 0.82 molecules of CO per molecule of CH_4 oxidized. The global CH_4 oxidation chain was estimated to produce, as a result, about 1.15 molecules of O_3 for each molecule of CH_4 destroyed.

C. Hydroxyl Radical Levels in the Troposphere

The hydroxyl radical is the most important reactive species in the troposphere. OH reacts with most trace species in the atmosphere, and its importance derives from both its high reactivity toward other molecules and its relatively high concentration. Were OH simply to react with other species and not in some manner be regenerated, its concentration would be far too low, in spite of its high reactivity, to play an important role in atmospheric chemistry. The key is that, when reacting with atmospheric trace species, OH is regenerated in catalytic cycles, leading to sustained concentrations on the order of 10^6 molecules/cm³ during daylight hours.

From tropospheric chemical mechanisms it is possible to estimate the atmospheric concentration of OH. Such calculations suggest a seasonally, diurnally, and globally averaged OH concentration ranging from 2×10^5 to 10^6 molecules/cm³. As noted earlier, highest OH levels are predicted in the tropics, where high humidities and strong radiative fluxes lead to a high rate of OH production from Reactions 2 and 5. In addition, OH levels are predicted to be about 20% higher in the Southern Hemisphere as a result of the large amounts of CO produced by human activities in the Northern Hemisphere that act to reduce OH through Reaction 18. Hydroxyl radical levels are about a factor of five higher over the continents than over the oceans, because of generation through hydrocarbon oxidation reactions involving continental anthropogenic and biogenic hydrocarbons.

The two principal routes of formation of OH in the troposphere are O_3 photolysis, Reactions 2 and 5, and reaction of HO_2 radicals with NO, Reaction 11. Under noontime conditions at 298 K, by assuming an O_3 mixing ratio of 50 ppb (1.2×10^{12} molecules/cm³), $[HO_2] = 10^8$ molecules/cm³, and NO of 10 ppb (2.45×10^{11} molecules/cm¹³), OH generation rates for the two pathways are:

$$O_3 + h\nu \longrightarrow 9 \times 10^7 \text{ molecules OH/cm}^3/\text{sec},$$

and

$$HO_2 \cdot + NO \longrightarrow 2.5 \times 10^8 \text{ molecules } OH/cm^{-3}/sec.$$

D. NO_x AND NO_y

Nitrogen oxides play a central role in the chemistry of the atmosphere. Of the estimated global sources of NO_x (about 44 Tg (N) per year) to the atmosphere, nearly two-thirds originate at the earth's surface from fossil fuel and biomass combustion. Although virtually all the NO_x is emitted as NO, it is rapidly converted to a variety of other species. The sum of NO_x ($NO + NO_2$) and associated products are collectively termed total reactive oxidized nitrogen and given the designation NO_y. NO_y includes NO, NO_2, HNO_3, $CH_3C(O)OONO_2$, PAN, N_2O_5, HNO_2, HNO_4, $RONO_2$, RO_2NO_2, and particulate NO_3^-. A complete understanding of the atmospheric NO_x cycle is hampered by the fact that many oxidation products are difficult to measure. Nitric oxide (NO) can be measured reliably down to about 5 ppt. In the marine boundary layer, mixing ratios are typically in the 3 to 10 ppt range, increasing with altitude. Significant latitudinal gradients in NO exist with largest mixing ratios in the upper troposphere at northern latitudes. Nitrogen dioxide (NO_2) can be detected down to about 10 ppt but its measurement is subject to more interferences than that of NO.

PAN, and other pernitrates, can constitute a substantial component of reactive nitrogen in some regions of the atmosphere. (We will discuss PAN chemistry shortly.) Substantial NO_y mixing ratios exist everywhere in the free troposphere. There are, however, sizable deficits in the budget of reactive nitrogen; measured species, mainly $NO_x + HNO_3 + PAN$, can account for 70 to 100% of NO_y but shortfalls are common.

The distribution of NO_y among its constituent species is of significant importance because this distribution is indicative of the chemical processing of the air mass that has occurred. Generally the predominant NO_y components are NO_x, PAN, and inorganic nitrate (HNO_3 plus particulate NO_3^-). The relative abundance of these species varies in a systematic way. NO_x is most abundant close to sources; PAN is most abundant in air masses in which there is active organic photochemistry. Inorganic nitrate is most abundant in more remote areas of the troposphere. Although organic nitrates, of the general form $RONO_2$, generally comprise less than 5% of total NO_y, their presence is an indicator of the extent of photochemical processing of an air mass (Roberts, 1990). These compounds are formed as products of hydrocarbon oxidation, usually at NO_x mixing ratios ≥ 500 ppt. Tropospheric loss rates of $RONO_2$ compounds are sufficiently slow that their concentration in an airmass is a good indicator of the degree of organic photochemistry that has occurred in a high NO_x environment over the preceding several days (Roberts *et al.*, 1996).

III. CHEMICAL LIFETIMES IN THE METHANE OXIDATION CHAIN

The lifetime of a species A, as a result of reaction with the hydroxyl radical,

$$A + OH \cdot \xrightarrow{k} products,$$

is

$$\tau_{OH}^A = \frac{1}{k[OH]}.$$

The lifetime as a result of photolysis,

$$A + h\nu \xrightarrow{j_A} product,$$

is just the inverse of the photolysis rate constant,

$$\tau_{h\nu}^A = \frac{1}{j_A}.$$

Up to this point, we have considered the tropospheric reactions of CH_4, HCHO, CO, and CH_3OOH. We can compare timescales for their reactions as in Table 2. Methane has a very long atmospheric lifetime; at the OH level assumed for the calculation in Table 2 it is about 3.4 years. Actually with a more realistic globally averaged OH concentration (i.e., about one-third of that assumed in Table 2) the CH_4 lifetime is about 10 years. Therefore, the ozone-generating and oxidation capacity of the troposphere depends, over a decadal timescale, on the background level of methane. Because of its one- to three-month lifetime, CO chemistry determines local HO_2 and OH levels and O_3 formation in different regions of the background troposphere.

Carbon monoxide is the most abundant readily oxidizable compound present in the global troposphere. A global background level of CO results from the methane oxidation chain. In most regions of the Northern Hemisphere, CO is substantially augmented by anthropogenic continental emissions. With its tropospheric lifetime of about one month, CO is an effective tracer of anthropogenic emissions, including combustion sources of NO_x. Measurements of O_3 and CO at distances the order of 1000 km or more from anthropogenic source regions show relatively large correlation between these two species (Roberts *et al.*, 1996; Fehsenfeld *et al.*, 1996; Wang *et al.*, 1996).

TABLE 2   Timescales for Reactions in the Methane Oxidation Chain

Reaction	OH rate constant (cm³/molecule/sec) @ 298 K	$\tau_{OH}{}^a$	$\tau_{h\nu}$
$CH_4 + OH \cdot \rightarrow$	6.3×10^{-15}	3.4 years	
$CO + OH \cdot \rightarrow$	1.5×10^{-13}	51 days	
$HCHO + OH \cdot \rightarrow$	9.37×10^{-12}	19.6 h	
$HCHO + h\nu \rightarrow$			4 h
$CH_3OOH + OH \rightarrow$	5.54×10^{-12}	33 h	
$CH_3OOH + h\nu \rightarrow$			3.5 days[b]

[a]Based on an assumed 12-h daytime mean OH concentration of 1.5×10^6 molecules/cm³.
[b]Based on absorption cross section of Vaghjiani and Ravishankara (1989) at solar zenith angle of 0°.

The lifetimes of HCHO resulting from photolysis and OH reaction are ~ 4 h and 1 day, respectively, leading to an overall lifetime of ~ 3 h for overhead sun conditions. The lifetime of methyl hydroperoxide in the troposphere resulting from photolysis and OH reaction is estimated to be 1 to 2 days. CH_3OOH is then a temporary sink of free radicals.

Methyl peroxynitrate, CH_3OONO_2, thermally dissociates back to its reactants with a lifetime of ~ 1 sec at 298 K and atmospheric pressure; this lifetime increases to ~ 2 days at the temperature and pressure conditions of the upper troposphere. Thus, CH_3OONO_2 can act as a temporary reservoir of CH_3O_2 radicals and NO_2 in the upper troposphere.

IV. NONMETHANE ORGANICS IN THE GLOBAL ATMOSPHERE

Although the chemistry of the troposphere is dominated by CH_4 and CO, a number of other organic species play important roles in global-scale chemistry. Various studies exist in which hydrocarbons and other organic species have been sampled in remote tropospheric air (see Chapter 6 of this volume). Selected data from one such study, PEM-West A, conducted in the northern Pacific during September and October 1991, are given in Table 3 (Gregory *et al.*, 1996).

Even though the Pacific Ocean is one of the few remaining regions of the

TABLE 3   Chemical Signatures of Aged Pacific Marine Air[a,b]

	Free troposphere (3–13 km)		Marine boundary layer (<0.5 km)	
	North	South	North	South
O_3, ppb	31.9	21.9	13.1	9.0
CO, ppb	90.4	72.6	85.5	67.8
CH_4, ppb	1724.9	1688.2	1713.8	1672.8
NO	15.3	10.3	3.3	4.6
NO_2	44.2	29.7	17.6	15.8
PAN	22.1	13.9	2.0	2.0
SO_2	86.0	61.0	66.0	36.0
H_2O_2	520.0	349.8	1061.1	545.9
CH_3OOH	292.0	373.5	1491.0	1826.0
HNO_3	65.0	7.0	65.0	17.0
HCOOH	121.0	45.0	191.0	86.0
CH_3COOH	242.0	429.0	173.0	176.0
Toluene	11.5	No data	13.5	No data
Ethane	621.5	428.0	434.0	360.0
Ethene	15.0	10.0	23.0	12.0
Propane	45.2	17.0	25.2	12.0
Propene	8.0	7.0	14.0	No data
Isobutane	9.0	<DL[c]	6.0	<DL
n-Butane	10.0	No data	7.0	No data
1-Butene	5.0	4.0	6.0	No data

[a]Gregory et al. (1996).
[b]Median values of measurements given in the table. Data obtained during September and October 1991 as part of the Pacific Exploratory Mission (PEM) over the western (west) Pacific, using the acronym PEM-West A. Aircraft sampling centered on the Northern Hemisphere western Pacific covering a latitude range of 0° to 40°N and a longitudinal range of about 115° to 180°E. The data in this table were obtained in air that had resided over the Pacific Ocean for at least 10 days. Two classifications of marine source air, North versus South, were made depending on whether the air mass sampled came from the Pacific Northern or Southern Hemisphere. All values in parts per trillion (ppt) unless noted otherwise.
[c]Below detection limit.

Northern Hemisphere that is relatively free of direct anthropogenic emissions, chemical signatures show that aged marine air that circulates around the semipermanent subtropical anticyclone located off the Asian continent is influenced by infusion of continental air with anthropogenic emissions. The air masses characterized as North in Table 3 show enhancements in some continental–anthropogenic source species as compared to aged marine air with a more southerly pathway.

A. CHEMISTRY OF NONMETHANE ORGANICS

1. Alkanes

The principal reaction of alkanes in the troposphere is with the OH radical proceeding by H atom abstraction from C—H bonds. Any H atom in the alkane molecule is subject to OH attack; generally the OH radical will tend to abstract the most weakly bound hydrogen atom in the molecule. The overall OH rate constant reflects the number of available H atoms and the strengths of the C—H bonds for each. For propane, $CH_3CH_2CH_3$, for example, 70% of the OH reaction occurs by H atom abstraction from the secondary carbon atom $(-CH_2-)$ and 30% from the $-CH_3$ groups. The predominant route for propane oxidation is

$$CH_3CH_2CH_3 + OH\cdot \longrightarrow CH_3\overset{\cdot}{C}HCH_3 + H_2O, \qquad (31)$$

$$CH_3\overset{\cdot}{C}HCH_3 + O_2 \xrightarrow[\text{fast}]{} CH_3CH(O_2\cdot)CH_3, \qquad (32)$$

$$CH_3CH(O_2\cdot)CH_3 + NO \longrightarrow NO_2 + CH_3CH(O\cdot)CH_3, \qquad (33)$$

and

$$CH_3CH(O\cdot)CH_3 + O_2 \xrightarrow[\text{fast}]{} \underset{\text{(acetone)}}{CH_3C(O)CH_3} + HO_2\cdot. \qquad (34)$$

By considering the subsequent reaction of HO_2 with NO, the net reaction representing atmospheric oxidation of propane,

$$CH_3CH_2CH_3 + OH\cdot + 2NO \longrightarrow 2NO_2 + CH_3C(O)CH_3 + OH\cdot, \qquad (35)$$

results in conversion of two molecules of NO to NO_2, production of one molecule of acetone, and regeneration of the OH radical. Conversion of two molecules of NO to NO_2 can be viewed as roughly tantamount to the generation of two molecules of ozone.

2. Alkenes (e.g., Ethene)

The principal reaction of alkenes under global tropospheric conditions is with the OH radical. In the case of alkenes, OH adds to the double bond instead of abstracting a hydrogen atom. The ethene–OH mechanism is

$$C_2H_4 + OH\cdot \longrightarrow HOCH_2CH_2\cdot, \qquad (36)$$

$$HOCH_2CH_2 \cdot\ +\ O_2 \xrightarrow{fast} HOCH_2CH_2O_2 \cdot, \tag{37}$$

$$HOCH_2CH_2O_2 \cdot\ +\ NO \longrightarrow NO_2 + HOCH_2CH_2O \cdot, \tag{38}$$

$$HOCH_2CH_2O \cdot\ +\ O_2 \longrightarrow 2HCHO + HO_2 \cdot. \tag{39}$$

By taking into account the subsequent fate of HCHO and assuming sufficient NO is present so that HO_2 preferentially reacts with NO, the atmospheric oxidation of each ethene molecule leads, on average, to the production of about 3.8 molecules of ozone (via Reactions 11, 36, and 37).

The atmosphere oxidation of propene by OH proceeds similarly to that of ethene. Because of the presence of three carbon atoms in the molecule, the number of reaction steps involved in propene oxidation is larger than that of ethene and more ozone molecules can be formed (on average 6.8 instead of 3.8). In the oxidation of propene, acetaldehyde (CH_3CHO) is formed as an intermediate.

3. Aldehydes

Formaldehyde is a first-generation product of the methane oxidation chain, Reaction 10. Formaldehyde photolyzes by Reactions 12a and 12b and reacts with OH by Reaction 13.

Acetaldehyde also photolyzes,

$$CH_3CHO + h\nu \longrightarrow CH_4 + CO \tag{40a}$$

$$\longrightarrow CH_3 \cdot\ +\ H\dot{C}O, \tag{40b}$$

and reacts with OH,

$$CH_3CHO + OH \cdot \longrightarrow CH_3\dot{C}O + H_2O. \tag{41}$$

The acetyl radical, CH_3CO, reacts with O_2,

$$CH_3\dot{C}O + O_2 \longrightarrow CH_3C(O)O_2 \cdot. \tag{42}$$

The peroxyacetyl radical, $CH_3C(O)O_2$, reacts with NO and NO_2,

$$CH_3C(O)O_2 \cdot\ +\ NO \longrightarrow NO_2 + CH_3C(O)O \cdot, \tag{43}$$

$$CH_3C(O)O_2 \cdot\ +\ NO_2 \rightleftarrows \underset{\text{(PAN)}}{CH_3C(O)O_2NO_2}, \tag{44,45}$$

and

$$CH_3C(O)O \cdot\ +\ O_2 \longrightarrow CH_3O_2 \cdot\ +\ CO_2. \tag{46}$$

4. Ketones

One of the most abundant reactive oxygenated species in the remote atmosphere is acetone, $CH_3C(O)CH_3$ (Singh $et\ al.$, 1994). Global background mixing ratios are in the range of 200 to 500 ppt. From atmospheric data and three-dimensional photochemical models, Singh $et\ al.$ (1994) estimated a global acetone source of 40 to 60 Tg per year, of which secondary formation from the atmospheric oxidation of hydrocarbons, principally propane (see Reaction 34), isobutane, and isobutene, was estimated to contribute 51%. The remainder is attributable to biomass burning (26%), direct biogenic emissions (21%), and primary anthropogenic emissions (3%).

Acetone photolyzes and reacts with OH radicals,

$$CH_3C(O)CH_3 + h\nu \longrightarrow CH_3\dot{C}(O) + CH_3\cdot \qquad (47)$$

and

$$CH_3C(O)CH_3 + OH\cdot \longrightarrow CH_3C(O)CH_2\cdot + H_2O. \qquad (48)$$

By incorporating the rapid O_2 addition to the free radicals, these two reactions can be written as

$$CH_3C(O)CH_3 + h\nu \xrightarrow{2O_2} CH_3C(O)O_2\cdot + CH_3O_2\cdot \qquad (49)$$

and

$$CH_3C(O)CH_3 + OH\cdot \xrightarrow{O_2} CH_3C(O)CH_2O_2\cdot + H_2O. \qquad (50)$$

We have already seen the reactions of the $CH_3C(O)O_2$ and CH_3O_2 radicals. The $CH_3C(O)CH_2O_2$ radical reacts as follows:

$$CH_3C(O)CH_2O_2\cdot + NO \longrightarrow CH_3C(O)CH_2O\cdot + NO_2 \qquad (51)$$

and

$$CH_3C(O)CH_2O\cdot + O_2 \xrightarrow{O_2} CH_3C(O)CHO + HO_2\cdot. \qquad (52)$$

Photolysis of acetone yields two HO_2 radicals and two HCHO molecules. About 30% of the HCHO molecules photolyze via the radical-forming path to yield two more HO_2 radicals. Thus, the yield from photolysis of a molecule of acetone is ~ 3.2 HO_x species, with subsequent formation of ozone when sufficient NO is present.

Acetone photolysis, which produces the $CH_3C(O)\cdot$ radical, is estimated to contribute 40 to 50 ppt of PAN in the middle and upper troposphere of the Northern Hemisphere. The lifetime of acetone is controlled mainly by photolysis. It varies globally between several days to more than a month. Acetone lifetime decreases with altitude and toward the equator. Thus, acetone transported to mid- and high latitudes can accumulate, especially during winter. Singh *et al.* (1995) estimated the average lifetime of acetone in the atmosphere to be 16 days. Comparison of measured concentrations of OH and HO_2 with those calculated based on production from H_2O vapor, O_3, and CH_4 showed that these sources were insufficient to explain observed radical concentrations in the upper troposphere (Wennberg *et al.*, 1998). Photolysis of acetone and other carbonyls may provide the needed source of HO_x.

5. Acids

Significant concentrations of formic, HCOOH, and acetic, CH_3COOH, acids exist in the troposphere (see Table 2 and Madronich *et al.*, 1990). Acetic acid is generally present in concentrations higher than those of formic acid (ratios of acetic to formic of roughly 2). Gas-phase production of acetic acid can occur by reaction of peroxyacetyl radicals with HO_2 and CH_3O_2 (Moortgat *et al.*, 1989; Madronich *et al.*, 1990; Wallington *et al.*, 1992)

$$CH_3C(O)O_2 \cdot + HO_2 \cdot \longrightarrow CH_3C(O)OH + O_3 \qquad (53a)$$
$$\longrightarrow CH_3C(O)OOH + O_2 \qquad (53b)$$

and

$$CH_3C(O)O_2 \cdot + CH_3O_2 \cdot \longrightarrow CH_3C(O)OH + O_2 + HCHO \qquad (54a)$$
$$\longrightarrow CH_3 \cdot + CO_2 + O_2 + CH_3O \cdot . \qquad (54b)$$

It has been estimated that Reaction 53 proceeds about 33% by the acetic acid channel 53a, with the remainder leading to peracetic acid, $CH_3C(O)OOH$, by channel 53b. Reaction 54 has been estimated to proceed in about equal proportions between channels 54a and 54b.

B. LIFETIMES OF TROPOSPHERIC ORGANICS

Estimated atmospheric lifetimes of selected nonmethane hydrocarbons are given in Table 4. The alkenes are relatively short lived in the atmosphere (lifetimes ≤ 2 days); as a result, these molecules persist only long enough to be observed relatively close to their sources of emission. Ethyne (acetylene) has a relatively long lifetime of 23 days, allowing it to persist sufficiently long to act as an atmospheric tracer for air masses that have been transported over long distances. In general, in the lower troposphere (≤ 2km) the behavior of the most reactive hydrocarbons (lifetime from OH oxidation ≤ 1 day) can be predicted based solely on photochemistry. Concentration levels of those with intermediate lifetimes (1 to 3 days) are controlled by both photochemistry

TABLE 4   Estimated Global-Average
Atmospheric Lifetime of Selected
Nonmethane Hydrocarbons[a]

Species	Lifetime (days)[b]
Ethane	78
Ethyne	23
Propane	18
Benzene	17
Isobutane	9
n-Butane	8
Isopentane	5
n-Pentane	5
n-Hexane	4
Toluene	3
Ethene	2
Propene	0.8
1-Butene	0.7
1-Pentene	0.7
Isoprene	0.2

[a]Blake et al. (1996).
[b]Estimated on the basis of OH rate constants (see Table 1 and Atkinson, 1994) relative to that of CH_3CCl_3. Lifetimes of hydrocarbons are calculated by comparison with the CH_3CCl_3 lifetime of 5.7 years (CH_3CCl_3 + OH rate constant, 0.01×10^{-12} cm^3 molecule/sec at 298 K, from DeMore et al., 1997) through ratios of OH rate constants. CH_3CCl_3 lifetime from Prinn et al. (1992).

and dynamic processes such as mixing; compounds with lifetimes greater than about 3 days are primarily influenced by mixing processes.

C. PEROXYACETYL NITRATE

The $CH_3C(O)O_2NO_2$ molecule, peroxyacetic nitric anhydride, known commonly as PAN, is an important atmospheric oxidation product of NO_x. Its precursor is the peroxyacetyl radical, $CH_3C(O)O_2$, one source of which is the acetaldehyde–OH Reaction 41. Another source is acetone photolysis, Reaction 47.

The principal atmospheric loss mechanism for PAN is thermal decomposition by Reaction 45 back to the peroxyacetyl radical and NO_2. The thermal decomposition rate constant for PAN, Reaction 45, is both temperature and pressure dependent. Over the temperature range 280 to 330 K,

$$k_{45} = \frac{k_0[M]}{1 + \frac{k_0[M]}{k_\infty}} F^{\{1 + (\log_{10} k_0[M]/k_\infty)^2\}^{-1}},$$

$k_0 = 4.9 \times 10^{-3} \exp(-12100/T)$ cm^3/molecule/sec,
$k_\infty = 4.0 \times 10^{16} \exp(-13600/T)$/sec,
$F = 0.3$.

At 298 K and 1 atm, $k_{45} = 5.2 \times 10^4$ sec. The lifetime of PAN is strongly temperature dependent and ranges from about 30 min at 298 K to about 8 h at 273 K. At temperatures of the upper troposphere, PAN lifetime from thermal decomposition is several months. PAN therefore acts as a reservoir for NO_x, a temporary one near the earth's surface and a considerably more permanent one in the upper troposphere.

V. OZONE FORMATION IN THE TROPOSPHERE

Identifying the tropospheric sources and sinks of ozone, together with their distribution with respect to latitude and altitude, is one of the predominant problems in atmospheric chemistry. Both transport and photochemistry play an important role in governing the tropospheric distribution of ozone.

Whereas there is abundant information on ground-level ozone concentrations in North America and Europe, there is less so in other portions of the globe. Relatively high concentrations of ozone (mixing ratios exceeding 50 ppb) are measured over large parts of rural areas of South America, Africa, and Southeast Asia during the dry season when biomass burning is practiced. Between 2 and 5×10^{12} kg of biomass carbon are burned each year in the tropics and subtropics (Crutzen and Andreae, 1990), leading to emissions of CO, hydrocarbons, and NO_x. By comparison, global fossil fuel combustion is estimated at about 5.5×10^{12} kg per year (Crutzen and Lawrence, 1997). In contrast, ozone mixing ratios in the cleanest regions of the globe, such as the tropical Pacific, are frequently less than 10 ppb.

A. Ozone Formation and Destruction

The rate-limiting reactions controlling net ozone formation can be expressed in terms of the difference between the local rates of O_3 production and destruction, $P(O_3)$,*

$$P(O_3) = \{k_{11}[HO_2] + k_9[CH_3O_2] + k_{34}[CH_3C(O)O_2]\}[NO]$$
$$- \{k_5[O(^1D)][H_2O] + k_{30}[HO_2][O_3]\}.$$

The generation terms assume that conversion of NO to NO_2 by a peroxy radical is tantamount to production of a molecule of O_3 because of the rapid photodissociation of NO_2, Reaction 27, followed immediately by Reaction 28. Thus $P(O_3)$ can be either positive or negative. The expression for $P(O_3)$ can be written as

$$P(O_3) = F(O_3) - D(O_3),$$

where $F(O_3)$ and $D(O_3)$ are the formation and destruction terms, respectively. When conditions are such that $P(O_3) > 0$, photochemical processes provide a net local source of O_3; when $P(O_3) > 0$, they constitute a net sink. Thus, the numerical value of $P(O_3)$, measured in molecules O_3 cm^3/sec, provides an estimate of the net effect of photochemistry on local levels of O_3. The sign and value of $P(O_3)$ depend on latitude and solar zenith angle (photodissociation rates), and water vapor and NO concentrations. We have already seen that, even at fixed solar intensity and water vapor level, there exists a critical NO concentration below which O_3 is destroyed and above which it is produced.

B. Ozone Formation and Destruction in the Remote Pacific Atmosphere

Davis *et al.* (1996) assessed $P(O_3)$ based on measurements made in the western North Pacific during the PEM-West A campaign carried out in 1991. They found two general trends: (1) at altitudes below 6 km, $P(O_3)$ was consistently < 0 with the largest negative values occurring within the first 3 to 4 km of the surface; and (2) above 7 km, $P(O_3)$ was found to be consistently >0, but the magnitude was strongly dependent on latitude. The altitude behavior of $P(O_3)$ was largely a result of a strong decrease of ozone destruction, $D(O_3)$,

---

*An additional minor sink for O_3 is

$$OH\cdot + O_3 \longrightarrow HO_2\cdot + O_2,$$

which can also be included in the expression for $P(O_3)$.

with increasing altitude. The dominant O_3 loss process at altitudes < 6 km is Reaction 5, $O(^1D) + H_2O$. Most of the remaining loss is a result of Reaction 30, $HO_2 + O_3$. The predominance of Reaction 5 at low altitudes is due to the high water vapor concentrations in the boundary layer. With respect to O_3 formation, at altitudes < 4 km the two major production processes are Reaction 11, $HO_2 + NO$, and Reaction 9, $CH_3O_2 + NO$. For altitudes > 4 km, Reaction 11 becomes the dominant process. At all altitudes in this area of the Pacific atmosphere, the contribution of nonmethane hydrocarbons to O_3 formation is no more than 11%. NO and peroxy radicals are the major chemical factors controlling $F(O_3)$. Water vapor and peroxy radical levels decrease with altitude, accompanied by a concomitant increase in NO and NO_x with altitude. Above 6 km, H_2O vapor levels are low and NO levels are sufficiently large such that $F(O_3) > D(O_3)$.

The critical NO concentration at which $P(O_3)$ changes sign was estimated by Davis et al. (1996) for the PEM-West A observations. Critical NO mixing ratios ranged from 6 to 17 ppt, with a median value of 11 ppt. Corresponding values of the critical NO_x mixing ratio ranged from 7 to 50 ppt, with a median value of 20 ppt. By considering Reaction 19, $OH + NO_2$, to be the major loss process for NO_x, NO_x lifetimes ranged from 1 to 1.5 days at low altitudes to 3 to 9 days at high altitudes (8 to 12 km). By contrast, high-altitude lifetimes for NO_2 were estimated to fall in the range of 1 to 2 days. The difference between the NO_x and NO_2 lifetimes reflects the large shift in the partitioning of NO_x toward NO at high altitude.

A quantity that is useful in characterizing the regions of photochemical ozone formation is the NO_x–O_3 chain length CL, defined as the number of O_3 molecules produced per NO_x molecule oxidized. By assuming conditions under which free-radical concentrations are at quasi-steady state and assuming that the only significant NO_x loss process is Reaction 19,

$$CL = \frac{F(O_3)}{k_{19}[OH][NO_2]}.$$

The numerator of the expression for CL is just the local rate of O_3 formation resulting from peroxy radical–NO reactions and the denominator is the rate of termination of NO_x into inactive status as HNO_3.[†] Thus, the CL is an estimate of the number of times a molecule of NO_x takes part in O_3 formation before it is removed. The effect of the NO_x level on the chain length can be seen since increasing NO_x promotes Reactions 9 and 11, and other $RO_2 +$ NO reactions, such as Reaction 43. Increasing NO_x also leads to lower levels

---

[†]In the upper free troposphere, because the lifetime of HNO_3 is much longer compared to that in the lower troposphere, HNO_3 may reside long enough to eventually be recycled back to NO_x by photolysis or reaction with OH.

of HO_x radicals through enhancement of Reaction 19. NO_x–O_3 chain lengths estimated from the PEM-West A data fall in a range roughly around 100.

A key issue in the chemistry of the global troposphere is the source of atmospheric NO_x. For the region of the Pacific we have been considering, it has been estimated that no more than about 25% of the high-altitude NO_x could be attributable to surface sources (Davis *et al.*, 1996; Singh *et al.*, 1996; Liu *et al.*, 1996). While possible high-altitude sources of NO_x include lightning, aircraft emissions, stratospheric intrusions, and NO_x from "recycled" NO_y, quantitatively evaluating such sources remains as one of the most important unsolved problems in atmospheric chemistry.

VI. CONCLUSIONS

The chemistry of the troposphere is driven by solar UV radiation of wavelengths shorter than about 310 nm. This radiation, acting on ozone, in the presence of water vapor, leads to the formation of hydroxyl (OH) radicals, which despite exceedingly low concentrations, oxidize virtually all organic compounds from natural and anthropogenic sources. In the background troposphere, about 75% of the OH radicals react with CO, the remainder reacting with CH_4. Although OH radicals react overwhelmingly with CO and CH_4 in the background troposphere, these reactions do not lead to the net removal of OH, but initiate reaction chains that regenerate OH. Many of the reaction chains involve nitrogen oxides (NO_x) as catalysts and some do not. The resulting chemistry is quite different in the two cases; for example, in the presence of sufficient NO, the oxidation of CO by OH leads to generation of O_3 without loss of OH. In NO-poor regions, the reaction of CO with OH leads to a net consumption of O_3, also without loss of OH. A "breakeven" NO concentration below which tropospheric chemistry is net O_3-depleting and above which it is net O_3-generating is an essential property of tropospheric chemistry.

REFERENCES

Atkinson, R. (1994). Gas-phase tropospheric chemistry of organic compounds. *J. Phys. Chem. Ref. Data,* Monograph, No. **2**, 1–216.

Blake, D. R., Chen, T. Y., Smith, T. W., Jr., Wang, C. J. L., Wingenter, O. W., Blake, N. J., and Rowland, F. S. (1996). Three-dimensional distribution of nonmethane hydrocarbons and halocarbons over the northwestern Pacific during the 1991 Pacific Exploratory Mission (PEM-West A). *J. Geophys. Res.* **101**, 1763–1778.

Crutzen, P. J., and Andreae, M. O. (1990). Biomass burning in the tropics—Impact on atmospheric chemistry and biogeochemical cycles. *Science.* **250**, 1669–1678.

Crutzen, P., and Lawrence, M. (1997). Ozone clouds over the Atlantic. *Nature.* **388**, 625–626.

Davis, D. D., Crawford, J., Chen, G., Chameides, W., Liu, S., Bradshaw, J., Sandholm, S., Sachse, G., Gregory, G., Anderson, B., Barrick, J., Bachmeier, A., Collins, J., Browell, E., Blake, D., Rowland, S., Kondo, Y., Singh, H., Talbot, R., Heikes, B., Merrill, J., Rodriguez, J., and Newell, R. E. (1996). Assessment of ozone photochemistry in the western North Pacific as inferred from PEM-West A observations during the fall of 1991. *J. Geophys. Res.* **101**, 2111–2134.

DeMore, W. G., Sander, S. P., Golden, D. M., Hampson, R. F., Kurylo, M. J., Howard, C. J., Ravishankara, A. R., Kolb, C. E., and Molina, M. J. (1997). Chemical Kinetics and Photochemical Data for Use in Stratospheric Modeling. Evaluation No. 12, Jet Propulsion Laboratory, Pasadena, CA.

Fehsenfeld, F. C., Trainer, M., Parrish, D. D., Volz-Thomas, A., and Penkett, S. (1996). North Atlantic regional experiment 1993 summer intensive: forward. *J. Geophys. Res.* **101**, 28869–28875.

Gregory, G. L., Bachmeier, A. S., Blake, D. R., Heikes, B. G., Thornton, D. C., Bandy, A. R., Bradshaw, J. D., and Kondo, Y. (1996). Chemical signatures of aged Pacific marine air: mixed layer and free troposphere as measured during PEM-West A. *J. Geophys. Res.* **101**, 1727–1742.

Hard, T. M., Chan, C. Y., Mehrabzadeh, A. A., and O'Brien, R. J. (1992). Diurnal HO_2 cycles at clean air and urban sites in the troposphere. *J. Geophys. Res.* **97**, 9785–9794.

Heikes, B. G., Lee, M., Bradshaw, J., Sandholm, S., Davis, D. D., Crawford, J., Rodriguez, J., Liu, S., McKeen, S., Thornton, D., Bandy, A., Gregory, G., Talbot, R., and Blake, D. (1996). Hydrogen peroxide and methylhydroperoxide distributions related to ozone and odd hydrogen over the North Pacific in the fall of 1991. *J. Geophys. Res.* **101**, 1891–1905.

Liu, S. C., McKeen, S. A., Hsie, E. Y., Lin, X., Kelly, K. K., Bradshaw, J. D., Sandholm, S. T., Browell, E. V., Gregory, G. L., Sachse, G. W., Bandy, A. R., Thornton, D. C., Blake, D. R., Rowland, F. S., Newell, R., Heikes, B. G., Singh, H., and Talbot, R. W. (1996). Model study of tropospheric trace species distributions during PEM-West A. *J. Geophys. Res.* **101**, 2073–2085.

Logan, J. A., Prather, M. J., Wofsy, S. C., and McElroy, M. B. (1981). Tropospheric chemistry: a global perspective. *J. Geophys. Res.* **86**, 7210–7254.

Madronich, S., and Calvert, J. G. (1990). Permutation reactions of organic peroxy radicals in the troposphere. *J. Geophys. Res.* **95**, 5697–5717.

Madronich, S., Chatfield, R. B., Calvert, J. G., Moortgat, G. K., Veyret, B., and Lesclaux, R. (1990). A photochemical origin of acetic acid in the troposphere. *Geophys. Res. Lett.* **17**, 2361–2364.

Mihelcic, D., Volz-Thomas, A., Pätz, H. W., Kley, D., and Mihelcic, M. (1990). Numerical analysis of ESR spectra from atmospheric samples. *J. Atmos. Chem.* **11**, 271–297.

Moortgat, G., Veyret, B., and Lesclaux, R. (1989). Kinetics of the reaction of HO_2 with $CH_3C(O)O_2$ in the temperature range 253–368 K. *Chem. Phys. Lett.* **160**, 443–447.

Prinn, R., Cunnold, D., Simmonds, P., Alyea, F., Boldi, R., Crawford, A., Fraser, P., Gutzler, D., Hartley, D., Rosen, R., and Rasmussen, R. (1992). Global average concentration and trend for hydroxyl radicals deduced from ALE/GAGE trichloroethane (methyl chloroform) data for 1978–1990. *J. Geophys. Res.* **97**, 2445–2461.

Roberts, J. M. (1990). The atmospheric chemistry of organic nitrates. *Atmos. Environ.* **24A**, 243–287.

Roberts, J. M., Parrish, D. D., Norton, R. B., Bertman, S. B., Holloway, J. S., Trainer, M., Fehsenfeld, F. C., Carroll, M. A., Albercook, G. M., Wang, T., and Forbes, G. (1996). Episodic removal of NO_y species from the marine boundary layer over the North Atlantic. *J. Geophys. Res.* **101**, 28947–28960.

Singh, H. B., Herlth, D., Kolyer, R., Salas, L., Bradshaw, J. D., Sandholm, S. T., Davis, D. D., Crawford, J., Kondo, Y., Koiki, M., Talbot, R., Gregory, G. L., Sachse, G. W., Browell, E.,

Blake, D. R., Rowland, F. S., Newell, R., Merrill, J., Heikes, B., Liu, S. C., Crutzen, P. J., and Kanakidou, M. (1996). Reactive nitrogen and ozone over the western Pacific: distribution, partitioning, and sources. *J. Geophys. Res.* **101**, 1793–1808.

Singh, H. B., Kanakidou, M., Crutzen, P. J., and Jacob, D. J. (1995). High concentrations and photochemical fate of oxygenated hydrocarbons in the global troposphere. *Nature* **378**, 50–54.

Singh, H. B., O'Hara, D., Herlth, D., Sachse, W., Blake, D. R., Bradshaw, J. D., Kanakidou, M., and Crutzen, P. J. (1994). Acetone in the atmosphere: distribution, sources, and sinks. *J. Geophys. Res.* **99**, 1805–1819.

Stockwell, W. R. (1995). On the HO_2-HO_2 reaction—its misapplication in atmospheric chemistry models, *J. Geophys. Res.* **100**, 11695–11698.

Tie, X., Kao, C.-Y., and Mroz, E. J. (1992). Net yield of OH, CO and O_3 from the oxidation of atmospheric methane, *Atmos. Environ.* **26A**, 125–136.

Vaghjiani, G. L., and Ravishankara, A. R. (1989). Absorption cross sections of CH_3OOH, H_2O_2 and D_2O_2 vapors between 210 and 365 nm at 297 K. *J. Geophys. Res.* **94**, 3487–3492.

Wallington, T. J., Dagaut, P., and Kurylo, M. J. (1992). Ultraviolet absorption cross sections and reaction kinetics and mechanisms for peroxy radicals in the gas phase. *Chem. Rev.* **92**, 667–710.

Wang, T., Carroll, M. A., Albercook, G. M., Owen, K. R., Duderstadt, K. A., Markevitch, A. N., Parrish, D. D., Holloway, J. S., Fehsenfeld, F. C., Forbes, G., and Ogren J. (1996). Ground-based measurements of NO_x and total reactive oxidized nitrogen (NO_y) at Sable Island, Nova Scotia during the NARE 1993 summer intensive. *J. Geophys. Res.* **101**, 28991–29004.

Wennberg, P. O., Hanisco, T. F., Jaeglé, L., Jacob, D. J., Hintsa, E. J., Lanzendorf, E. J., Anderson, J. G., Gao, R.-S., Keim, E. R., Donnelly, S. G., Del Negro, L. A., Fahey, D. W., McKeen, S. A., Salawitch, R. J., Webster, C. R., May, R. D., Herman, R. L., Proffitt, M. H., Margitan, J. J., Atlas, E. L., Schauffler, S. M., Flocke, F., McElroy, C. T., and Bui, T. P. (1998). Hydrogen radicals, nitrogen radicals, and the production of O_3 in the upper troposphere. *Science* **279**, 49–53.

INDEX